Edexcel GCSE Maths

Foundation Practice Book

Use and apply standard techniques

Kath Hipkiss

Contents

How to use this book

Welcome to Collins *Edexcel GCSE Maths Foundation Practice Book*. This book follows the structure of the Collins *Edexcel GCSE Maths 4th edition Foundation Student Book*, so is ideal to use alongside it.

Colour-coded questions

Know what level of difficulty you are working at with questions ranging from more accessible (green), through intermediate (blue) to more challenging (pink).

Hints and tips

These are provided where extra guidance can save you time or help you out.

Use of calculators

Questions when you could use a calculator are marked with a calculator icon.

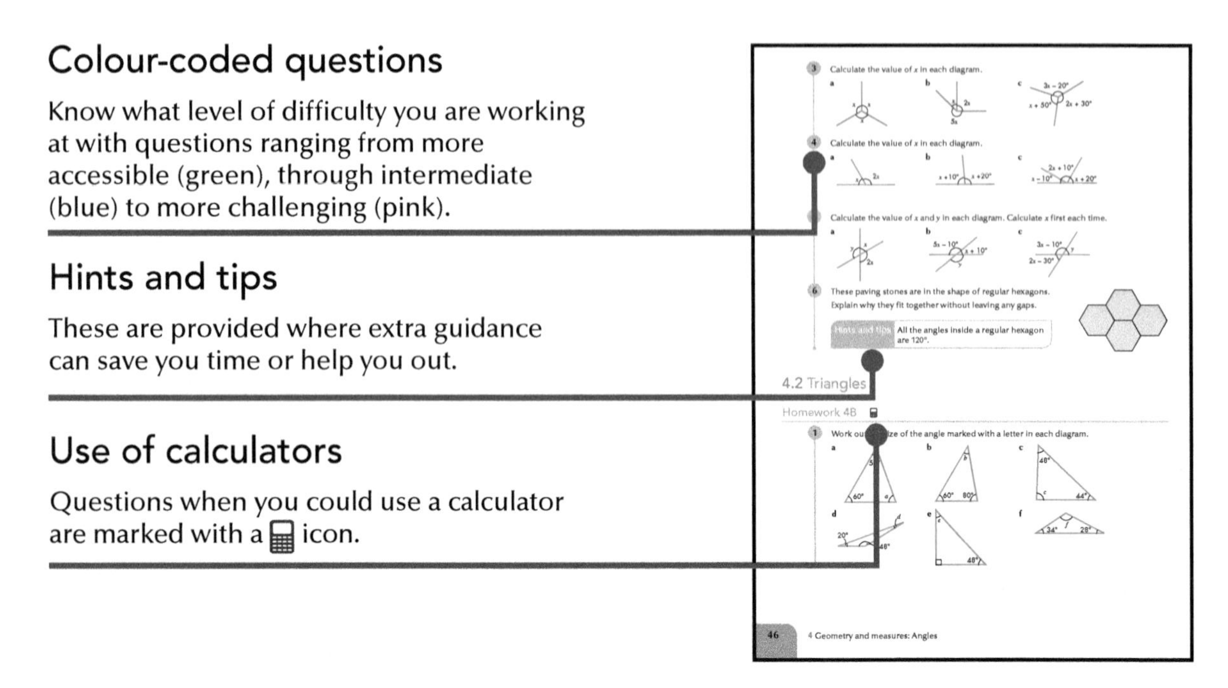

3 Calculate the value of x in each diagram.

4 Calculate the value of x in each diagram.

Calculate the value of x and y in each diagram. Calculate x first each time.

6 These paving stones are in the shape of regular hexagons. Explain why they fit together without leaving any gaps.

Hints and tips: All the angles inside a regular hexagon are 120°.

4.2 Triangles

Homework 4B

1 Work out the size of the angle marked with a letter in each diagram.

46 4 Geometry and measures: Angles

Examples

Understand the topic before you start the exercise by reading the examples in blue boxes.

These take you through questions step by step for certain topics where examples can provide additional support.

Answers

The answers are provided online at **www.collins.co.uk/gcsemaths4eanswers**.

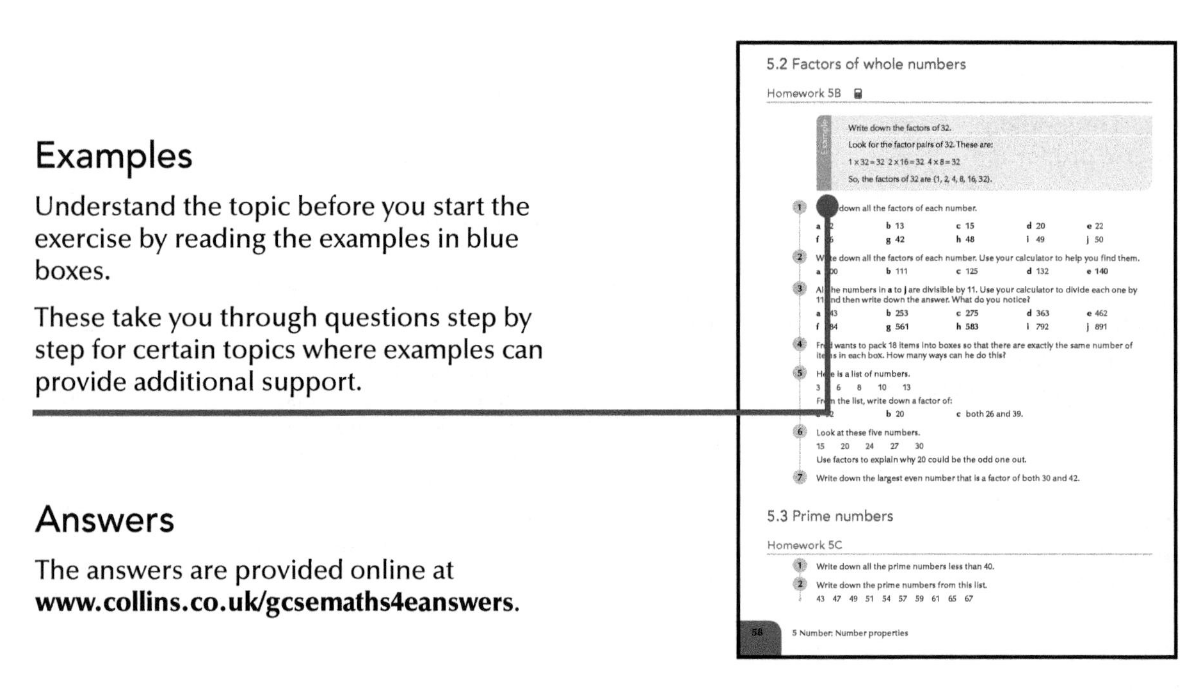

5.2 Factors of whole numbers

Homework 5B

Example: Write down the factors of 32. Look for the factor pairs of 32. These are: 1 × 32 = 32 2 × 16 = 32 4 × 8 = 32. So, the factors of 32 are {1, 2, 4, 8, 16, 32}.

6 Look at these five numbers. 15 20 24 27 30. Use factors to explain why 20 could be the odd one out.

7 Write down the largest even number that is a factor of both 30 and 42.

5.3 Prime numbers

Homework 5C

1 Write down all the prime numbers less than 40.

2 Write down the prime numbers from this list. 43 47 49 51 54 57 59 61 65 67

58 5 Number: Number properties

1 Number: Basic number

1.1 Place value and ordering numbers

Homework 1A 

1 Write down the value of each underlined digit.

a 576 **b** 374 **c** 689 **d** 4785 **e** 3007

f 7608 **g** 3542 **h** 12 745 **i** 87 409 **j** 7 777 777

2 Write each of the following in words.

a 7245 **b** 9072 **c** 29 450 **d** 2 760 000 **e** 5 800 000

3 Write each of the following in digits.

a Eight thousand and five hundred

b Forty two thousand and forty two

c Six million

d Five million and five

4 Write these numbers in order, starting with the *smallest*.

a 31, 20, 14, 22, 8, 25, 30, 12

b 159, 155, 176, 167, 170, 168, 151, 172

c 2100, 2070, 2002, 1990, 2010, 1998, 2000, 2092

5 Write these numbers in order, starting with the *largest*.

a 49, 62, 75, 57, 50, 72

b 988, 1052, 999, 1010, 980, 1007

c 4567, 4765, 4675, 4576, 4657, 4756

6 Here are the distances from Emily's house to five seaside towns.

Skegness (86 miles)

Rhyl (115 miles)

Great Yarmouth (166 miles)

Scarborough (80 miles)

Blackpool (105 miles)

a Which town is the furthest from Emily's house?

b Which town is the nearest to Emily's house?

7 a Write down as many four-digit numbers as you can using the digits 5, 7, 8 and 9. Use each digit once only.

b Which of your numbers is the smallest?

c Which of your numbers is the largest?

8 Write down, in order of size and starting with the largest, all the two-digit numbers that can be made using 2, 4 and 6. You can repeat the digits.

9 Copy each of these sentences, writing the numbers in words.

a The diameter of the Earth at the equator is 12 756 kilometres.

b The Moon is approximately 238 000 miles from the Earth.

c The greatest distance of the Earth from the Sun is 94 600 000 miles.

10 Using each of the digits 1, 5, 6 and 9, make a four-digit even number greater than eight thousand.

11 Write each set of numbers in order, starting with the *smallest*.

a 20, –10, –30, 12, –5, 5, –28, –13

b –2, 0, –1, 1.6, 1.1 , 2, 1 , –2.9, –1.1

c –6, 5, 26, –13, 1, –1, 0, –12

d –1.3, 1.8, $-\frac{1}{2}$, 2, –6, $2\frac{3}{4}$, 0, 3.1, –4

12 Calculate the temperature, when:

a 6 °C increases by 9 °C

b 7 °C falls by 3 °C

c 16 °C falls by 15 °C

d –2 °C increases by 4 °C

e –6 °C falls by 8 °C

f 11 °C falls by 4 °C

g –9 °C falls by 12 °C

h –3 °C increases by 2 °C

i 11°C falls by 12 °C

j –4 °C falls by 5 °C.

13 What is the difference between each pair of temperatures?

a 4 °C and 7 °C

b 13 °C and 23 °C

c 6 °C and 4 °C

d –2 °C and 2 °C

e –7 °C and –3 °C

f –1 °C and 0 °C

g 8 °C and 2 °C

h 5 °C and –2 °C

i 9 °C and –3 °C

j –2 °C and –4 °C

1.2 Order of operations and BIDMAS

Homework 1B

1 Work out each of these.

a $3 \times 4 + 7 =$ **b** $8 + 2 \times 4 =$ **c** $12 \div 3 + 4 =$ **d** $10 - 8 \div 2 =$

e $7 + 2 - 3 =$ **f** $5 \times 4 - 8 =$ **g** $9 + 10 \div 5 =$ **h** $11 - 9 \div 1 =$

i $12 \div 1 - 6 =$ **j** $4 + 4 \times 4 =$ **k** $10 \div 2 + 8 =$ **l** $6 \times 3 - 5 =$

2 Work out each of these.

Remember to work out the brackets first.

a $3 \times (2 + 4) =$ **b** $12 \div (4 + 2) =$ **c** $(4 + 6) \div 5 =$

d $(10 - 6) + 5 =$ **e** $3 \times 9 \div 3 =$ **f** $5 + 4 \times 2 =$

g $(5 + 3) \div 2 =$ **h** $5 \div 1 \times 4 =$ **i** $(7 - 4) \times (1 + 4) =$

j $(7 + 5) \div (6 - 3) =$ **k** $(8 - 2) \div (2 + 1) =$ **l** $15 \div (15 - 12) =$

3 Copy each number sentence and put in brackets to make the statement correct, where necessary.

a $4 \times 5 - 1 = 16$ **b** $8 \div 2 + 4 = 8$ **c** $8 - 3 \times 4 = 20$ **d** $12 - 5 \times 2 = 2$

e $3 \times 3 + 2 = 15$ **f** $12 \div 2 + 1 = 4$ **g** $9 \times 6 \div 3 = 18$ **h** $20 - 8 + 5 = 7$

i $6 + 4 \div 2 = 5$ **j** $16 \div 4 \div 2 = 8$ **k** $20 \div 2 + 2 = 12$ **l** $5 \times 3 - 5 = 10$

4 Jo says that $8 - 3 \times 2$ is equal to 10.

Is she correct? Explain your answer.

5 Add ÷, ×, + or – signs, and brackets where necessary, to make each number sentence true.

a 2 ... 5 ... 10 = 0 **b** 10 ... 2 ... 5 = 1 **c** 10 ... 5 ... 2 = 3

d 10 ... 2 ... 5 = 4 **e** 10 ... 5 ... 2 = 7 **f** 5 ... 10 ... 2 = 10

g 10 ... 5 ... 2 = 13 **h** 5 ... 10 ... 2 = 17 **i** 10 ... 2 ... 5 = 20

j 5 ... 10 ... 2 = 25 **k** 2 ... 2 ... 2 = 2

6 Amanda worked out $3 + 4 \times 5$ and got the answer 35. Andrew worked out $3 + 4 \times 5$ and got the answer 23. Explain why they got different answers.

7 Explain how to work out this calculation.

$7 + 2 \times 6$

8 Here is a list of numbers, some signs and one pair of brackets.

2 5 6 42 + × = ( )

Use *all* of them to make a correct calculation.

9 Here is a list of numbers, some signs and one pair of brackets.

1 3 5 8 – ÷ = ( )

Use *all* of them to make a correct calculation.

10 Jon has a piece of wood that is 8 m long.

He wants to use his calculator to work out how much wood will be left when he cuts off three pieces, each of length 1.2 m.

Which calculation(s) would give him the correct answer?

i $8 - 1.2 - 1.2 - 1.2$

ii $8 - 1.2 \times 3$

iii $8 - 1.2 + 1.2 + 1.2$

1.3 The four rules

Homework 1C 

1 Copy and complete each addition.

a $\begin{array}{r} 75 \\ +\ 23 \\ \hline \end{array}$ **b** $\begin{array}{r} 245 \\ +\ 156 \\ \hline \end{array}$ **c** $\begin{array}{r} 307 \\ +\ 293 \\ \hline \end{array}$ **d** $\begin{array}{r} 4158 \\ +\ 3951 \\ \hline \end{array}$ **e** $\begin{array}{r} 4289 \\ 532 \\ +\ 96 \\ \hline \end{array}$

2 Work out each addition.

a $25 + 89 + 12$ **b** $211 + 385 + 46$ **c** $125 + 88 + 720$

d $478 + 207 + 300$ **e** $1275 + 3245 + 524$

3 Copy and complete each subtraction.

a $\begin{array}{r} 354 \\ -\ 120 \\ \hline \end{array}$ **b** $\begin{array}{r} 651 \\ -\ 128 \\ \hline \end{array}$ **c** $\begin{array}{r} 785 \\ -\ 207 \\ \hline \end{array}$ **d** $\begin{array}{r} 450 \\ -\ 178 \\ \hline \end{array}$ **e** $\begin{array}{r} 5421 \\ -\ 2568 \\ \hline \end{array}$

4 Work out these subtractions.

a $386 - 296$ **b** $709 - 518$ **c** $452 - 386$

d $800 - 258$ **e** $7208 - 1564$

5 The train from Brighton to London takes 68 minutes.

The train from London to Birmingham takes 85 minutes.

a I travel from Brighton to London and then London to Birmingham. How long does my journey take altogether, if there is a 30-minute wait between trains in London?

b How much longer is the train journey from London to Birmingham than from Brighton to London?

6 Michael is checking the addition of two numbers.

His answer is 917.

One of the numbers is 482.

What should the other number be?

7 Copy each of these and fill in the missing digits.

a $\begin{array}{r} 4\ 5 \\ +3\square \\ \hline \square 7 \end{array}$ **b** $\begin{array}{r} \square 7 \\ +4\square \\ \hline 9\ 2 \end{array}$ **c** $\begin{array}{r} 3\square 4 \\ +2\ 8\ 6 \\ \hline \square 4\square \end{array}$ **d** $\begin{array}{r} \square\square\square \\ +2\ 8\ 7 \\ \hline 5\ 5\ 5 \end{array}$

8 Copy each of these and fill in the missing digits.

a
```
  7 5
 −1□
 ───
  □3
```

b
```
  3 2□
 −1□4
 ────
  1 8 2
```

c
```
  5 8 3
 −□□□
 ─────
  1 3 5
```

d
```
  □□□
 −2 4 8
 ─────
  3 7 4
```

9 Copy and complete the following additions and subtractions.

a 88.21 − 81.33 =

b 26.74 + 41.21 =

c 77.83 − 66.16 =

d 73.44 + 29.27 =

e 51.84 + 21.97 =

f 89.23 − 35.91 =

g 82.28 + 33.29 =

h 21.31 + 34.35 =

i 99.95 − 17.49 =

j 49.82 − 38.24 =

10 Copy and complete these multiplications.

a $\begin{array}{r} 24 \\ \times\ 3 \\ \hline \end{array}$ **b** $\begin{array}{r} 38 \\ \times\ 4 \\ \hline \end{array}$ **c** $\begin{array}{r} 124 \\ \times\ 5 \\ \hline \end{array}$ **d** $\begin{array}{r} 408 \\ \times\ 6 \\ \hline \end{array}$ **e** $\begin{array}{r} 359 \\ \times\ 8 \\ \hline \end{array}$

11 Set each multiplication in columns and complete it.

a 21 × 5 **b** 37 × 7 **c** 203 × 9 **d** 4 × 876 **e** 6 × 3214

f 2.1 × 3 **g** 3.7 × 4 **h** 20.3 × 6 **i** 2.14 × 1.6 **j** 0.6 × 32.15

12 Copy and complete each division.

a 684 ÷ 2 **b** 525 ÷ 3 **c** 804 ÷ 4 **d** 7260 ÷ 5 **e** 2560 ÷ 8

13 Jake, Tomas and Theo are footballers.

The manager of their club has offered them a bonus of £5 for every goal they score.

Jake scores 15 goals.

Tomas scores 12 goals.

Theo scores 20 goals.

a How many goals do they score altogether?

b How much does each footballer receive as a bonus?

14 The footballers in question **13** are given a goals target for the following season.

The manager wants the total of the goals target to add up to 55, with Theo having the highest target and Tomas the lowest.

Copy and complete the table to show how this can be done.

	Goals Target
Jake	
Tomas	
Theo	
Total	**55**

15 Answer each of these questions by completing a suitable multiplication.

a How many people could seven 55-seater coaches hold?

b Adam buys seven postcards at 23p each. How much does he spend in pounds?

c Nails are packed in boxes of 144. How many nails are there in five boxes?

d Eight people book a holiday, costing £784 each. What is the total cost?

e There are 1760 yards in a mile. How many yards are there in 6 miles?

16 Answer each of these questions by completing a suitable division.

a There are 288 students in eight forms in Year 10. There are the same number of students in each form. How many is this?

b Phil jogs 7 miles every morning. How many days will it take him to cover a total distance of 441 miles?

c In a supermarket, cans of cola are sold in packs of six. If there are 750 cans on the shelf, how many packs are there?

d Sandra's wages for a month were £2060. Assuming there are four weeks in a month, how much does she earn per week?

e Tickets for a charity disco were sold at £5 each. The total sales were £1710. How many tickets were sold?

17 Give your answers to these divisions as decimals.

a $17.0 \div 2 =$ **b** $29.0 \div 4 =$ **c** $58 \div 8 =$ **d** $34 \div 5 =$ **e** $38 \div 4 =$

f $933 \div 6 =$ **g** $188 \div 8 =$ **h** $90 \div 6 =$ **i** $84 \div 7 =$ **j** $273 \div 6 =$

Homework 1D

1 Use a thermometer scale to help you work out the answers to these calculations.

a $-5 + 7 =$ **b** $-4 + 8 =$ **c** $-3 + 6 =$

d $-4 + 7 =$ **e** $-7 + 4 =$ **f** $-11 + 10 =$

2 Copy and complete these calculations *without* using a thermometer scale.

a $10 - 14 =$ **b** $14 - 15 =$ **c** $17 - 15 =$

d $-48 + 78 =$ **e** $-14 + 18 =$ **f** $-12 + 19 =$

3 Work these out.

a $-102 + 15 - 47 =$ **b** $78 - 104 + 48 =$ **c** $-15 - 17 + 41 =$

d $-11 - 11 + 22 =$ **e** $-15 + 59 - 75 =$ **f** $11 + 16 - 27 =$

4 At 6:00 am the temperature in Lincoln was −2 °C. By 2:00 pm it had risen by 14 °C. What was the temperature in Lincoln at 2:00 pm?

5 I have £129 in my bank account. If I spend £251, what is the balance of my bank account?

6 On the same day, the average temperature in Canada was −21 °C and the average temperature in Mexico was 41 °C. Calculate the difference in temperature between these two countries.

Homework 1E

1 Work out each of these without a calculator. Then use a calculator to check your answers.

a $2 + (-7)$ **b** $5 + (-6)$ **c** $4 + (-11)$

d $8 + (-18)$ **e** $9 + (-11)$ **f** $11 + (-19)$

2 Write down the answer to each calculation without a calculator. Then use a calculator to check your answers.

a $-8 - (9)$ **b** $-4 + (-5)$ **c** $-14 + (-7)$

d $-13 + (-7)$ **e** $-8 - (-6)$ **f** $-7 - (-4)$

3 Work out the answer to each calculation without a calculator. Then use a calculator to check your answers.

a $-7 + (-4) - 9$ **b** $-4 + 5 - 18$ **c** $-5 + 17 - (-16)$

d $-8 + 19 - (-17)$ **e** $9 - 11 + 4$ **f** $8 - 17 - (-21)$

4 Write down the number missing from the box to make each number sentence true.

a $66 + \square = -11$ **b** $51 + \square = -34$ **c** $12 + \square = -65$ **d** $17 + \square = -12$

e $53 + \square = -19$ **f** $56 - \square = -10$ **g** $27 - \square = -13$ **h** $33 - \square = -9$

i $50 - \square = -1$ **j** $8 - \square = -7$

Homework 1F

1 Write down the answers to these.

a -5×8 **b** -7×-4 **c** -8×7

d 7×-9 **e** 9×-4 **f** 13×-13

2 Work out the answers.

a $-36 \div -3$ **b** $-48 \div -12$ **c** $-64 \div 4$

d $144 \div -24$ **e** $132 \div -11$ **f** $56 \div -8$

3 Work out the answers.

a $(-6) \times 3$ **b** $(-7) \times (-4)$ **c** $(-6) \div 2$

d $49 \div (-7)$ **e** $80 \div (-4)$ **f** $(-12) \div (-3)$

g $(-8) \times (-3)$ **h** $45 \div (-9)$ **i** $(-6) \times 10$

j $(-2) \times (-5)$ **k** $66 \div (-3)$ **l** $148 \div (-4)$

4 Write down the number missing from the box to make each number sentence true.

a $\square \times (-6) = 12$ **b** $(-40) \div \square = 5$

c $3 \times \square = -18$ **d** $(-81) \div \square = -9$

e $7 \times \square = 21$ **f** $(-12) \div \square = 3$

g $(-2) \times \square = 14$ **h** $(-2) \times \square \times (-3) = -24$

5 Copy and complete the multiplication grid.

×	–2	2	6
–3			
–7			
8			

6 Work these out.

a $(-4)^2$ **b** $(5-7)^2$ **c** $(-5)^2 \times (-2)^2$

d $(-3 \times -4)^2$ **e** $(11-13)^2$ **f** $(-7)^2 - (-3)^2$

Homework 1G 

1 Use your preferred written method to complete these multiplications.

a 24 × 82 **b** 44 × 18 **c** 28 × 47 **d** 83 × 84

e 86 × 54 **f** 121 × 57 **g** 216 × 67 **h** 143 × 34

i 286 × 47 **j** 354 × 86

2 Use your preferred written method to complete these multiplications.

a 56 × 21 **b** 45 × 57 **c** 56 × 78

d 12 × 34 **e** 78 × 910 **f** 23 × 57

3 Use a range of written methods to work these out. Try to use the grid method, the column method and the partition method at least once each.

a 2504 × 123 **b** 6514 × 847 **c** 1478 × 963

d 1591 × 357 **e** 1245 × 365 **f** 4895 × 362

4 There are 19 stickers on a sheet. A teacher buys 87 sheets of stickers. How many stickers is this?

5 26 baguette trays can fit inside an industrial oven. Each baguette tray can hold 12 baguettes. How many baguettes can be baked in the oven at the same time?

6 Billy has 18 complete books of stamps. Each book holds 232 stamps. How many stamps is this?

Homework 1H 

1 Solve these by long division.

a 286 ÷ 13 **b** 646 ÷ 19 **c** 1053 ÷ 27 **d** 744 ÷ 31 **e** 864 ÷ 18

2 Mrs English has 360 new pencils to share between 20 students. How many pencils will each student get?

3 **a** Cinema tickets cost £9.00 each. If you have £48, how many of your friends can you take to the cinema with you?

b At the cinema, you notice that the 504 seats are divided into seven equal sections. How many seats are there in each section?

4 288 new maths textbooks have been delivered to school. Each class needs 32 books. How many classes will get the new books?

5 Jennifer bought 48 teddy bears at £9.55 each.

a Work out the total amount she paid.

Jennifer sold all the teddy bears for the same price. The total revenue was £696.

b Work out the price at which Jennifer sold each teddy bear.

6 Nick fills his van with large wooden crates. The mass of each crate is 69 kg. The van can safely hold 990 kg. Work out the greatest number of crates he should pack in the van.

7 Mario makes pizzas. On Wednesday, he made 36 pizzas. He sold them for £2.45 each.

a Work out the total amount he charged for 36 pizzas.

Marco delivers pizzas. He is paid 65p for each pizza he delivers. Last Thursday he was paid £27.30 for delivering pizzas.

b How many pizzas did Marco deliver?

Homework 1I

1 Work out each of these.

a 2.4×5.6 **b** 7.4×5.07 **c** 5×4.37

d 3.6×5.47 **e** 0.7×6.82 **f** 6.7×4.92

g 0.7×7.27 **h** 6.3×3.48 **i** 5.4×2.81

j 2×6.16 **k** 3.2×1.21 **l** 5.5×0.92

m 3.2×1.07 **n** 5.3×1.53 **o** 6.8×4.9

2 Solve these multiplications. Remember to keep your decimal points in line in your working out.

a 16.94×45.18 **b** 66.96×89.61 **c** 10.41×34.46

d 63.39×81.42 **e** 36.42×96.64 **f** 96.85×45.94

g 13.41×19.88 **h** 72.28×22.38 **i** 25.74×82.98

j 84.13×63.51

3 Karen bought 11 cans of food costing £1.99 each and 14 cans of food costing £1.28 each. What was the total cost?

4 Lewis bought 19 boxes of nails costing £4.58 and 7 boxes of screws costing £2.83. How much change did he receive from £110?

5 Zack's regular wage is £11.29 per hour. He is paid £15.68 per hour for overtime. He was paid for 148.5 regular hours and 27.25 hours of overtime last month. Calculate his total pay.

2 Geometry and measures: Measures and scale drawings

2.1 Systems of measurement

Homework 2A

1. Decide which metric unit you would most likely use to measure each of the following amounts.
   - **a** The height of your best friend
   - **b** The distance from school to your home
   - **c** The thickness of a CD
   - **d** The mass of your maths teacher
   - **e** The amount of water in a lake
   - **f** The mass of a slice of bread
   - **g** The length of a double-decker bus
   - **h** The mass of a kitten
2. Estimate the approximate metric length, mass or capacity of each of the following.
   - **a** The length and mass of this book
   - **b** The length of the road you live on
   - **c** The capacity of a bottle of wine
   - **d** The length, width and weight of a door
   - **e** The diameter of a £1 coin, and its weight.
   - **f** The distance from your school to the Houses of Parliament (London)
3. The distance from Sheffield to Tintagel is shown on a website as 473 kilometres. Why is this unit used instead of metres?
4. Sarah makes a living cleaning windows of houses. She has three ladders: a 2-metre, a 4-metre and a 6-metre ladder. Which one should she use to clean the upper windows of a two-storey house? Give a reason for your choice.

Hints and tips

Length	10 mm = 1 cm 1000 mm = 100 cm = 1 m 1000 m = 1 km
Weight	1000 g = 1 kg 1000 kg = 1 t
Capacity	10 ml = 1 cl 1000 ml = 100 cl = 1 litre
Volume	1000 litres = 1 m^3 1 ml = 1 cl^3

5 Copy and complete these statements.

a 155 cm = ____ m
b 95 mm = ____ cm
c 780 mm = ____ m
d 3100 m = ____ km
e 310 cm = ____ m
f 3050 mm = ____ m
g 156 mm = ____ cm
h 2180 m = ____ km
i 1070 mm = ____ m
j 1324 cm = ____ m
k 175 m = ____ km
l 83 mm = ____ m
m 620 mm = ____ cm
n 2130 cm = ____ m
o 5120 m = ____ km
p 8150 g = ____ kg
q 2300 kg = ____ t
r 32 ml = ____ cl
s 1360 ml = ____ l
t 580 cl = ____ l
u 950 kg = ____ t

6 Copy and complete these statements.

a 120 g = ____ kg
b 150 ml = ____ l
c 350 cl = ____ l
d 540 ml = ____ cl
e 2060 kg = ____ t
f 7500 ml = ____ l
g 3800 g = ____ kg
h 605 cl = ____ l
i 15 ml = ____ l
j 6300 l = ____ m^3
k 45 ml = ____ cm^3
l 2350 l = ____ m^3
m 720 l = ____ m^3
n 8.2 m = ____ cm
o 71 km = ____ m
p 8.6 m = ____ mm
q 15.6 cm = ____ mm
r 0.83 m = ____ cm
s 5.15 km = ____ m
t 1.85 cm = ____ mm
u 2.75 m = ____ cm

7 Alesha wants to put up six shelves, each 65 cm long. She finds that she can buy planks of the correct width in three different lengths: 1200 mm, 1800 mm and 2400 mm. What combination of planks should she buy to have the least amount of waste wood?

8 How many square centimetres are there in a square kilometre?

9 Could you pour all the water from a full 2-litre bottle into a container with a volume of 101 cm^3?

Explain how you know.

Homework 2B

Hints and tips		
	Length	12 inches = 1 foot, 3 feet = 1 yard, 1760 yards = 1 mile
	Weight	16 ounces = 1 pound, 14 pounds = 1 stone, 2240 pounds = 1 ton
	Capacity	8 pints = 1 gallon

1 Copy and complete these statements.

a 5 feet = ___ inches
b 5 yards = ___ feet
c 3 miles = ___ yards
d 6 pounds = ___ ounces
e 5 stones = ___ pounds
f 2 tons = ___ pounds
g 4 gallons = ___ pints
h 7 feet = ___ inches
i 2 yards = ___ inches
j 11 yards = ___ feet
k 5 pounds = ___ ounces
l 39 feet = ___ yards
m 2 stones = ___ ounces
n 4400 yards = ___ miles
o 12 gallons = ___ pints
p 2 miles = ___ feet
q 84 inches = ___ feet
r 48 ounces = ___ pounds
s 21 feet = ___ yards
t 22 400 pounds = ___ tons
u 2 miles = ___ inches
v 256 ounces = ___ pounds
w 80 pints = ___ gallons
x 280 pounds = ___ stones
y 31 680 feet = ___ miles
z 2 tons = ___ ounces

2 Work out how many square feet are in a square mile.

3 Running tracks in the UK and the USA used to be 400 yards long. How many times would you need to run around the track in a six-mile race?

4 1 pound is approximately 450 grams.
Explain how you know that 1 tonne is lighter than 1 ton.

2.2 Conversion factors

Homework 2C

Hints and tips		
	Length	**Weight**
	1 metre ≈ 39 inches	2.2 pounds ≈ 1 kilogram
	1 foot ≈ 30 centimetres	**Capacity**
	1 foot ≈ 12 inches	1 litre ≈ 1.75 pints
	1 kilometre ≈ 1.62 miles	1 gallon ≈ 4.5 litres

1 Change each of these masses to pounds.

a 6 kg **b** 8 kg **c** 15 kg **d** 32 kg **e** 45 kg

2 Change each of these masses to kilograms. (Give each answer to one decimal place.)

a 10 lb **b** 18 lb **c** 25 lb **d** 40 lb **e** 56 lb

3 Change each of these capacities to pints.

a 2 litres **b** 8 litres **c** 25 litres **d** 60 litres **e** 75 litres

4 Change each of these capacities to litres. (Give each answer to the nearest litre.)

a 7 pints **b** 20 pints **c** 35 pints **d** 42 pints **e** 100 pints

5 Change each of these distances to kilometres.

a 20 miles **b** 30 miles **c** 50 miles **d** 65 miles **e** 120 miles

6 Change each of these distances to miles.

a 16 km **b** 24 km **c** 40 km **d** 72 km **e** 300 km

7 Change each of these capacities to litres.

a 5 gallons **b** 12 gallons **c** 27 gallons **d** 50 gallons **e** 72 gallons

8 Change each of these capacities to gallons.

a 18 litres **b** 45 litres **c** 72 litres **d** 270 litres **e** 900 litres

9 Change each of these distances to inches.

a 2 m **b** 5 m **c** 8 m **d** 10 m **e** 12 m

10 Change each of these distances to centimetres.

a 3 ft **b** 5 ft **c** 7 ft **d** 10 ft **e** 30 ft

11 Change each of these distances to metres. (Give your answer to one decimal place.)

a 48 inches **b** 52 inches **c** 60 inches **d** 75 inches **e** 100 inches

12 While on holiday in Iceland, I saw a road sign that said 'Blue Lagoon 26 km'. I was in a coach travelling on a road with a speed limit of 40 km/h.

a Approximately how many miles was I from the Blue Lagoon?

b What was the speed limit in miles per hour?

c If the coach travelled at this top speed, how many minutes would it take us to get to the Blue Lagoon?

13 Tom was cycling in the Alps. He planned to cycle 60 km each day. Because of the steep terrain, his average speed was only 11.5 mph.

How long would he expect it to take him to cycle 60 km with no stops?

14 How many cubic inches are in a five gallon drum?

2.3 Scale drawings

Homework 2D

1 The diagram shows the floor plan of a kitchen. The scale is 1 cm to 30 cm.

Work space
Cooker
Work space
Fridge
Sink unit
Door
Cupboards
Door

a State the actual dimensions of:

i the sink unit

ii the cooker

iii the fridge

iv the cupboards.

b Calculate the actual total area of work space.

2 The sketch shows a ladder leaning against a wall.

The bottom of the ladder is 1 m away from the wall and it reaches 4 m up the wall.

a Make a scale drawing to show the position of the ladder. Use a scale of 4 cm to 1 m.

b Use your scale drawing to work out the actual length of the ladder.

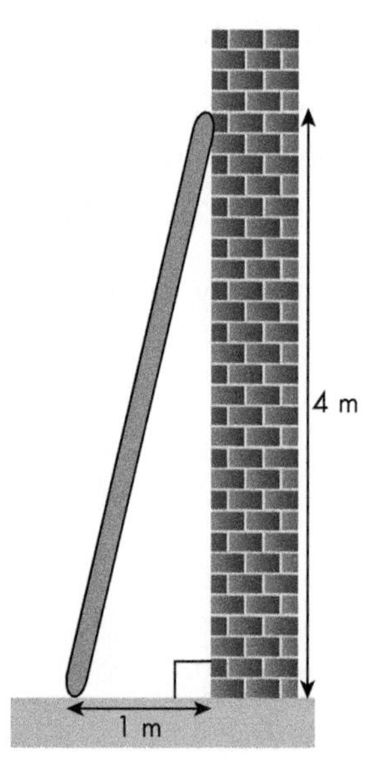

3 The map below is drawn to a scale of 1 cm to 2 km.

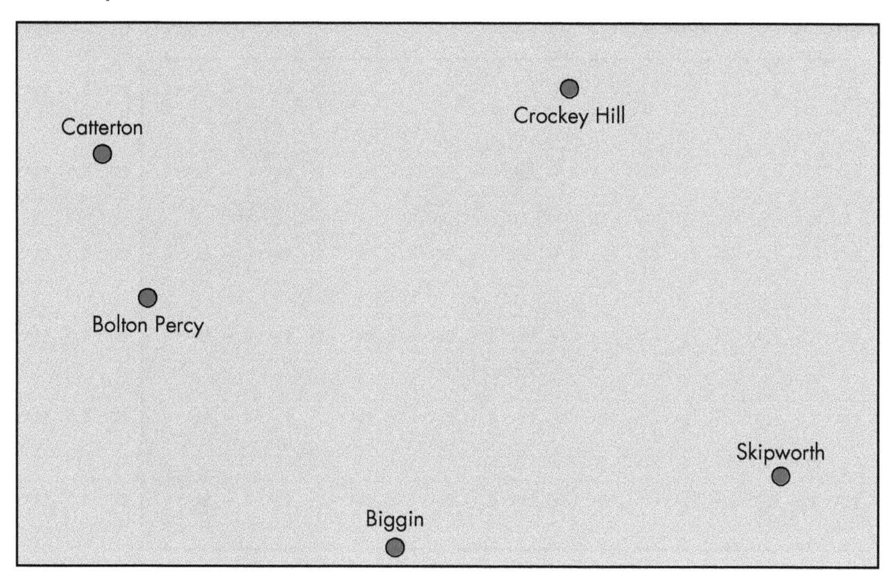

Work out the actual distance between:

a Biggin and Skipworth

b Bolton Percy and Crockey Hill

c Skipworth and Catterton

d Crockey Hill and Biggin

e Catterton and Bolton Percy.

4 This diagram shows a farmer's sketch of one of his fields.

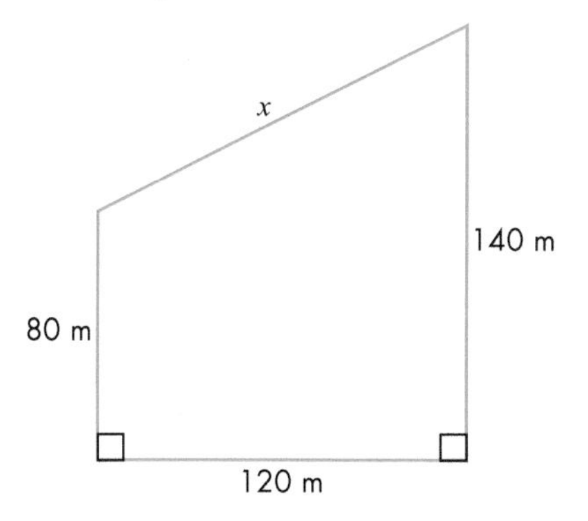

a Make a scale drawing of the field. Use the scale 1 cm represents 20 m.

b The farmer wants to build a wall along the side marked x on the diagram. Each metre length of wall uses 60 bricks.

Work out the number of bricks the farmer will need.

5 The map below shows the position of four fells in the Lake District. The map is drawn to a scale of 1 : 150 000.

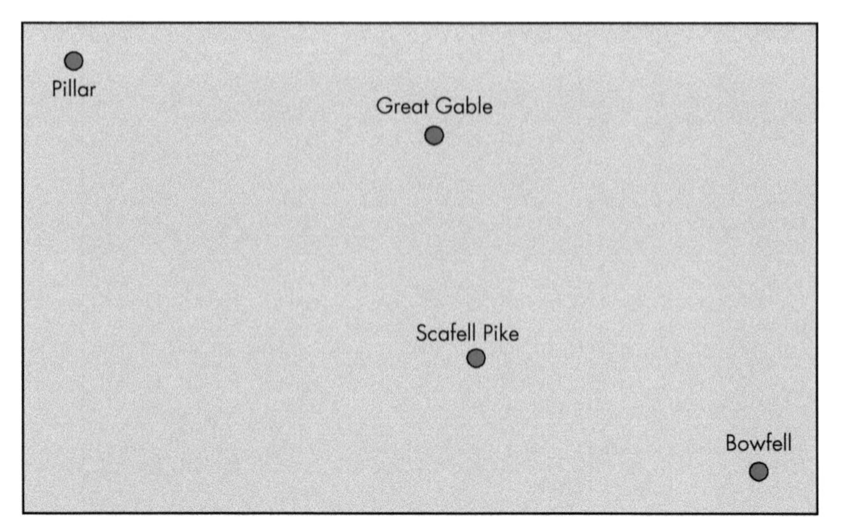

State these direct distances to the nearest kilometre.

a Scafell Pike to Great Gable

b Scafell Pike to Pillar

c Great Gable to Pillar

d Pillar to Bowfell

e Bowfell to Great Gable

6 Here is a scale drawing of a ferry crossing a river from port A to port B.

The width of the river is 400 m.

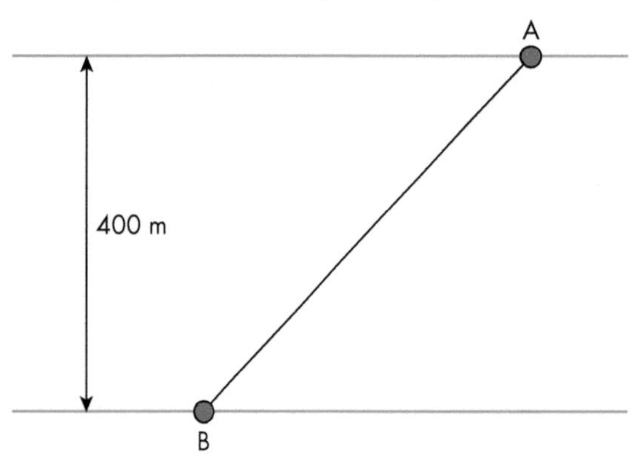

a Which of the following is the correct scale for the drawing?

1 : 1000   1 : 10 000   1 : 40 000   1 : 100 000

b What is the actual distance from port A to port B?

Homework 2E

1 The man in the pictures is 1.8 m tall.

Use this to estimate the height of:

a a door in a house

b a double decker bus.

2 Estimate the weight of:

a an average-sized man

b a car

c an egg.

3 The car in the pictures is 2.4 m long.

Use this to estimate the length of:

a an articulated lorry

b a football pitch.

4 Estimate the capacity of:

a a coffee cup

b a cylinder in a vacuum cleaner

c a small paddling pool.

2.4 Nets

Homework 2F

1 Four nets are shown below. Copy the nets that would make a cube.

a

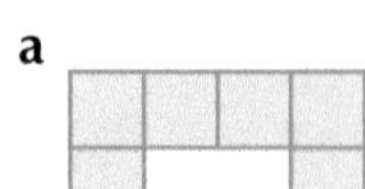

b

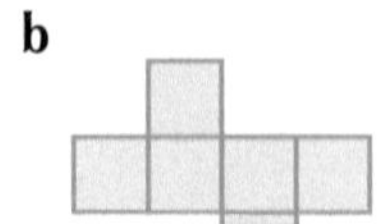

c

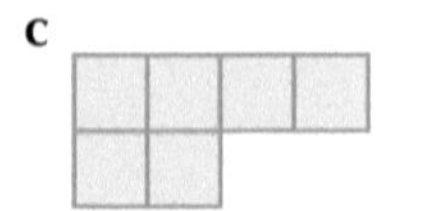

d

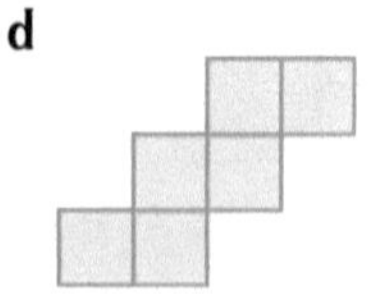

2 Draw, on squared paper, an accurate net for each cuboid.

a

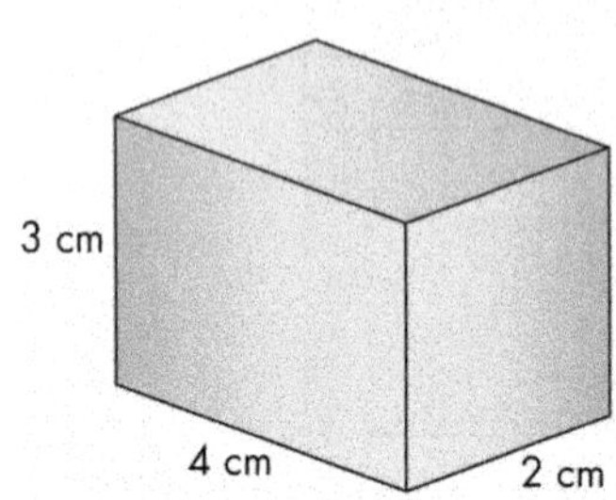

b

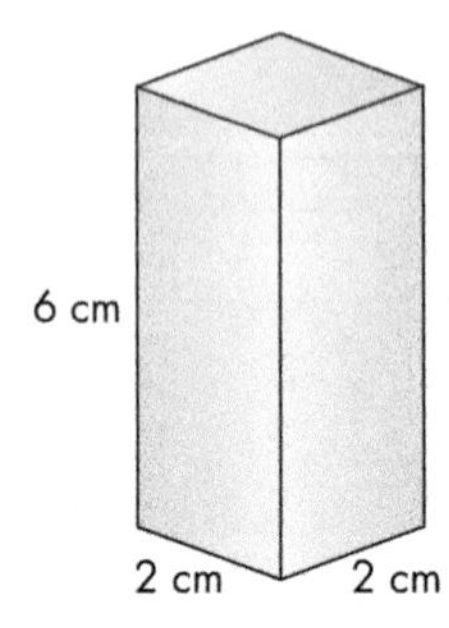

3 Draw, on squared paper, an accurate net for this triangular prism.

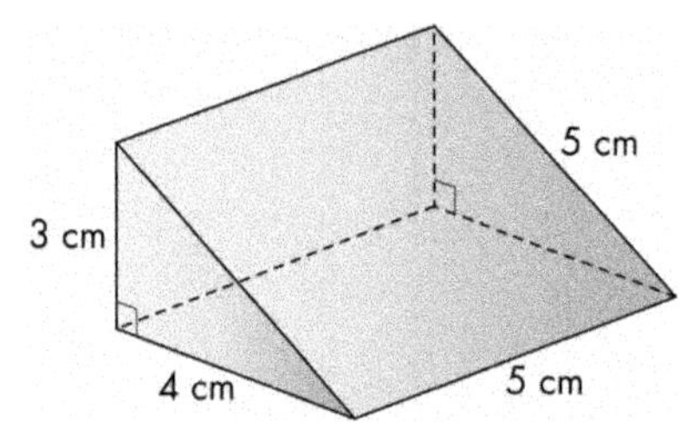

4 The diagram shows a sketch of a square-based pyramid.

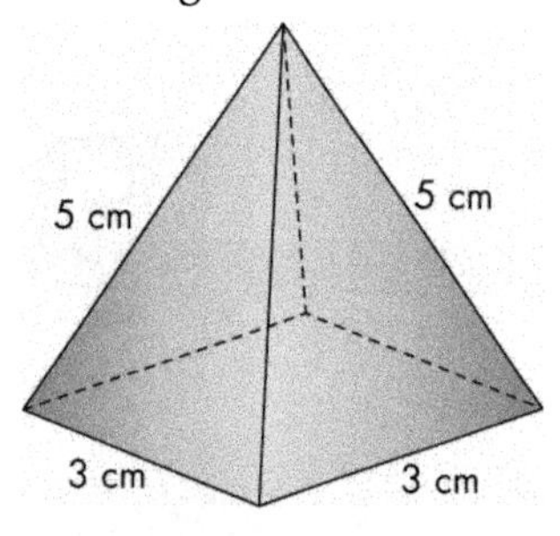

a Write down the number of:

i vertices

ii edges

iii faces

a square-based pyramid has.

b Draw a sketch of a net for this pyramid.

5 Paul is making this dice out of card.

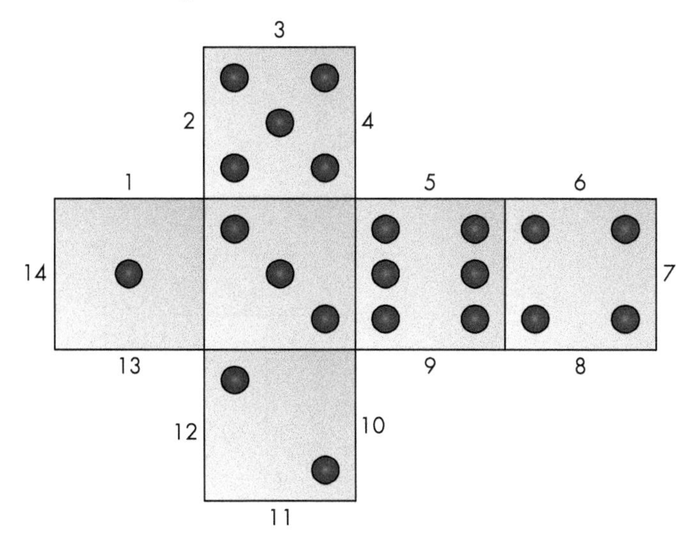

For each pair of edges that join together, such as edge 1 and edge 2, he will need to add a tab. List the other pairs of edges that join together.

6 The diagram shows a regular tetrahedron. Its four faces are all equilateral triangles.

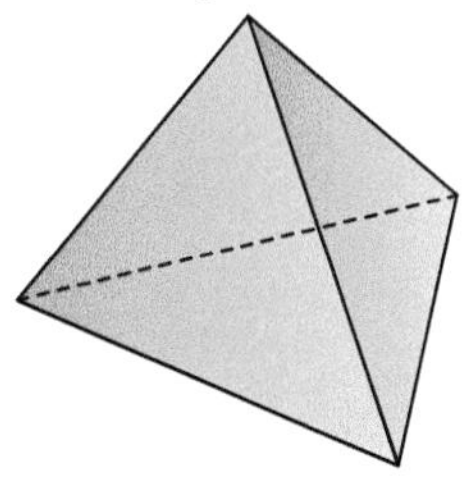

A tetrahedron has two nets. Draw them both.

7 Zoë has five shapes:

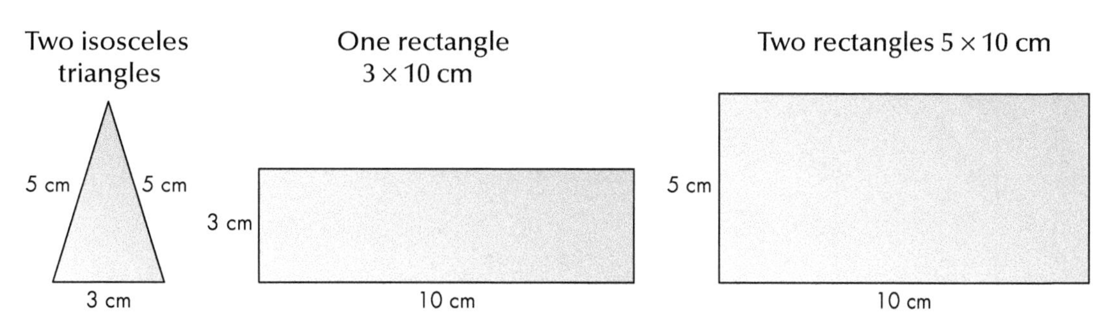

Draw a sketch to show how she can put the five shapes together to make a net of a triangular prism.

2.5 Using an isometric grid

Homework 2G

1 Draw each of these 3D shapes on an isometric grid.

a

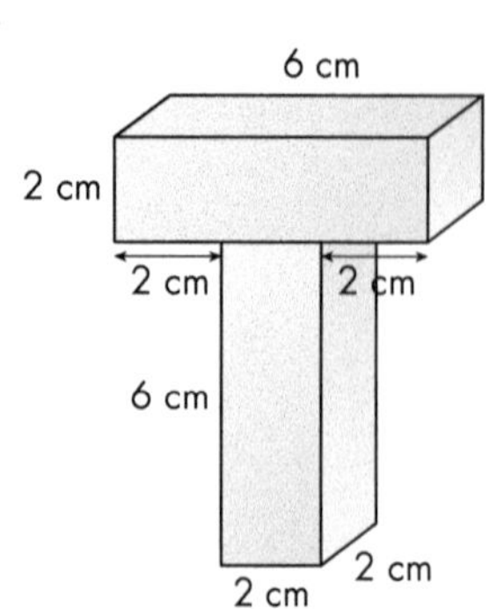

b

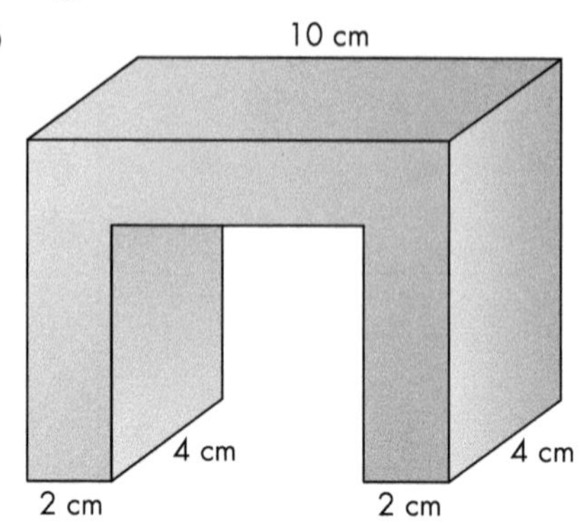

2 For each shape, draw on squared paper:

i the plan

ii the front elevation

iii the side elevation.

a

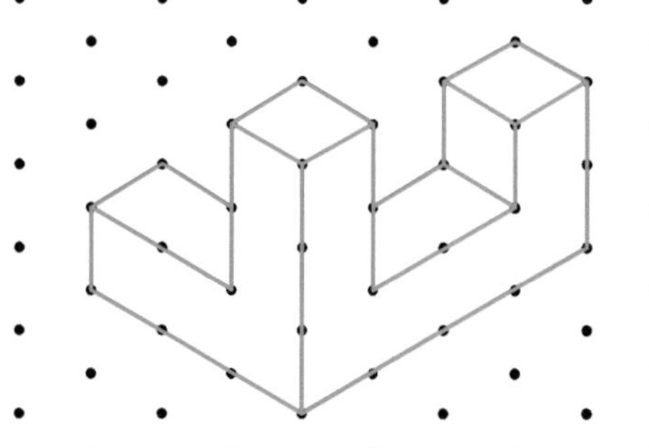

b

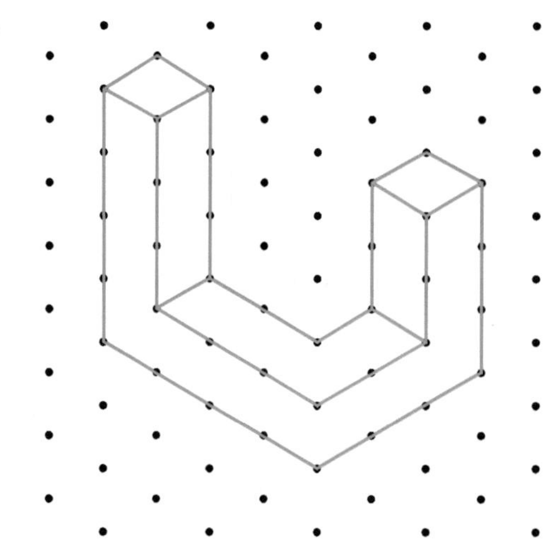

3 Here are three views of a 3D shape.

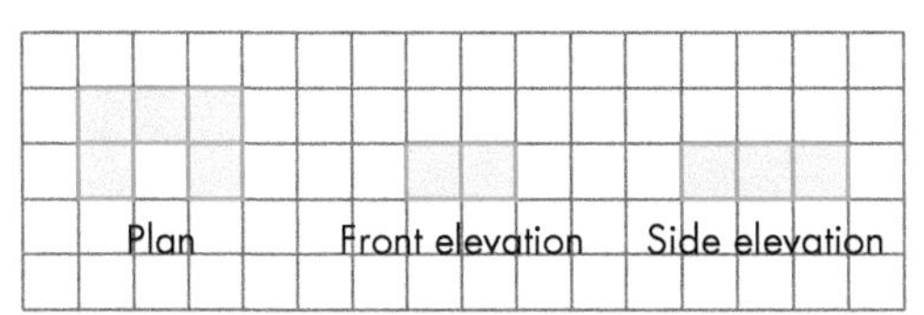

Draw the 3D shape on an isometric grid.

4 This solid shape is made from cubes.

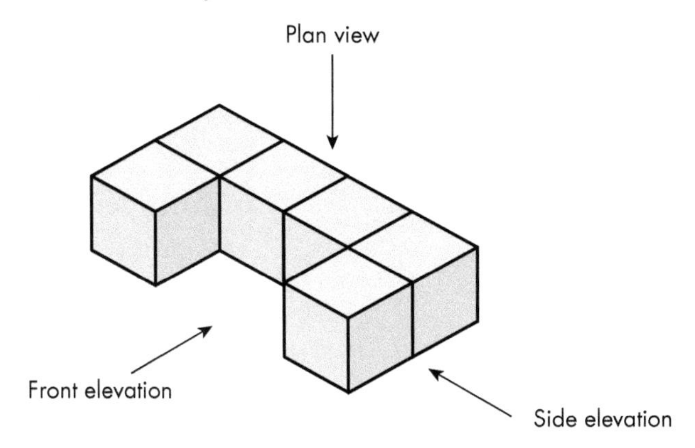

Here are some diagrams of the shape.

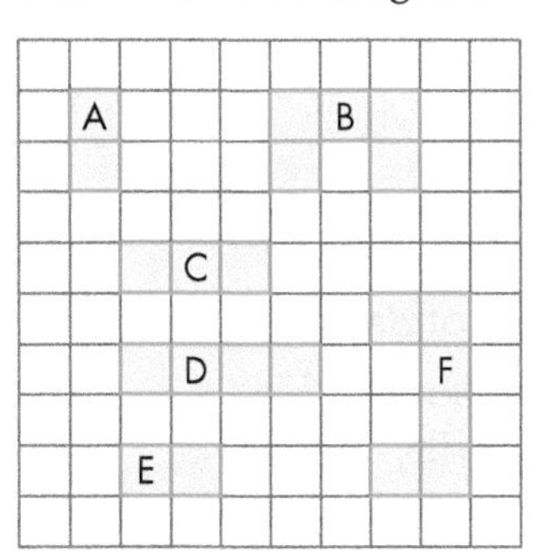

a Which diagram shows the plan view?

b Which diagram shows the front elevation?

3 Statistics: Charts, tables and averages

3.1 Frequency tables

Homework 3A

1 For each set of data:

i draw a frequency table

ii write down the most frequent value

iii write down the total number of values.

a 2, 4, 6, 8, 7, 3, 8, 7, 9, 8, 4, 2, 7, 8, 2, 7, 8, 3, 5, 6, 8

b 9, 7, 5, 2, 1, 8, 9, 7, 2, 8, 6, 7, 4, 4, 2, 5, 6, 8

c 9, 7, 1, 3, 2, 1, 3, 2, 6, 7, 8, 6, 3, 4, 2, 4, 7, 6, 8

d 8, 8, 7, 7, 3, 2, 4, 3, 7, 9, 3, 2, 3, 6, 6, 9, 7

e 3, 4, 5, 6, 6, 7, 8, 3, 6, 5, 2, 4, 3, 6

2 Draw a sensible grouped frequency table for each set of data.

a The age of the oldest sibling of a class of Year 7 students:
12, 13, 12, 18, 12, 15, 24, 27, 22, 23, 17, 18, 18, 20, 26, 18, 26, 18, 17, 27

b The grades of a group of violin students:
8, 3, 1, 2, 8, 3, 1, 2, 7, 8, 3, 1, 7, 6, 5, 4, 5, 5, 6, 7, 8

c The number of times a group of students have been abroad on holiday:
8, 4, 6, 2, 4, 4, 6, 9, 3, 5, 6, 3, 15, 3, 0, 11, 6, 12

d The age a group of work colleagues first rented their own home:
31, 18, 20, 21, 27, 24, 21, 22, 23, 21, 25, 23, 25

3.2 Statistical diagrams

Homework 3B

1 The pictogram shows the number of copies of *The Times* sold by a newsagent in a particular week.

Copies of *The Times* sold		Total
Monday	▬ ▬ ▬	12
Tuesday	▬ ▬ ▬ ▬	
Wednesday	▬ ▬ ▪	
Thursday	▬ ▬ ▬ ▬	
Friday		
Saturday		

a How many newspapers does the symbol ▬ represent?

b Copy the pictogram and complete the totals for Tuesday, Wednesday and Thursday. Remember to include a key.

c The newsagent sold 15 copies on Friday and 22 copies on Saturday.

Use this information to complete the pictogram.

2 The pictogram shows the amount of sunshine in five English holiday resorts on one day in August.

Hours of sunshine in five English resorts

Blackpool	Brighton	Scarborough	Skegness	Torbay
☼ ☼ ☼	☼ ◖	☼ ☼ ☼	☼ ☼	☼ ☼ ☼ ◖

Key ☼ represents 3 hours

a Write down the number of hours of sunshine for each resort.

b Great Yarmouth had $5\frac{1}{2}$ hours of sunshine on the same day.

Explain why this would be difficult to show on this pictogram.

3 The pictogram shows the number of call-outs received by five taxi drivers one evening.

Number of call-outs

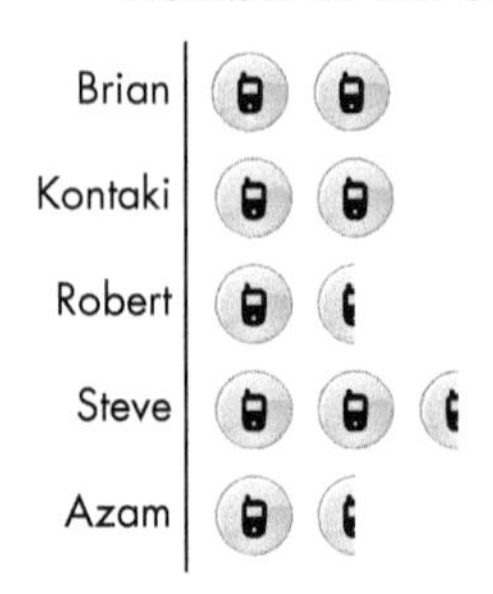

Key represents 10 call-outs

a How many call-outs did each taxi driver have?

b Explain why the symbol used in this pictogram is not really suitable.

c Joanne had 16 call-outs on the same evening. Draw a pictogram that will accurately show the call-outs for the six taxi drivers.

4 Rachel recorded the number of people travelling in each car that passed the end of her street on a particular morning.

Number of people in each car	Frequency
1	30
2	19
3	12
4	5
5 or more	1

Draw a pictogram to illustrate her data.

5 Form 5 did a survey of the number of emails received by all class members to their school email account.

Number of emails received by class members to school email account

Form teacher	
Boys	
Girls	
Teaching assistant	

Key represents 20 emails

a How many emails did:

i the form teacher receive

ii the boys receive?

b The girls in the class received 110 emails. Show this on a copy of the pictogram.

c The teaching assistant received 13 emails. Can this be accurately shown on the pictogram? Give a reason for your answer.

6 Jay asked all his classmates what types of shows their parents had seen at the local theatre in the last six months. His data is shown in the frequency table.

Frequency of theatre visits by genre

	Frequency
Musicals	128
Comedy	48
Drama	80

Draw a pictogram to show this information. Think carefully about how many people each symbol should represent.

7 Alfie is drawing a pictogram to show where students in his school would prefer to hold their activity week. He is using data in this frequency table.

Students' votes for activity week destinations

England	405
Ireland	85
Wales	115
Scotland	325

Explain why choosing one symbol to represent five people is not sensible.

8 Linda asked a sample of people: 'What is your favourite soap opera?'

The bar chart shows their replies.

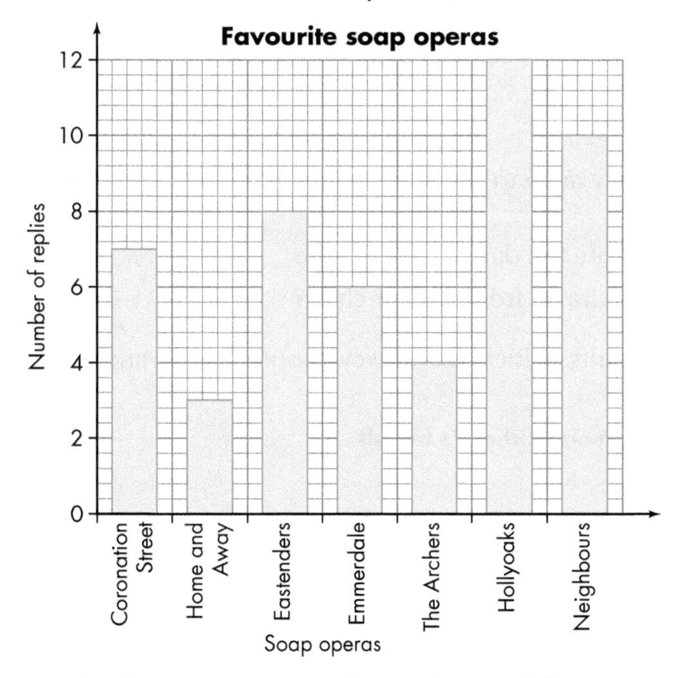

a Which soap opera was chosen by six of the respondents?

b How many people were in Linda's sample?

c Linda collected the data from all her friends in Year 10 at school. Is this a good way to collect data? Give two reasons to support your answer.

9 The bar chart shows the shoe sizes of all the students in form 10KE.

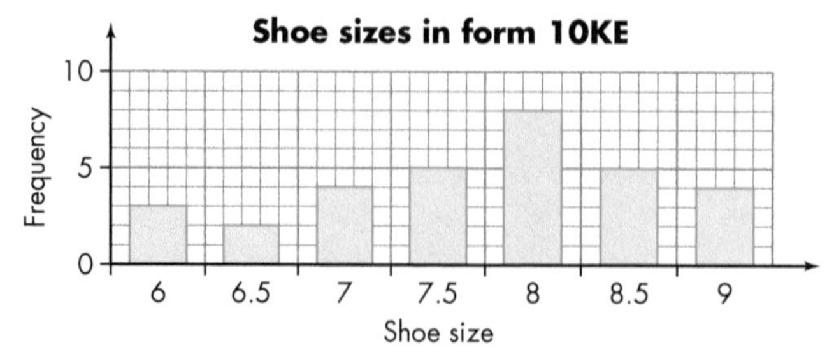

a How many students wear size $7\frac{1}{2}$ shoes?

b How many students are in Form 10KE?

c What is the most common shoe size?

d Can you tell how many boys were in the survey? Explain your answer.

10 The table shows the lowest and highest marks six students gained in a series of mental arithmetic tests.

	Rana	Ben	Chris	Dave	Emma	Ade
Lowest mark	7	11	10	10	15	9
Highest mark	11	12	12	13	16	14

Draw a dual bar chart to illustrate the data.

11 A surgery recorded the times, to the nearest minute, that patients had to wait before seeing a doctor during one morning clinic.

5	12	14	24	32	7	12	35	23	27	13	6
28	4	20	13	40	5	2	11	16	31	10	26
25	30	29	9	12	27	13	20	24	11	14	38

a Draw a grouped frequency table to show the waiting times of the patients. Use the class intervals 1–10, 11–20, …

b Draw a bar chart to illustrate the data.

c What conclusions can be drawn from the bar chart?

12 The bar chart shows the results of Richard's survey about which brand of crisps his friends preferred.

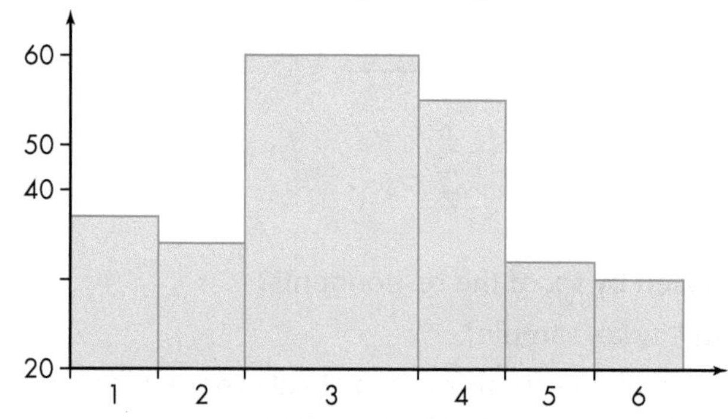

His bar chart is misleading.

a Explain how he could improve it.

b Redraw it, including all your suggested improvements.

13 This table shows the number of accidents involving buses that occurred in Redlow over a six-year period.

Year	2009	2010	2011	2012	2013	2014
Number of accidents	13	17	14	18	13	19

a Draw a pictogram to show this data.

b Draw a bar chart to show this data.

c Evie is writing an article for the local newspaper to show that bus drivers need to drive more carefully. Which diagram should she use? Explain your answer.

14 The bar chart shows the results of 45 students in a mental maths test.

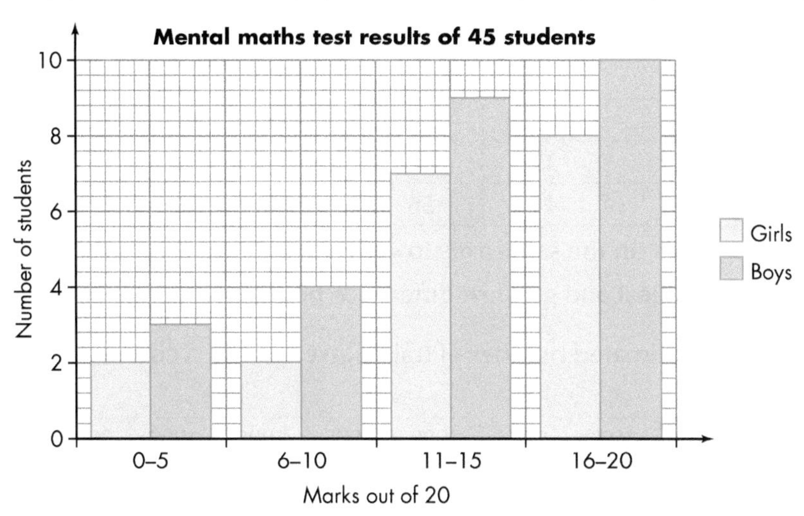

a What is the mean score for boys and girls?

b Why is the bar chart misleading?

15 The bar chart shows the average annual rainfall of two major cities, one in England and one in Wales.

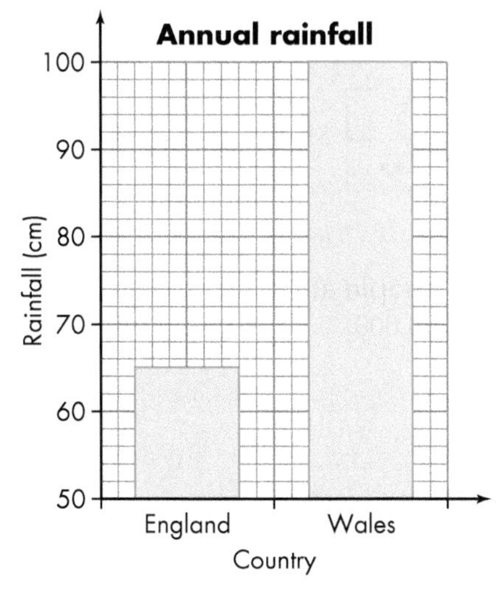

Elwyn says, 'There is more than three times as much rain in Wales as there is in England.'

Is Elwyn correct?

Give reasons to explain your answer.

3.3 Line graphs

Homework 3C

1 The table shows the temperature in Cuzco, Peru over a 24-hour period.

Time	00:00	04:00	08:00	12:00	16:00	20:00	00:00
Temperature (°C)	1	–4	6	15	21	9	–1

a Draw a line graph for the data.

b From your graph estimate the temperature at 18:00.

2 The table shows the value, to the nearest million pounds, of a country's imports and exports.

Year	2004	2005	2006	2007	2008	2009
Imports	22	35	48	51	62	55
Exports	35	41	56	53	63	58

a Draw two line graphs on the same axes to show the country's imports and exports.

b Write down the smallest and greatest difference between the imports and exports.

3 The table shows the estimated number of train passengers in a country at five-yearly intervals.

Year	1970	1975	1980	1985	1990	1995	2000	2005
Passengers (thousands)	210	310	450	570	590	650	690	770

a Draw a line graph to show this data.

b From your graph, estimate the number of passengers in 2010.

c In which five-year period did the number of train passengers increase the most?

d Comment on the trend in the numbers of train passengers. Suggest one possible reason to explain this trend.

4 The height of a baby giraffe is measured at the end of each week.

Week	1	2	3	4	5
Height (cm)	110	160	200	220	235

Estimate the height of the giraffe after six weeks.

5 When plotting a graph to show the number of people attending cricket matches at Headingley, Kevin decided to start his graph at 18 000.

Explain why he might have made this decision.

6 The line graph shows the monthly average exchange rate of £1 in Japanese Yen over a six-month period.

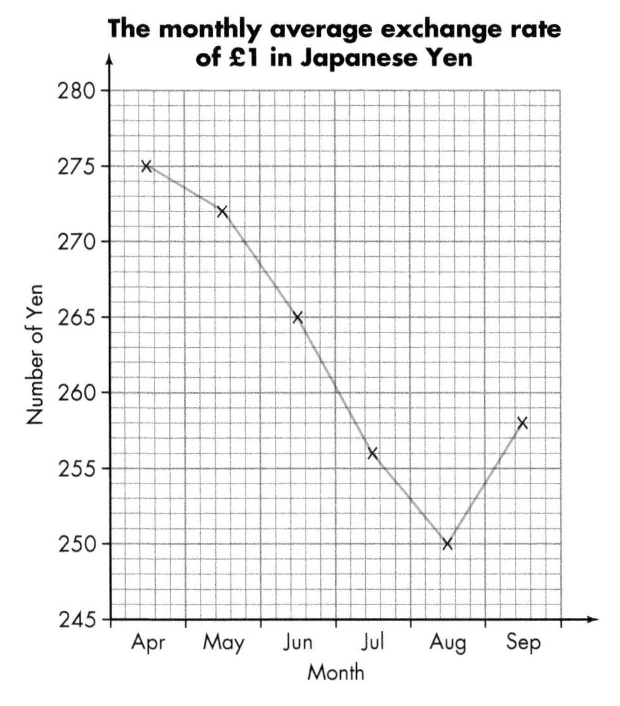

a In which month was the exchange rate lowest? What was that value?

b By how much did the exchange rate fall between April and August?

c Between which two months did the exchange rate fall the most?

d Mr Hargreaves changed £200 into Yen during July. How many Yen did he receive?

3.4 Statistical averages

Homework 3D

Example

The MOde is the number or numbers that occur Most Often.

2, 4, 6, 8, 7, 3, 8, 7, 9, 8, 4, 2, 7, 8, 2, 7, 8, 3, 5, 6, 8

Number	Frequency
2	3
3	2
4	2
5	1
6	2
7	4
8	6
9	1

The number that occurs most often is 8 as it has the highest frequency.

1 Write down the mode for each set of data.

a 3, 1, 2, 5, 6, 4, 1, 5, 1, 3, 6, 1, 4, 2, 3, 2, 4, 2, 4, 2, 6, 5

b 17, 11, 12, 15, 11, 13, 18, 14, 17, 15, 13, 15, 16, 14

c 110, 10, 101, 10, 111, 110, 11, 101, 11, 111, 11, 101, 101, 111

d 1, −3, 3, 2, −1, 1, −3, −2, 3, −1, 2, 1, −1, 1, 2

e $7, 6\frac{1}{2}, 6, 7\frac{1}{2}, 8, 5\frac{1}{2}, 6\frac{1}{2}, 6, 7, 6\frac{1}{2}, 7, 6\frac{1}{2}, 6, 7\frac{1}{2}$

2 Write down the modal category for each set of data.

a I, A, E, U, A, O, A, E, U, A, I, A, E, I, E, O, E, I, E, O

b ITV, C4, BBC1, C5, BBC2, C4, BBC1, C5, ITV, C4, BBC1, C4, ITV

c ↑, →, ↑, ←, ↓, →, ←, ↑, ←, →, ↓, ←, ←, ↑, →, ↓

d ♥, ♣, ♦, ♣, ♠, ♥, ♣, ♦, ♣, ♦, ♥, ♠

e ¥, €, £, €, $, £, ¥, €, £, $, €, £, $, €

3 The table shows the number of days each week that Bethan travelled to Manchester on business.

Days	0	1	2	3	4	5
Number of weeks	17	2	4	13	15	1

Explain how you would work out the median number of days per week that Bethan travelled to Manchester.

4 The diagram shows the number of eggs laid by a farmer's hens in five days.

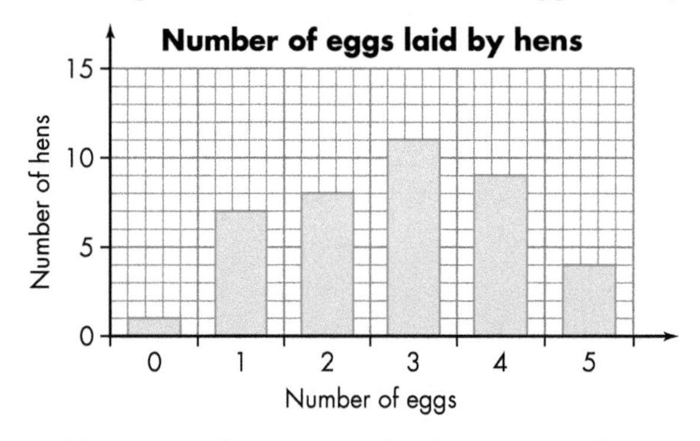

a How many hens does the farmer have?

b What is the mode of the number of eggs laid?

c How many eggs were laid altogether?

5 What is the mode of these numbers?

5	7	1	3	7	8	7	9	2	3	1	6
5	1	4	6	8	3	9	3	2	2	4	5
7	5	8	7	3	7	8	2	3	5	6	6
3	4	3	1	2	3	5	6	5	4	4	

6 This table shows favourite pets of students in form 10E.

	Dog	Cat	Rabbit	Guinea pig	Other
Boys	8	1	3	2	0
Girls	3	3	7	3	1

a How many students are in form 10E?

b What is the modal pet for:

i boys

ii girls

iii the whole form?

c After two new students join the form, the modal pet for the whole form is rabbit.

Which of the following statements are true?

i Both students like cats.

ii Both students like rabbits.

iii One student likes rabbits; the other likes cats.

iv You cannot tell what pets they like.

7 The manufacturers of cars in an office car park were recorded:

Ford, Peugeot, Austin, Honda, Ford, Austin, Peugeot, Ford, Peugeot, Austin, Ford, Toyota, Honda, Honda, Peugeot, Ford, Honda, Austin, Honda, Peugeot, Toyota, Toyota, Toyota, Austin, Toyota

What is the problem with identifying the mode of this data?

8 The grouped frequency table shows the number of marks Class 11BD obtained in a test.

Number of marks	Frequency
1–5	1
6–10	2
11–15	4
16–20	8
21–25	10
26–30	5

a How many students are in Class 11BD?

b Write down the modal group.

c Leo looked at the table and said that at least one person got full marks. Explain why he could be wrong.

9 The data shows the times, to the nearest minute, that 30 shoppers had to wait in the queue at a supermarket checkout.

1	3	8	12	7	4	0	9	10	15	8	1	2	7	4
2	4	7	1	0	5	4	8	4	10	7	5	4	1	5

a Copy and complete the grouped frequency table.

Time in minutes	Tally	Frequency
0–3		
4–7		
8–11		
12–15		

b Draw a bar chart to illustrate the data.

c How many shoppers had to wait more than 7 minutes?

d Write down the modal class.

e How could the supermarket manager decrease the waiting time of the shoppers?

Homework 3E

Example

The median is the middle value in an ordered list.

To find the median for this list:

2, 3, 5, 6, 1, 2, 3, 4, 5, 4, 6

First write the list in numerical order.

1, 2, 2, 3, 3, 4, 4, 5, 5, 6, 6

Then find the middle number.

There are 11 numbers in the list, so the middle of the list is the sixth number. Therefore, the median is 4.

1 Write down the median of each set of data.

a 18, 12, 15, 19, 13, 16, 10, 14, 17, 20, 11

b 22, 28, 42, 37, 26, 51, 30, 34, 43

c 1, −3, 0, 2, −4, 3, −1, 2, 0, 1, −2

d 12, 4, 16, 12, 14, 8, 10, 4, 6, 14

e 1.7, 2.1, 1.1, 2.7, 1.3, 0.9, 1.5, 1.8, 2.3, 1.4

2 The masses of 11 men in a local rugby team are shown below.

81 kg, 85 kg, 82 kg, 71 kg, 62 kg, 63 kg, 62 kg, 64 kg, 70 kg, 87 kg, 74 kg

a Work out the median mass.

b Write down the modal mass.

c Which is the better average to use? Give a reason for your answer.

3 The bar chart shows the scores obtained in 20 throws of a dice.

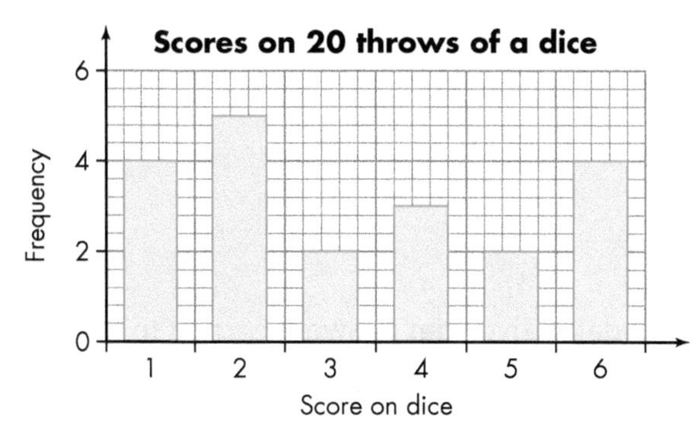

a Write down the modal score.

b Work out the median score. Remember to consider all the scores.

c Do you think that the dice is biased? Explain your answer.

4 Look at this list of numbers.

3 3 4 7 9 10 11 14 14 15 19

a Write down four more numbers to make the median 11.

b Write down six more numbers to make the median 11.

c What is the least number of values you must add to the list to make the median 3?

5 Explain why the median is not a good average for this data set.

5 g 7 g 10 g 200 g 4 kg

6 **a** Write down a list of seven numbers that has a median of 10 and a mode of 20.

b Write down a list of eight numbers that has a median of 10 and a mode of 20.

7 21 students take a science test. Their results, out of 100, are shown below.

45, 62, 27, 77, 40, 55, 80, 87, 49, 57, 35, 52, 59, 78, 48, 67, 43, 68, 38, 72, 81

a What is the median?

b Is the median a good average for this data?

Homework 3F

The mean of a set of data is the sum of all the values in the set divided by the total number of values in the set.

That is, $\text{mean} = \dfrac{\text{sum of all values}}{\text{total number of values}}$

To calculate the mean of: 4, 8, 7, 5, 9, 4, 8, 3

Work out the sum of all the values: $4 + 8 + 7 + 5 + 9 + 4 + 8 + 3 = 48$

Work out the total number of values: 8

Use the formula: $48 \div 8 = 6$

1 Work out the mean for each set of data.

a 4, 2, 5, 8, 6, 4, 2, 3, 5, 1

b 21, 25, 27, 20, 23, 26, 28, 22

c 324, 423, 342, 234, 432, 243

d 2.5, 3.6, 3.1, 4.2, 3.5, 2.9

e 1, 4, 3, 0, 1, 2, 5, 0, 2, 4, 2, 0

2 Calculate the mean for each set of data. Give your answers correct to one decimal place.

a 17, 24, 18, 32, 16, 28, 20

b 92, 101, 98, 102, 95, 104, 99, 96, 103

c 9.8, 9.3, 10.1, 8.7, 11.8, 10.5, 8.5

d 202, 212, 220, 102, 112, 201, 222

e 4, 2, –1, 0, 1, –3, 5, 0, –1, 4, –2, 1

3 A group of eight people took part in a marathon to raise money for charity. Their finishing times were:

2 hours 40 minutes, 3 hours 6 minutes, 2 hours 50 minutes, 3 hours 25 minutes,

4 hours 32 minutes, 3 hours 47 minutes, 2 hours 46 minutes, 3 hours 18 minutes

Calculate their mean time in hours and minutes.

4 The monthly wages of 11 full-time restaurant staff are:

£820, £520, £860, £2000, £800, £1600, £760, £810, £620, £570, £650

a Work out their median wage.

b Calculate their mean wage.

c How many of the staff earn more than:

i the median wage **ii** the mean wage?

d Which is the better average to use? Give a reason for your answer.

5 Maiden End scored 30 goals in 10 games of football.

What is the least number of goals they must score in their next match to increase their average score?

6 The table shows five couples' scores in a dance competition.

	Kath & Brian	Tom & Helen	Joe & Nik	Azan & Phyllis	David & Hannah
Tango	10	6	4	8	6
Salsa	6	8	3	8	6
Ballroom	8	4	4	8	8

a Kath claims that the salsa was the hardest dance. Work out the mean score for each dance. Is Kath correct?

b Which couple obtained the score closest to the mean in all three dances?

c How many couples were above average in all three dances?

7 Two families took part in a tug o' war competition.

Key family	Charlton family
Brian weighed 58 kg	David weighed 60 kg
Ann weighed 32 kg	Hannah weighed 56 kg
Steve weighed 49 kg	Pete weighed 42 kg
Alison weighed 39 kg	Barbara weighed 76 kg
Jill weighed 64 kg	Chris weighed 71 kg
Holly weighed 75 kg	Julie weighed 39 kg
Albert weighed 52 kg	George weighed 22 kg

Each family had to choose four members with a mean weight between 45 kg and 50 kg.

Choose two possible teams, one from each family.

8 The table shows the percentage scores of 10 students in Paper 1 and Paper 2 of their GCSE Mathematics examination.

	Ann	Bridget	Carole	Daniel	Edwin	Fay	George	Hannah	Imman	Joseph
Paper 1	72	61	43	92	56	62	73	56	38	67
Paper 2	81	57	49	85	62	61	70	66	48	51

a Calculate the mean mark for Paper 1.

b Calculate the mean mark for Paper 2.

c Which student scored marks closest to the mean on both papers?

d How many students were above the mean mark on both papers?

9 The number of runs scored by a cricketer in seven innings were: 48, 32, 0, 62, 11, 21, 43

a Calculate the mean number of runs per innings.

b After eight innings, her mean score increased to 33 runs per innings.

How many runs did she score in her eighth innings?

Homework 3G

1 **a** Work out the mode, the median and the mean for each set of data.

i 6, 4, 5, 6, 2, 3, 2, 4, 5, 6, 1

ii 14, 15, 15, 16, 15, 15, 14, 16, 15, 16, 15

iii 31, 34, 33, 32, 46, 29, 30, 32, 31, 32, 33

b Decide on the best average to represent each set of data. Give a reason for each answer.

2 A supermarket sells oranges in bags of 10.

The masses of each orange in a selected bag are shown below.

134 g, 135 g, 142 g, 153 g, 156 g, 132 g, 135 g, 140 g, 148 g, 155 g

a Work out:

i the mode **ii** the median **iii** the mean mass.

b The supermarket wants to state the average mass of the oranges. Which of the three averages would you advise the supermarket to use? Explain why.

3 Three players were hoping to be chosen for the hockey team.

The table shows the goals they scored in each of the last few games they played.

Adam	4, 2, 3, 2, 3, 2, 2
Faisal	4, 2, 4, 6, 2
Maya	4, 0, 4, 0, 1

The teacher said she would select the players with the best average scores. By which average would each player choose to be selected?

4 The masses, in kilograms, of players in a school football team are listed below.

68, 72, 74, 68, 71, 78, 53, 67, 72, 77, 70

a What is the median mass of the team?

b Calculate the mean mass of the team.

c Which average is the better one to use? Explain why.

5 Jez is a member of a local quiz team. The number of points he has scored in the last eight weeks are listed below.

62, 58, 24, 47, 64, 52, 60, 65

a What is the median for the number of points he scored over the eight weeks?

b Work out the mean for the number of points he scored over the eight weeks.

c The team captain wanted to know the average for each member of the team.

Which average would Jez use? Give a reason for your answer.

6 A class of students took a test.

When talking about the results, the teacher said the average mark was 32. One of the students said it was 28.

Explain how they could both be correct.

Homework 3H

Example

The range for a set of data is the difference between the highest value and the lowest value in the set.

Rachel's marks in 10 mental arithmetic tests were 4, 4, 7, 6, 6, 5, 7, 6, 9, 6.
Her mean mark is $60 \div 10 = 6$ marks, and her range is $9 - 4 = 5$ marks.

Robert's marks in the same tests were 6, 7, 6, 8, 5, 6, 5, 6, 5, 6.
His mean mark is $60 \div 10 = 6$ marks, and his range is $8 - 5 = 3$ marks.

Although the means are the same, Robert has a smaller range. This shows that Robert's results are amore consistent.

1 Work out the range for each set of data.

a 23, 18, 27, 14, 25, 19, 20, 26, 17, 24

b 92, 89, 101, 96, 100, 96, 102, 88, 99, 95

c 14, 30, 44, 25, 36, 27, 15, 42, 27, 12, 40, 31, 34, 24

d 3.2, 4.8, 5.7, 3.1, 3.8, 4.9, 5.8, 3.5, 5.6, 3.7

e 5, –4, 0, 2, –5, –1, 4, –3, 2, 2, 0, 1, –4, 0, –2

2 The table shows the ages of a group of students on an outdoor activity course.

Age	14	15	16	17	18	19
Number of students	2	3	8	5	6	1

a How many students were on the course?

b Write down the modal age of the students.

c What is the range of their ages?

d Draw a bar chart to illustrate the data.

3 A travel brochure shows the average monthly temperatures, in °F, for the island of Crete.

Month	April	May	June	July	August	September	October
Temperature (°F)	68	74	78	83	82	75	72

a Calculate the mean of these temperatures.

b Write down the range of these temperatures.

c The mean temperature for the island of Corfu was 77°F and the range was 20°F. Compare the temperatures for the two islands.

4 The table shows the daily attendance of three forms of 30 students over a week.

	Monday	Tuesday	Wednesday	Thursday	Friday
Form 10KG	25	25	26	27	27
Form 10RH	22	23	30	26	24
Form 10PB	24	29	28	25	29

a Calculate the mean attendance for each form.

b Write down the range for the attendance of each form.

c Which form had:

i the best attendance

ii the most consistent attendance?

Give reasons for your answers.

5 The table shows the amounts taken by a sandwich shop over a three-week period.

	Mon	Tue	Wed	Thurs	Fri
Week 1	£139	£190	£30	£219	£343
Week 2	£132	£188	£19	£203	£339
Week 3	£151	£194	£43	£212	£299

a Calculate the mean amount taken each week.

b Work out the range for each week.

c What can you say about the total amounts taken for each of the three weeks?

6 Look at this list of children, showing their current ages and weights.

Name	Age (years)	Weight (kg)
Olly	12	40
Elinor	7	23
Latham	11	36
Aimee	13	45
Chelsie	6	20
Kemunto	7	23
Kai-Yan	6	18
Anna	5	15
Zoe	12	38
Kesia	10	32

a Write down an age range of 4 which includes four children.

b Write down the smallest range of weights that includes three children.

7 The age range of a football team is 1 year.

What type of football team would you expect this to be? Explain your answer.

8 **a** Write down three numbers that have a range of 3 *and* a mean of 3.

b Write down three numbers that have a range of 3, a median of 3 *and* a mean of 3.

3.5 Stem-and-leaf diagrams

Homework 3I

1 The stem-and-leaf diagram shows the number of TVs a retailer sold each day over a three-week period.

TVs sold each day over a three-week period

Stem	Leaf
1	2 8 9
2	0 2 4 4 4 4 5 7 8 8 9
3	1 2 4 8

Key 1 | 2 represents 12 TVs

a What is the greatest number of TVs the retailer sold in one day?

b What is the most common number of TVs sold daily?

c What is the difference between the greatest number and the least number of TVs sold?

2 The stem-and-leaf diagram shows the ages of a group of people waiting for a train.

The ages of a group of people waiting for a train

1	6 8 9
2	4 7 8 8
3	0 2 4 5 6
4	2 5 5 6 8
5	0 4 8

Key 1 | 6 represents an age of 16

a How many people were waiting for a train?

b What is the age of the youngest person?

c What is the difference in age between the youngest person and oldest person?

3 The speeds, in miles per hour, of 30 vehicles travelling on a motorway are shown below.

62 45 70 58 68 70 75 80 72 65 40 55 65 72 38

70 75 68 50 48 65 60 68 72 70 45 68 69 68 60

a Show the data on an ordered stem-and-leaf diagram. (Remember to include a key.)

b What is the most common speed?

c What is the range of the speeds?

4 A teacher has given her class a spelling test. She wants to compare the boys' results with the girls' results so she lists the data in the back-to-back stem-and-leaf diagram below.

Spelling test results

Boys		Girls
5 3 2 1	1	4 8 8
8 8 5 1	2	1 1 2 7 7 7
6 5 5 5	3	0 0 4

Key Boys: 2 | 1 represents 12 correct
Girls: 1 | 4 represents 14 correct

a What was the highest score for the boys?

b What was the lowest score for the girls?

c What was the most common score for:

i boys **ii** girls?

d Overall, who did better in the test: boys or girls? Give a reason for your answer.

5 The following data is gathered about the scores from an IQ test in Form 10EF.

Boys: 117 105 128 132 110 108 123 114 128 110

Girls: 122 137 113 118 131 129 104 120 117 134

Use a diagram to represent this information so that the two sets of scores can be compared.

6 Godwin was asked to create a stem-and-leaf diagram from some numerical data, but he said, 'It is impossible to do this sensibly!'

Give an example of 10 items of numerical data that could not sensibly be represented in a stem-and-leaf diagram.

4 Geometry and measures: Angles

4.1 Angle facts

Homework 4A

Example

Work out value of x in the diagram.

These angles are around a point and add up to 360°.

So $x + x + 40 + 2x - 20 = 360°$

$4x + 20 = 360°$

$4x = 340°$

$x = 85°$

1 Work out the size of the angle marked x in each diagram.

a

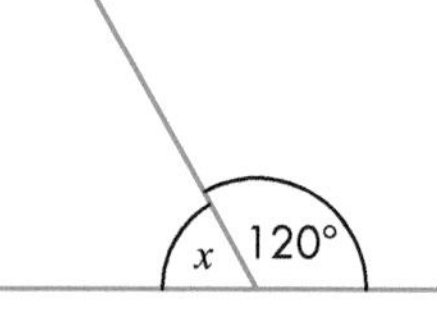

b

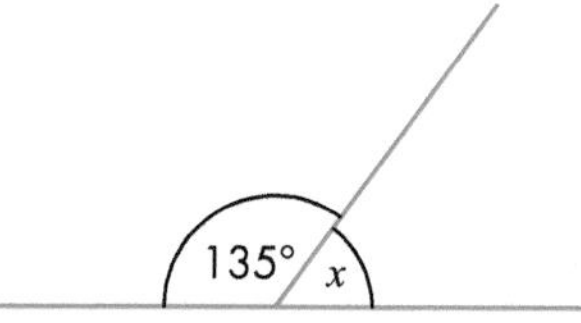

c

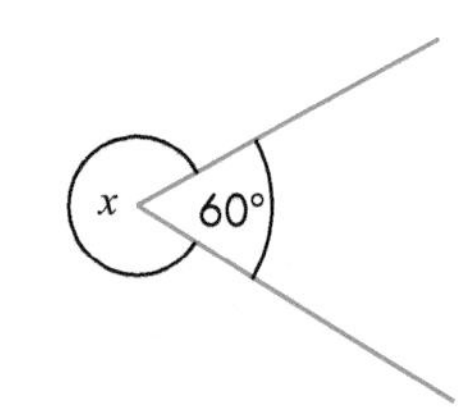

d

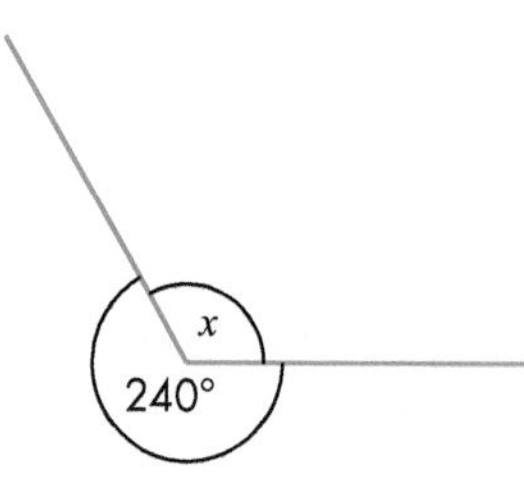

e

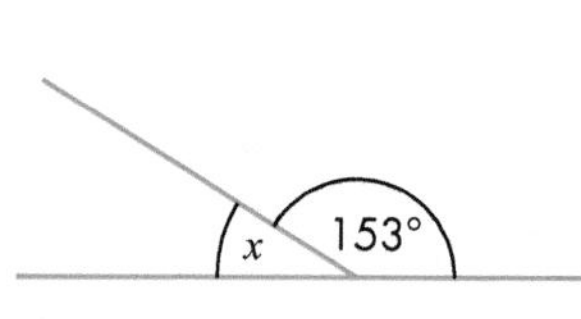

f

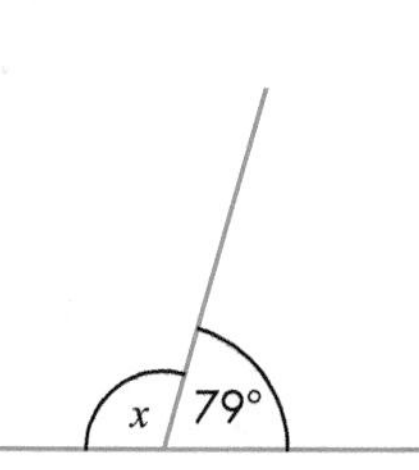

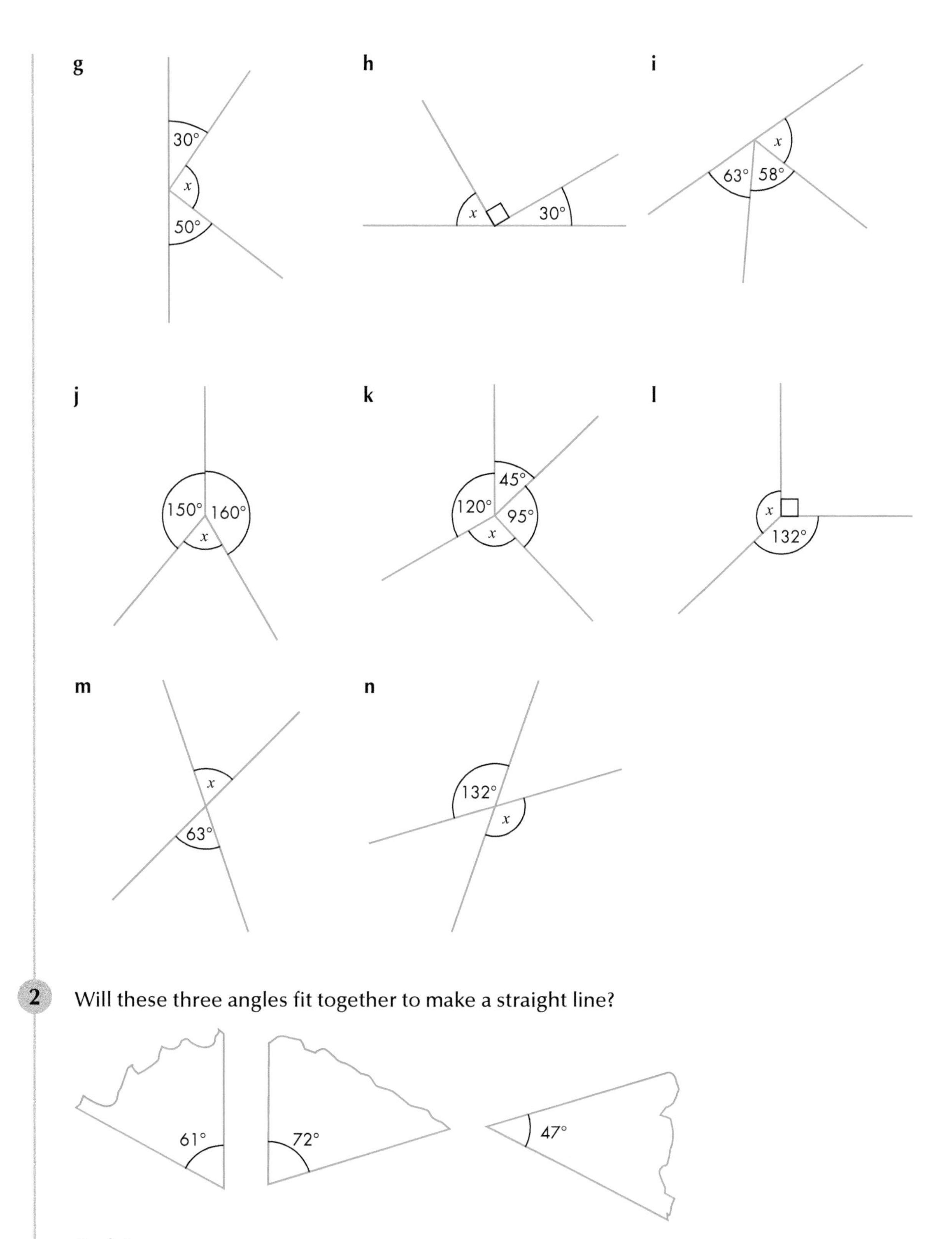

2 Will these three angles fit together to make a straight line?

Explain your answer.

3 Calculate the value of x in each diagram.

a

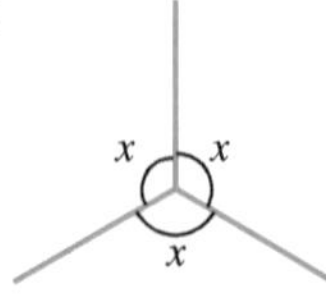

b

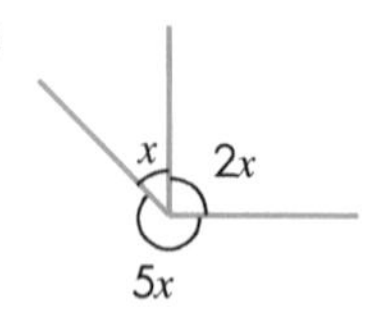

c

3x – 20°
x + 50°
2x + 30°

4 Calculate the value of x in each diagram.

a

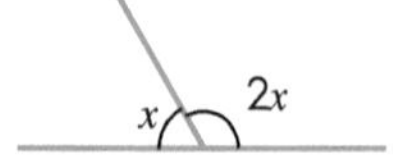

b

x +10°
x +20°

c

2x + 10°
x – 10°
x + 20°

5 Calculate the value of x and y in each diagram. Calculate x first each time.

a

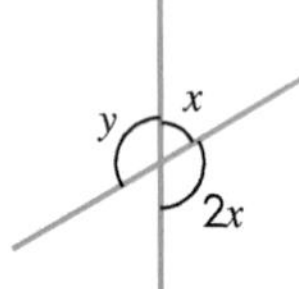

b

5x – 10°
x + 10°
y

c

3x – 10°
y
2x – 30°

6 These paving stones are in the shape of regular hexagons. Explain why they fit together without leaving any gaps.

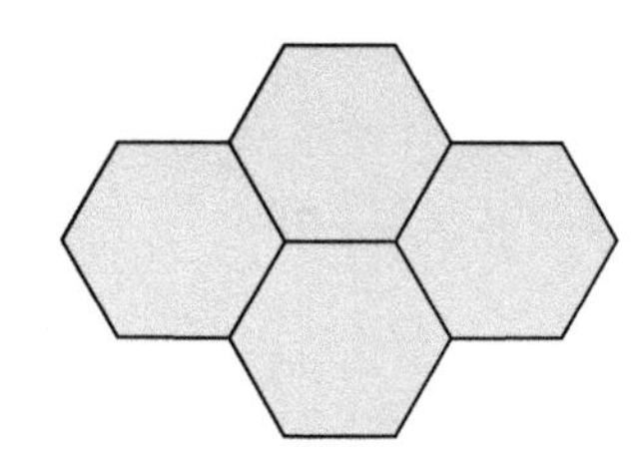

Hints and tips All the angles inside a regular hexagon are 120°.

4.2 Triangles

Homework 4B 

1 Work out the size of the angle marked with a letter in each diagram.

a

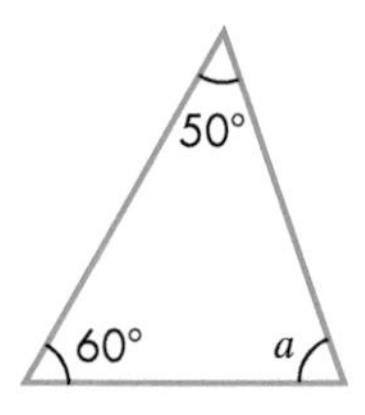

b

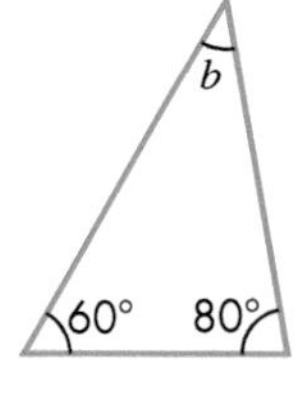

c

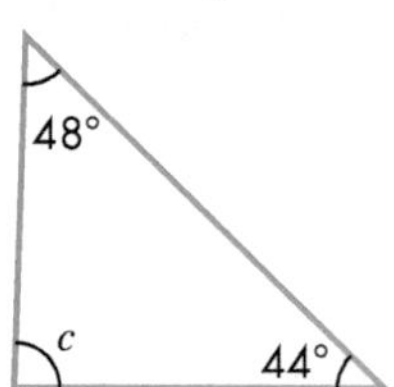

d

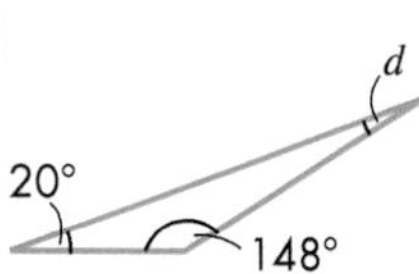

e

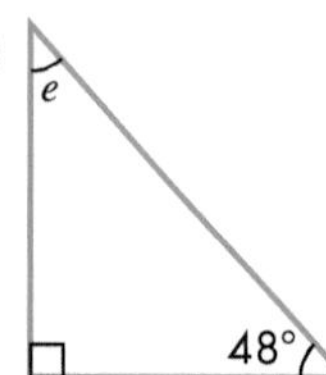

f

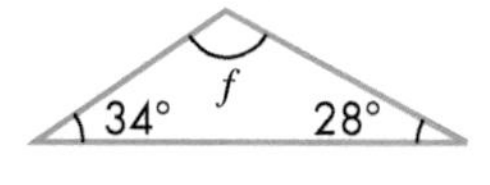

2 Which of these sets of angles form the three interior angles of a triangle? How do you know?

a 15°, 85° and 80° **b** 40°, 60° and 90° **c** 25°, 25° and 110°

d 40°, 40° and 100° **e** 32°, 37° and 111° **f** 61°, 59° and 70°

3 Each set of angles form the three interior angles of a triangle. Work out the value of the angle given by a letter in each set.

a 40°, 70° and a° **b** 60°, 60° and b° **c** 80°, 90° and c°

d 65°, 72° and d° **e** 130°, 45° and e° **f** 112°, 27° and f°

4 All the interior angles of a particular triangle are the same.

a What size is each angle?

b What is the name of this type of triangle?

c What is special about the sides of this triangle?

5 In the triangle on the right, two of the angles are the same.

a Work out the size of the angles marked x.

b What is the name of this type of triangle?

c What is special about the sides AB and AC of this triangle?

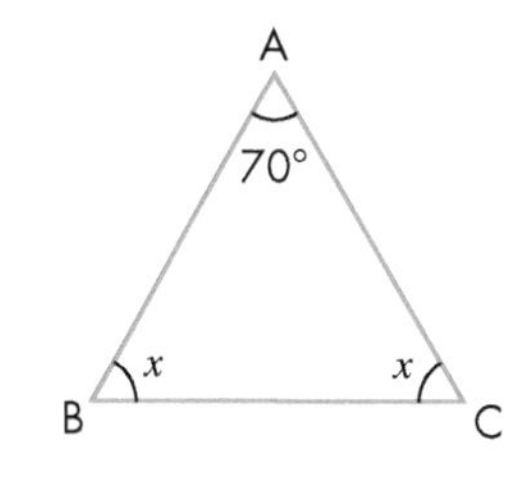

6 In the triangle on the right, the size of the angle at C is twice the size of the angle at A. Work out the size of the angles x and y.

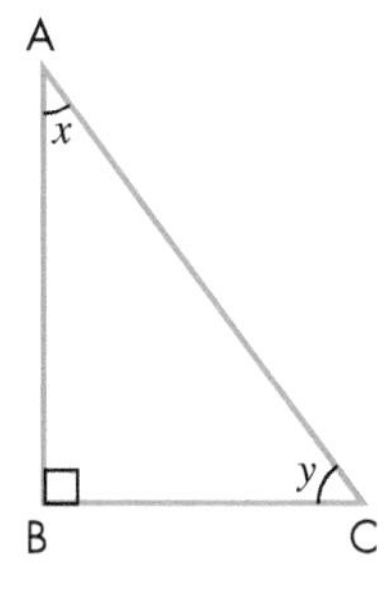

7 Work out the size of the angle marked with a letter in each diagram.

a

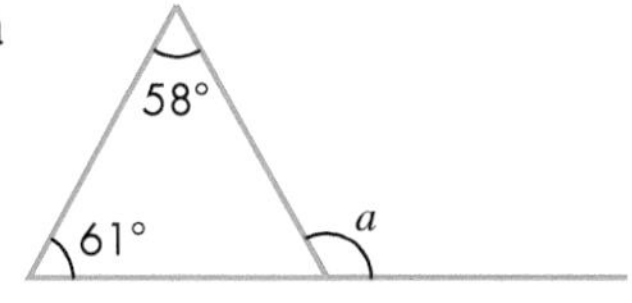

b

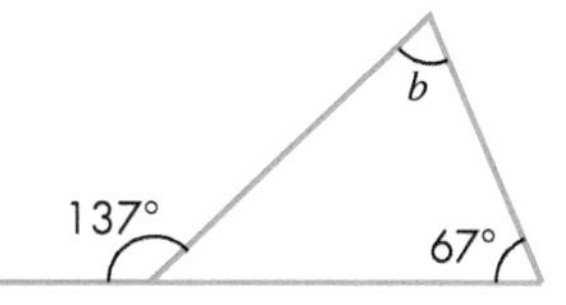

8 The diagram shows a guy rope attached to a mast on a marquee.

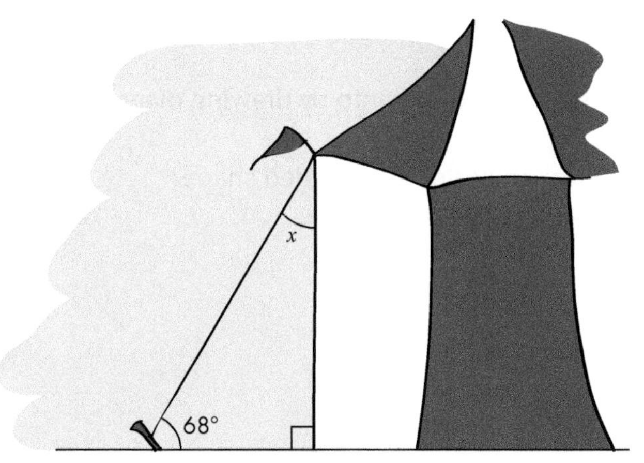

Work out the size of angle x marked on the diagram.

9 Here are five statements about triangles. Some are true and some are false.

A A triangle can have three acute angles.

B A triangle can have two acute angles and one obtuse angle.

C A triangle can have one acute angle and two obtuse angles.

D A triangle can have two acute angles and one right angle.

E A triangle can have one acute angle and two right angles.

If a statement is true, draw a sketch of a possible triangle.

If a statement is false, explain why.

10 Explain why angle a is 25 degrees.

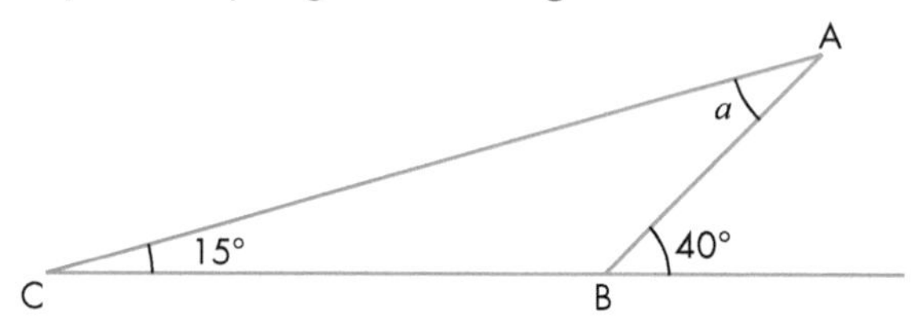

4.3 Angles in a polygon

Homework 4C

1 A regular octagon has eight sides.

a How many triangles can this shape be split into by drawing diagonals from one vertex?

b What is the sum of the interior angles of the octagon?

c What is the size of one interior angle?

2 A regular shape has 12 sides.

a How many triangles can this shape be split into by drawing diagonals from one vertex?

b What is the sum of the interior angles of the 12-sided shape?

c What is the size of one interior angle?

3 A regular shape has 30 sides.

a How many triangles can this shape be split into by drawing diagonals from one vertex?

b What is the sum of the interior angles of the 30-sided shape?

c What is the size of one interior angle?

Homework 4D

1 Work out the size of the angle marked with a letter in each quadrilateral.

a

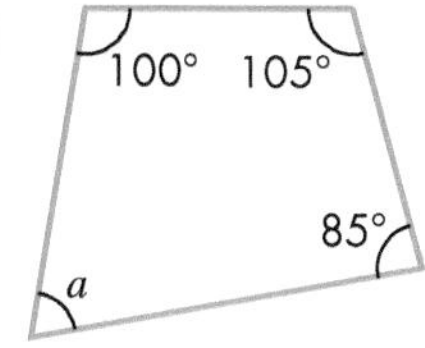

b

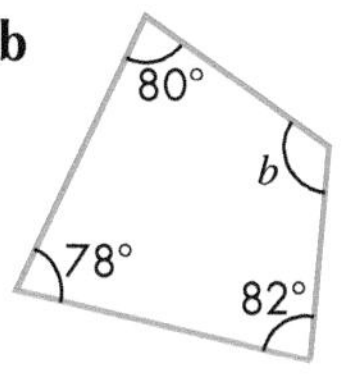

c

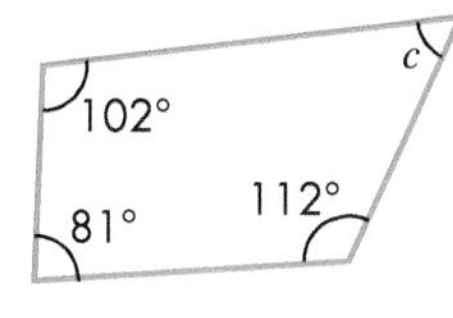

d

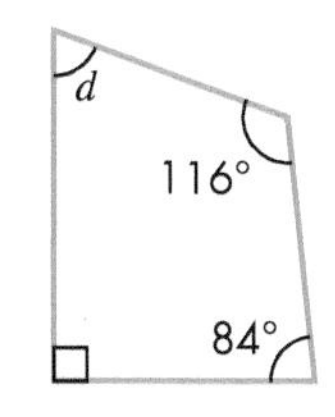

e

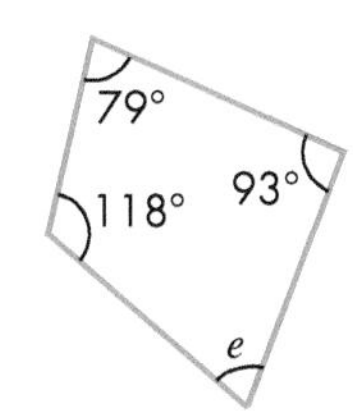

f

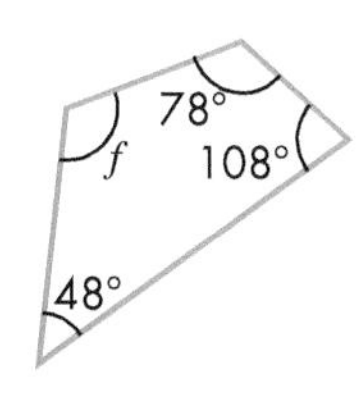

2 Which of these sets of angles form the four interior angles of a quadrilateral? How do you know?

a 125°, 65°, 70° and 90°

b 100°, 60°, 70° and 130°

c 85°, 95°, 85° and 95°

d 120°, 120°, 70° and 60°

e 112°, 68°, 32° and 138°

f 151°, 102°, 73° and 34°

3 Each set of angles form the four interior angles of a quadrilateral. Calculate the value of the angle given by a letter in each set.

a 110°, 90°, 70° and a°

b 100°, 100°, 80° and b°

c 60°, 60°, 160° and c°

d 135°, 122°, 57° and d°

e 125°, 142°, 63° and e°

f 102°, 72°, 49° and f°

4 For this quadrilateral:

a Work out the size of angle x.

b What type of angle is x?

c What is the special name of a quadrilateral like this?

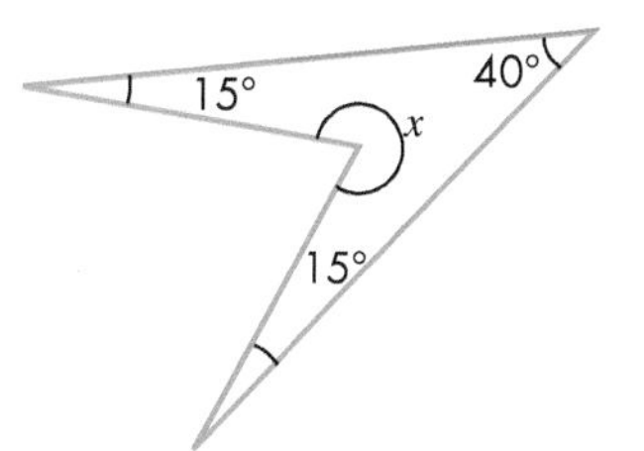

5 **a** Draw a diagram to show why the sum of the interior angles of any pentagon is 540°.

b Work out the size of the angle x in the pentagon.

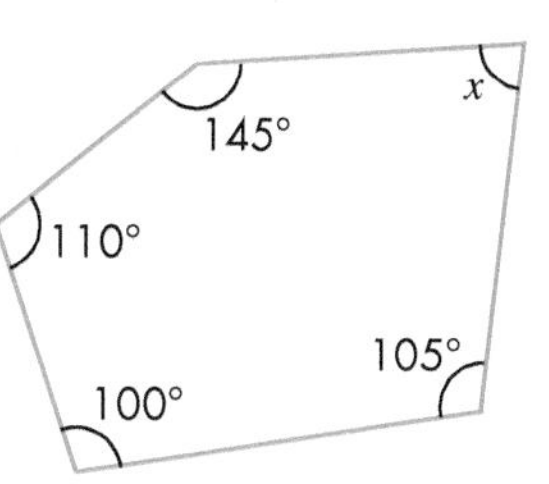

6 Calculate the size of the angle marked with a letter in each polygon.

a

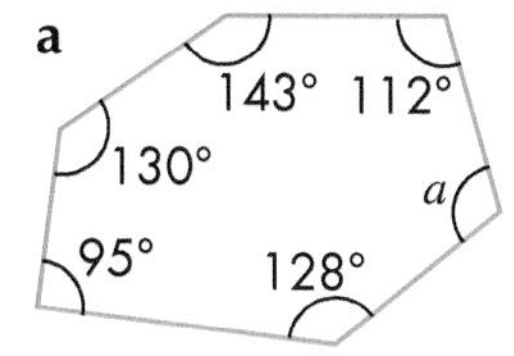

b

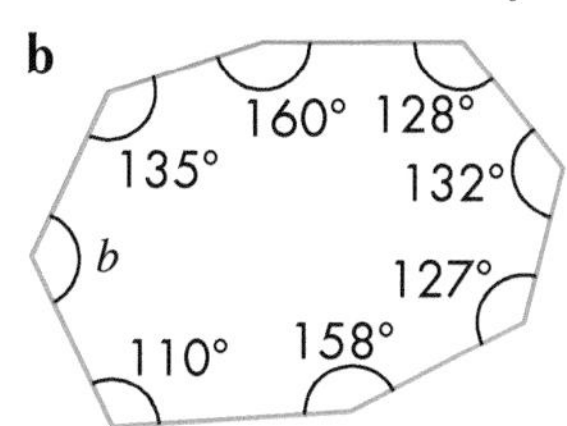

7 Jamal is cutting metal from a rectangular sheet to make this sign.

He needs to cut the two angles marked x accurately. What is the size of each one?

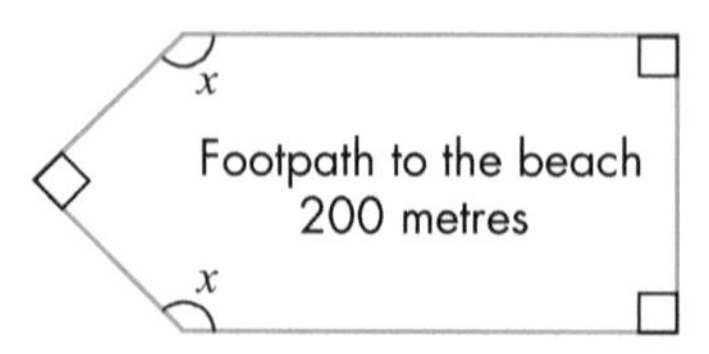

8 Work out the value of x in the diagram.

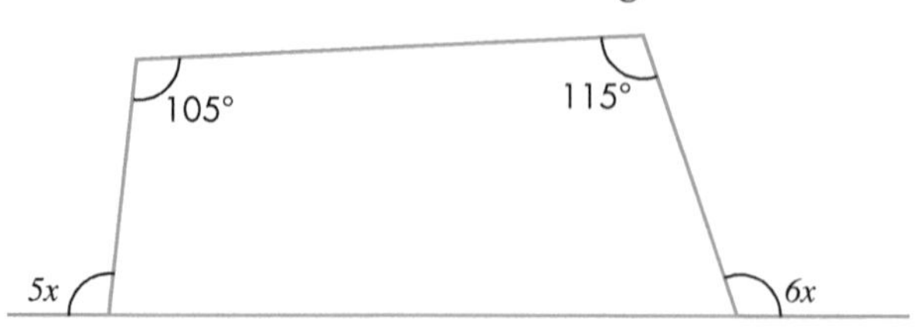

9 The diagram shows a quadrilateral.

Paul says that the size of angle x is 52°.

Explain why Paul is wrong.

Work out the correct value for x.

4.4 Regular polygons

Homework 4E

1 For each regular polygon below, work out the size of the interior angle x and the exterior angle y.

a

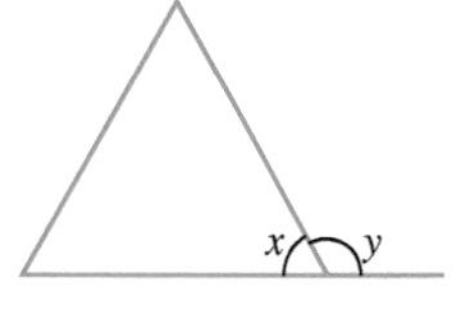

b

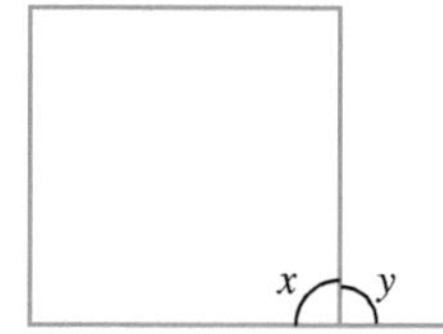

c

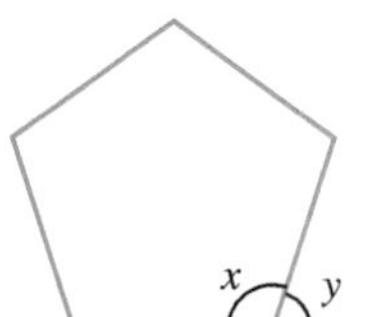

d

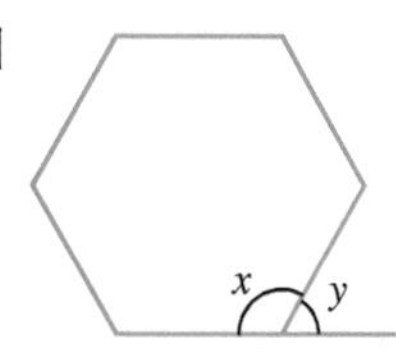

e

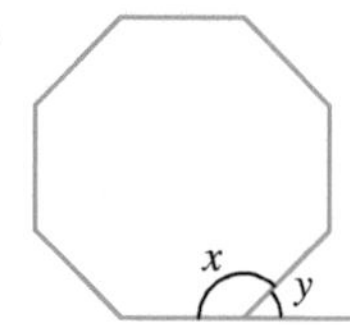

2 Calculate the number of sides of the regular polygon with an exterior angle of:

a 20° **b** 30° **c** 18° **d** 4°

3 Calculate the number of sides of the regular polygon with an interior angle of:

a 135° **b** 165° **c** 170° **d** 156°

4 What is the name of the regular polygon whose interior angle is treble its exterior angle?

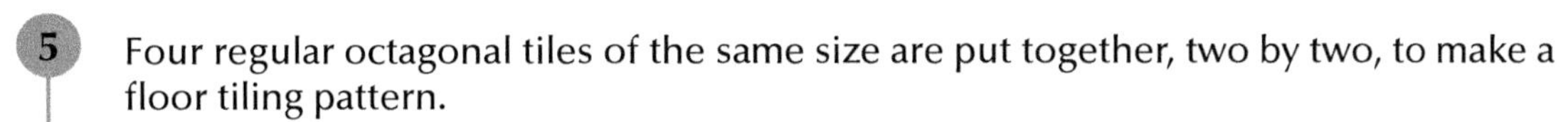

5. Four regular octagonal tiles of the same size are put together, two by two, to make a floor tiling pattern.

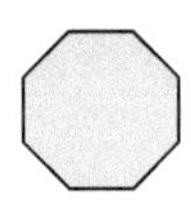 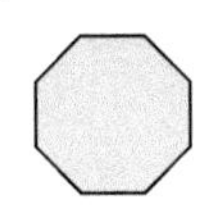 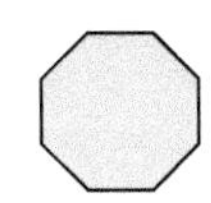 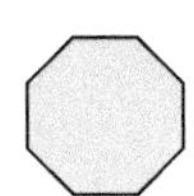

What is the shape of the tile that is required to fill the gap?

6. ABCDE is a regular pentagon.

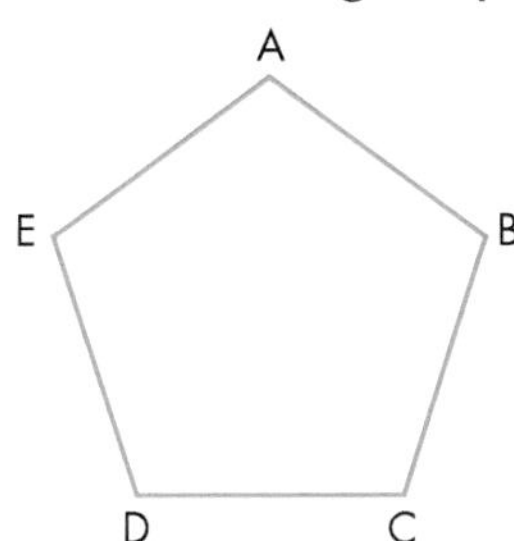

Work out the size of angle ADE. Give reasons for your answer.

7. Which of the following statements are true for a regular hexagon?

A The size of each interior angle is 60°.

B The size of each interior angle is 120°.

C The size of each exterior angle is 60°.

D The size of each exterior angle is 240°.

4.5 Angles in parallel lines

Homework 4F

1. State the sizes of the lettered angles in each diagram.

a 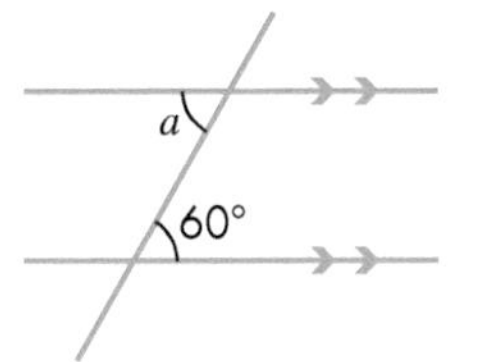

b 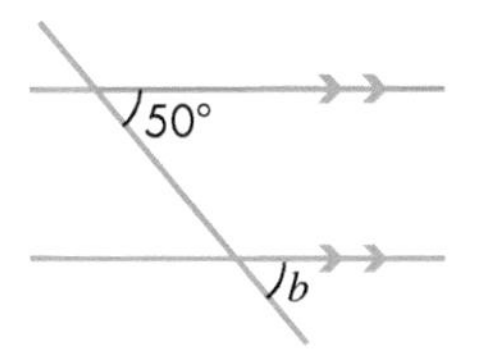

c 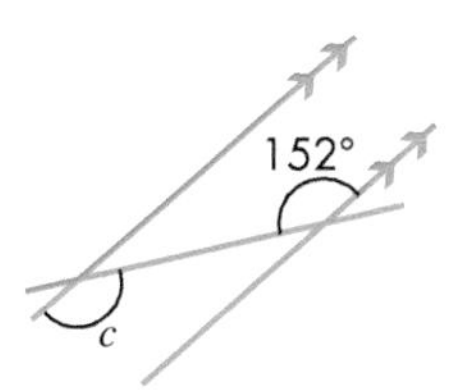

d 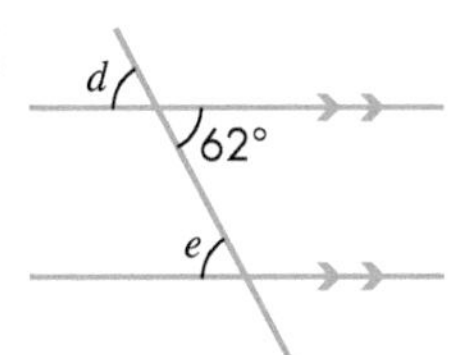

e 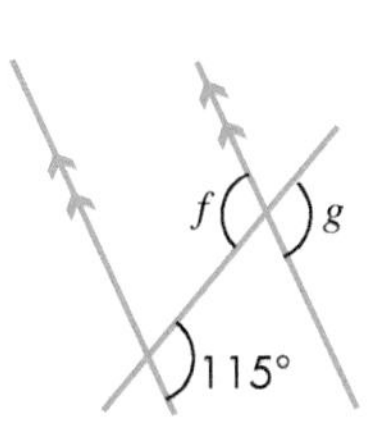

f 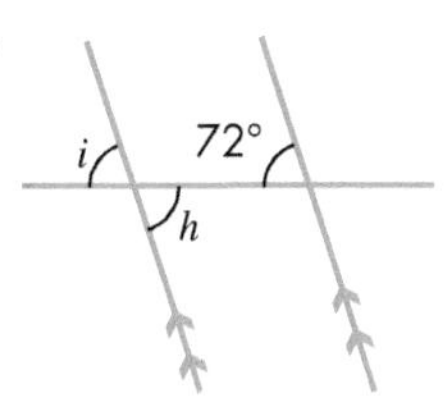

2. State the sizes of the lettered angles in each diagram. Give a reason for your answers.

a 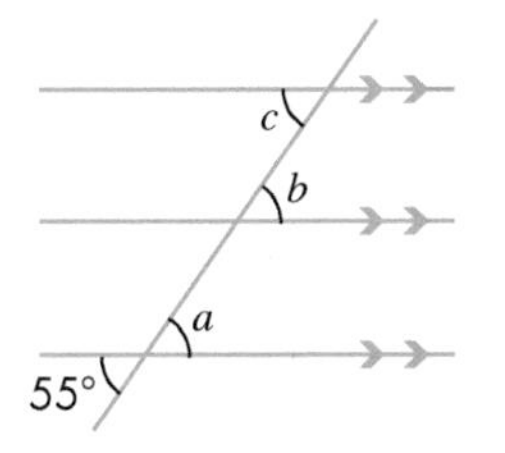

b 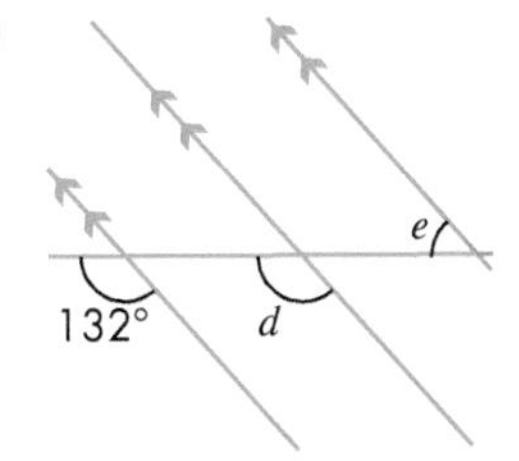

c 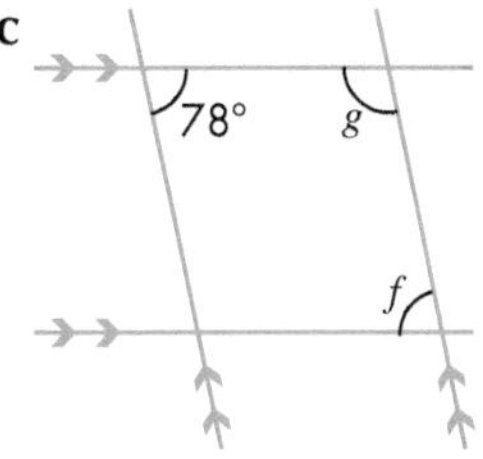

3 State the size of the lettered angles in these diagrams.

a

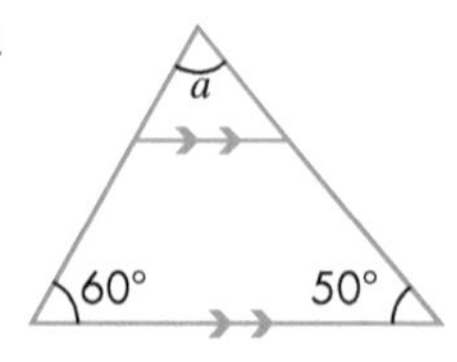

b

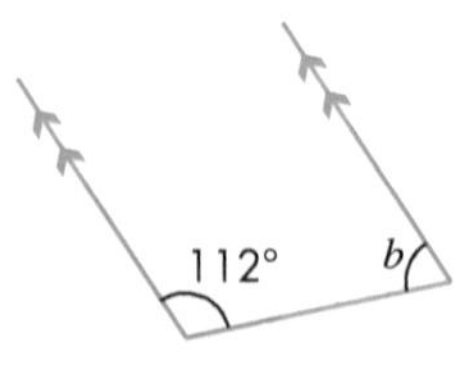

4 Calculate the values of x and y in these diagrams.

a

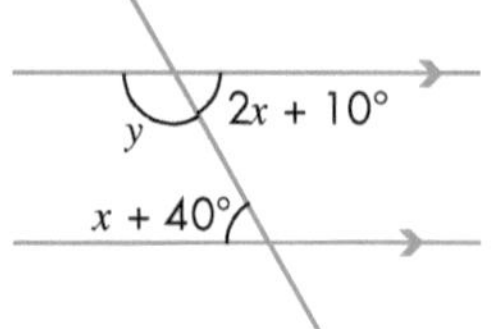

b

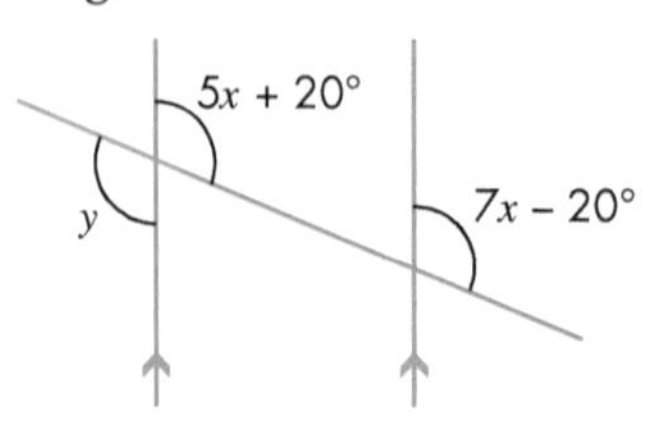

5 ABC is an isosceles triangle with angle ABC = 52°.

XY is parallel to BC.

Work out the size of angle BAC. Describe clearly how you calculated your answer.

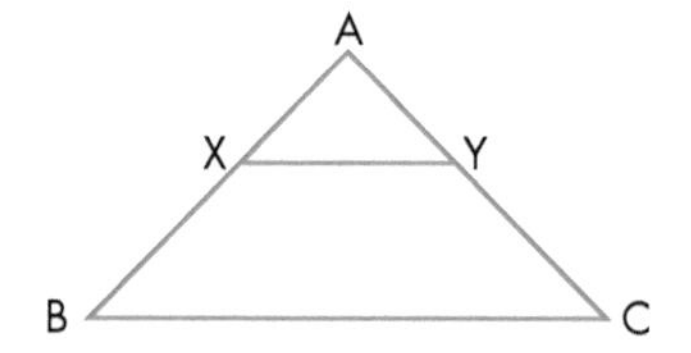

6 Work out the size of angle r in terms of p and q.

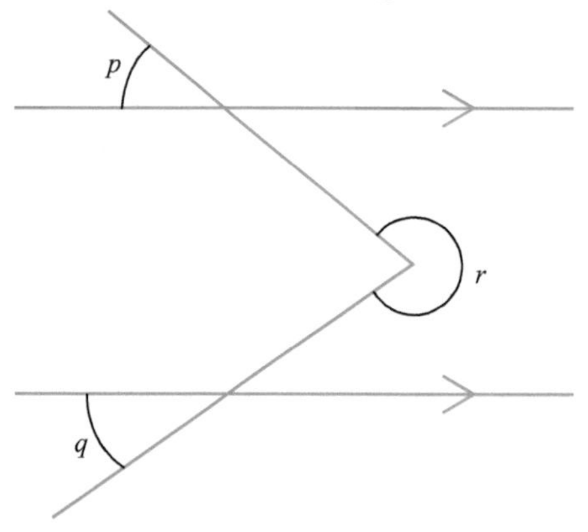

Hints and tips Draw a third parallel line that passes through angle r.

7 In the diagram, AB is parallel to CD.

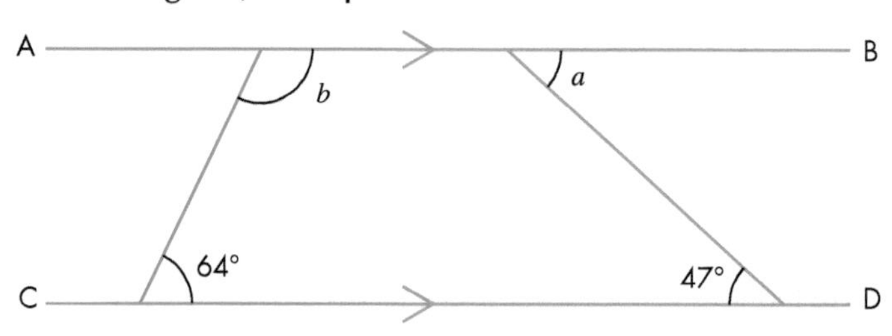

Explain why $a + b = 163°$.

4.6 Special quadrilaterals

Homework 4G

1 Calculate the sizes of the lettered angles in each trapezium.

a

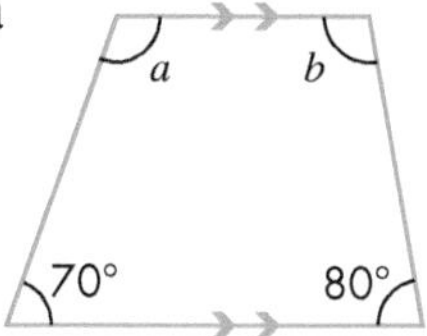

b

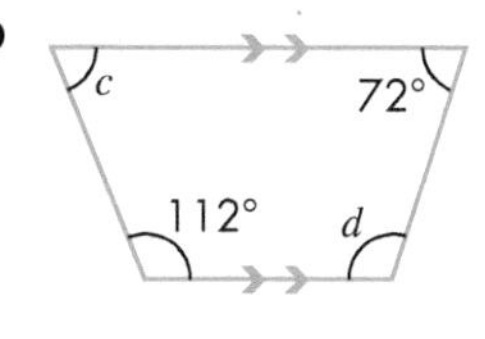

c

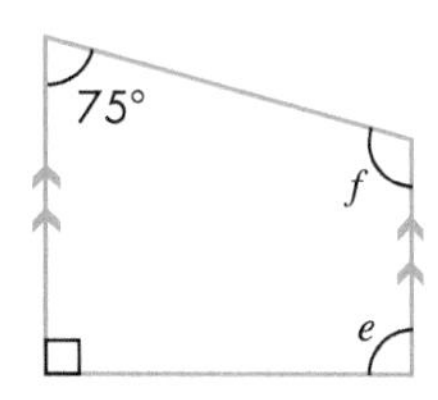

2 Calculate the sizes of the lettered angles in each parallelogram.

a

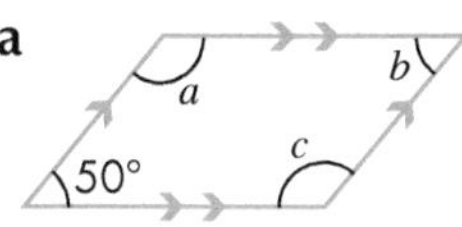

b

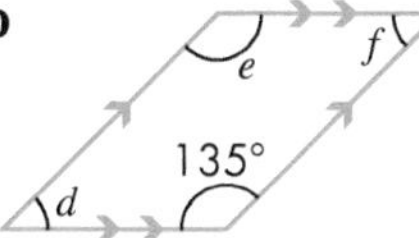

c

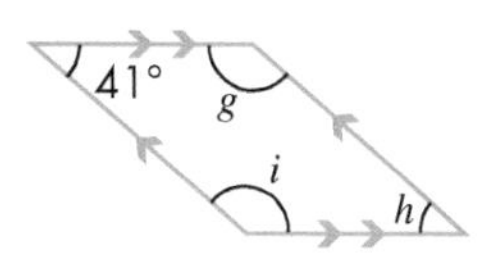

3 Calculate the sizes of the lettered angles in each kite.

a

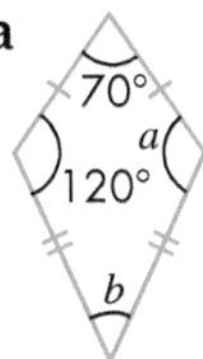

b

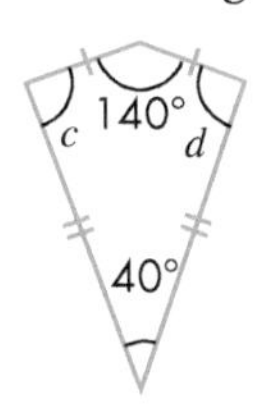

c

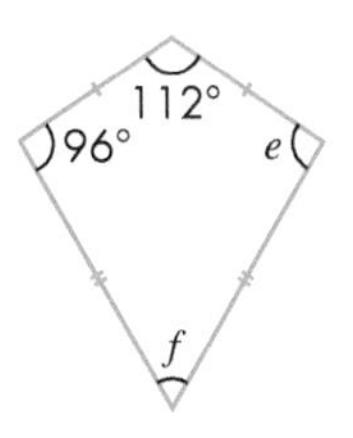

4 Calculate the sizes of the lettered angles in each rhombus.

a

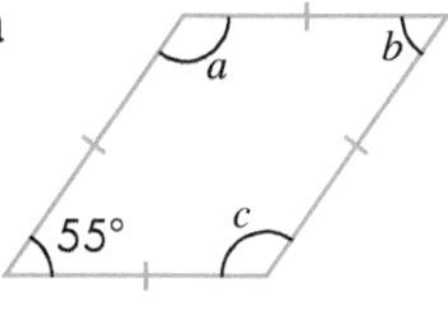

b

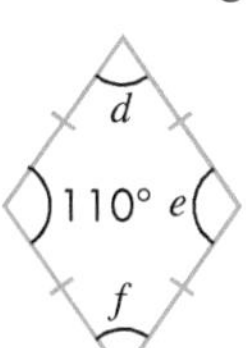

c

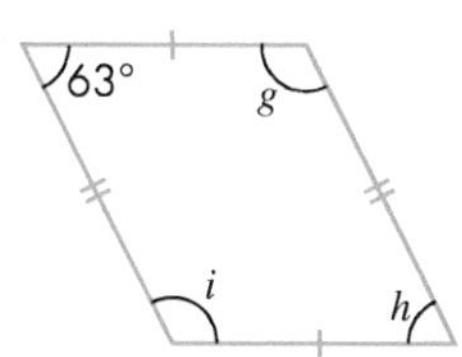

5 The diagram shows the side wall of a barn. The owner wants the angle between the roof and the horizontal to be at least 20°, so that rain will run off quickly. What can you say about the size of angles A and D?

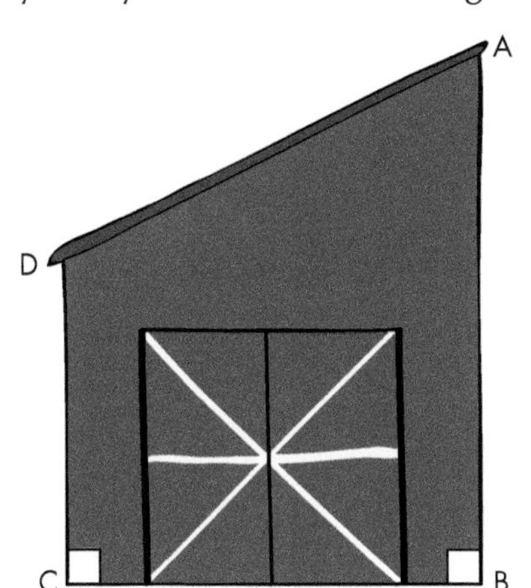

The diagram shows a parallelogram ABCD.

AC is a diagonal.

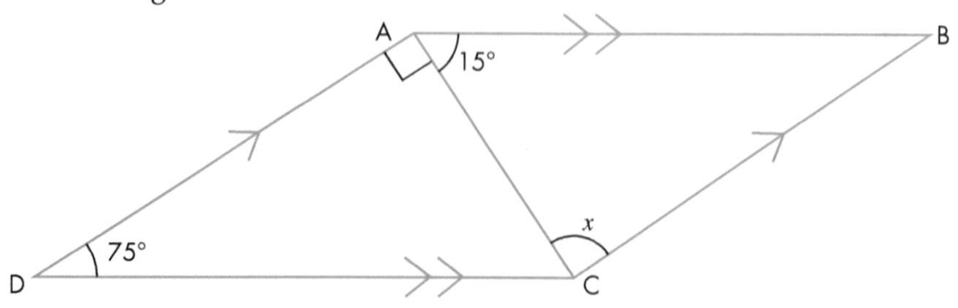

a Show that angle x is 90°.

b Calculate the size of angle BCD.

Explain two differences between the trapezium and the parallelogram.

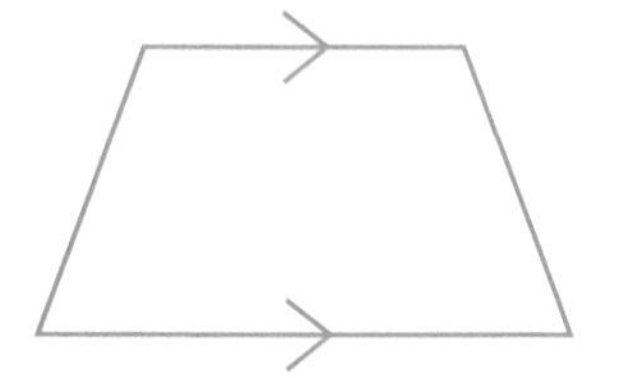

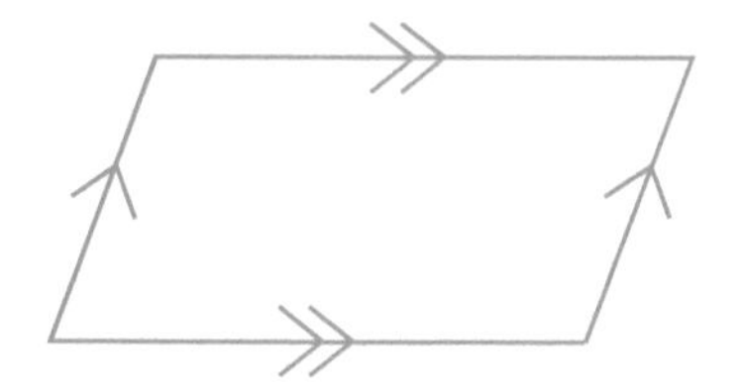

4.7 Bearings

Homework 4H

a Write down the bearing of B from A.

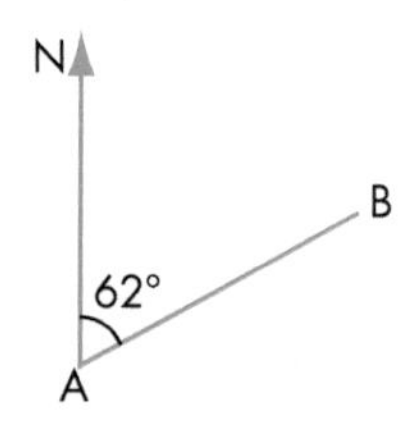

b Write down the bearing of D from C.

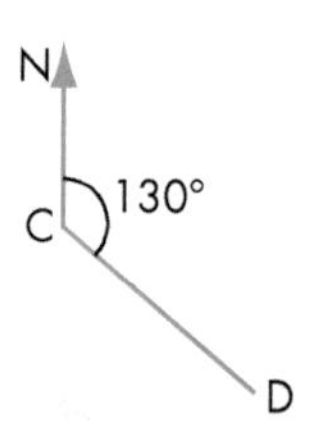

c Write down the bearing of F from E.

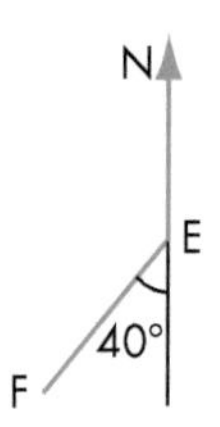

d Write down the bearing of H from G.

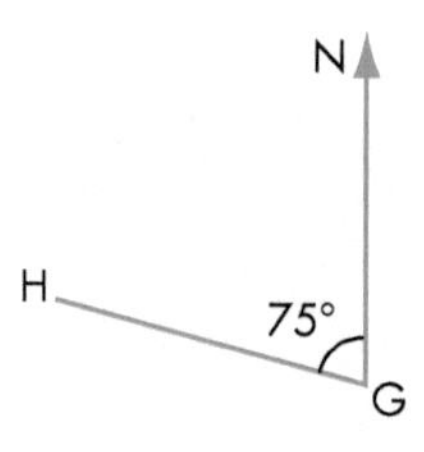

2 Below is a map of Britain.

By measuring angles, write down the bearings of:

a London from Edinburgh

b London from Cardiff

c Edinburgh from Cardiff

d Cardiff from London.

3 The diagram shows an aircraft's flight path from London to Paris.

a Write down the three-figure bearing of Paris from London.

b Work out the actual distance from London to Paris.

c The aircraft flies directly back to London.

What is the three-figure bearing of London from Paris?

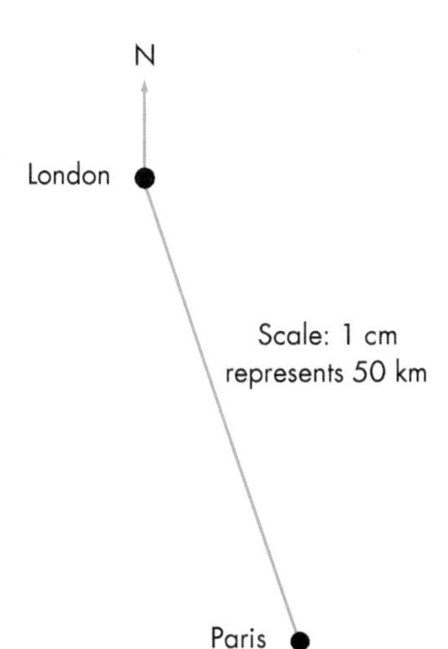

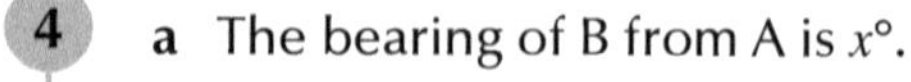

4 **a** The bearing of B from A is $x°$.

What is the bearing of A from B?

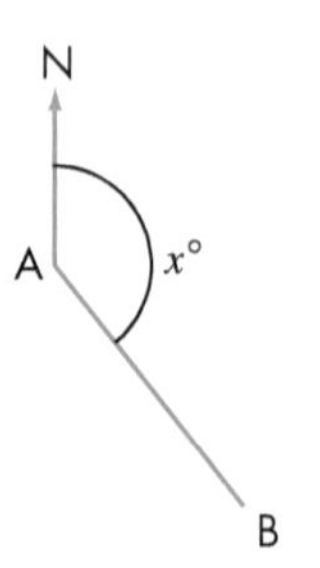

b The bearing of Y from X is $y°$.

What is the bearing of X from Y?

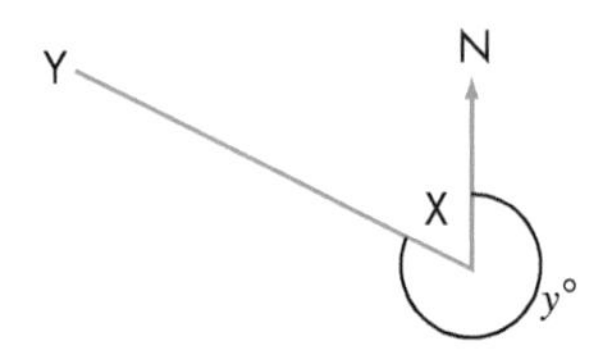

5 Town B is 40 km from Town A and on a bearing of 050°.

Town C is 60 km from Town A and on a bearing of 300°.

Make a scale drawing to work out the bearing of Town B from Town C.

6 A ship is sailing at a bearing of 324° when it receives an order to change course to sail due east. Through how many degrees should it turn?

7 Three towns, A, B and C, form an equilateral triangle. B is due north of A. The bearing of C from A is 060°. What is the bearing of C from B?

5 Number: Number properties

5.1 Multiples of whole numbers

Homework 5A

1 Write out the first five multiples of each number.

a 4 **b** 6 **c** 8 **d** 12 **e** 15

Hints and tips Remember, the first multiple is the number itself.

2 Look at the list of numbers below.

28 19 36 43 64 53 77 66 56 60 15 29 61 45 51

Write down the numbers from the list that are multiples of:

a 4 **b** 5 **c** 8 **d** 11.

3 Look at the list of numbers below.

225 252 361 297 162 363 161 289 224 205 312 378 315 182 369

Write down the numbers from the list that are multiples of:

a 7 **b** 9 **c** 12.

4 Write down the biggest number less than 200 that is a multiple of:

a 2 **b** 4 **c** 5 **d** 8 **e** 9.

5 Write down the smallest multiple of 3 that is greater than:

a 10 **b** 100 **c** 1000 **d** 10 000 **e** 1 000 000 000.

6 Kaja is packing sweets into bags of 12. She has 96 sweets. Will all the bags be full? Give a reason for your answer.

7 48 people are at a wedding reception. The tables are arranged so that the same number of people sit at each table.

How many people sit at each table? Give two possible answers.

8 Here is a list of numbers.

4 9 10 12 14 20

From the list, write down a multiple of:

a 7 **b** 6 **c** both 4 and 5.

9 Write down the lowest odd number that is a multiple of 9 and a multiple of 15.

5.2 Factors of whole numbers

Homework 5B

Example

Write down the factors of 32.

Look for the factor pairs of 32. These are:

$1 \times 32 = 32$ $2 \times 16 = 32$ $4 \times 8 = 32$

So, the factors of 32 are {1, 2, 4, 8, 16, 32}.

1 Write down all the factors of each number.

a 12 **b** 13 **c** 15 **d** 20 **e** 22

f 36 **g** 42 **h** 48 **i** 49 **j** 50

2 Write down all the factors of each number. Use your calculator to help you find them.

a 100 **b** 111 **c** 125 **d** 132 **e** 140

3 All the numbers in **a** to **j** are divisible by 11. Use your calculator to divide each one by 11 and then write down the answer. What do you notice?

a 143 **b** 253 **c** 275 **d** 363 **e** 462

f 484 **g** 561 **h** 583 **i** 792 **j** 891

4 Fred wants to pack 18 items into boxes so that there are exactly the same number of items in each box. How many ways can he do this?

5 Here is a list of numbers.

3 6 8 10 13

From the list, write down a factor of:

a 32 **b** 20 **c** both 26 and 39.

6 Look at these five numbers.

15 20 24 27 30

Use factors to explain why 20 could be the odd one out.

7 Write down the largest even number that is a factor of both 30 and 42.

5.3 Prime numbers

Homework 5C

1 Write down all the prime numbers less than 40.

2 Write down the prime numbers from this list.

43 47 49 51 54 57 59 61 65 67

3 This is a number pattern to generate odd numbers.

Line 1 $2 - 1 = 1$

Line 2 $2 \times 2 - 1 = 3$

Line 3 $2 \times 2 \times 2 - 1 = 7$

a Work out the next three lines of the pattern.

b Which lines have answers that are prime numbers?

4 Use the rules for recognising multiples to decide which of these numbers are not prime.

39 41 51 71 123

5 When two different prime numbers are multiplied together, the answer is 91. What are the two prime numbers?

6 **a** Write down two prime numbers with a difference of 6.

b Write down two more prime numbers with a difference of 6.

7 A mechanic has a set of 23 spanners.

He is trying to put them in a toolbox so that he has the same number of spanners in each part of his box.

Is this possible?

Give a reason for your answer.

5.4 Prime factors, LCM and HCF

Homework 5D

1 Copy and complete these factor trees and prime factorisations.

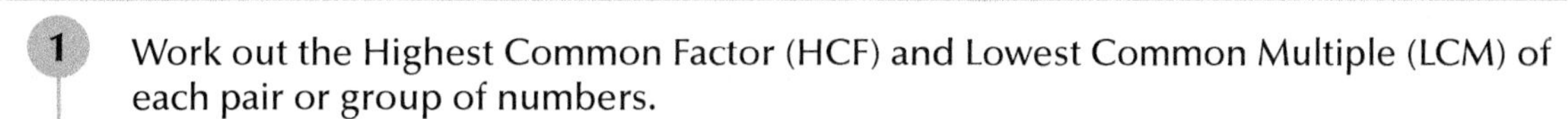

2 **a** Write 12 as a product of its prime factors.

b Write 144 as a product of its prime factors.

c Compare your answers to parts **a** and **b**.

Homework 5E

1 Work out the Highest Common Factor (HCF) and Lowest Common Multiple (LCM) of each pair or group of numbers.

a 20 and 45 **b** 42 and 70 **c** 120 and 130

d 36 and 40 **e** 120, 160 and 180

2 Two comets both passed Earth in 1992. One orbits the Earth once every 75 years. The other has an orbital period of 21 years. In which year will they next pass Earth together?

3 A large rectangular patio measures 24 ft by 120 ft. What are the dimensions of the largest square tile that can be used to pave the patio without cutting?

4 Grace visits Hazel every 4 days. Miriam visits Hazel every 5 days. If they both visited today, how many days will it be before they visit on the same day again?

5.5 Square numbers

Homework 5F

1 Use your calculator to work out these values.

a 5^2 **b** 15^2 **c** 25^2 **d** 35^2 **e** 45^2

f 55^2 **g** 65^2 **h** 75^2 **i** 85^2 **j** 95^2

Describe any patterns you notice.

2 **a** Write down the value of 11^2.

b Estimate the value of 10.5^2.

3 How much do 15 rulers at 15 pence each cost?

4 A builder buys 60 bricks for 60 pence each.

She has £40. How many extra bricks can she afford to buy?

5 In a warehouse, books are stored on shelves in piles of 20.

How many books are on two shelves, if there are 10 piles of books on each shelf?

Homework 5G

1 Write down the first five multiples of each number.

a 5 **b** 7 **c** 16 **d** 25 **e** 30

Hints and tips Remember: the first multiple is the number itself.

2 Write down all the factors of each of these numbers.

a 18 **b** 25 **c** 28 **d** 35 **e** 40

3 Write down the first three numbers that are multiples of both numbers in each pair.

a 2 and 5 **b** 3 and 4 **c** 5 and 6 **d** 4 and 6 **e** 8 and 10

4 In a prize draw, raffle tickets are numbered from 1 to 100.

A prize is given if a ticket drawn is a multiple of 10 or a multiple of 15. Which ticket holders will receive two prizes?

5 Here is a number pattern using square numbers.

$1^2 - 0^2 = 1$

$2^2 - 1^2 = 3$

$3^2 - 2^2 = 5$

$4^2 - 3^2 = 7$

a Write down the next three lines in the pattern.

b What do you think the answer to $21^2 - 20^2$ is? Explain your answer.

6 From the list of numbers below, write down the:

a prime numbers **b** square numbers.

4 6 7 10 13 16 21 23 25 28 34 37 40 49 50

7 Here are four numbers.

3 12 25 36

Copy and complete the table by writing the numbers in the correct cells.

	Square number	Factor of 24
Odd number		
Multiple of 6		

8 Use the four numbers on these cards to make a square number.

0 1 2 4

5.6 Square roots

Homework 5H

1 Write down the positive square root of each number.

a 64 **b** 25 **c** 49 **d** 81 **e** 16

f 36 **g** 100 **h** 121 **i** 144 **j** 400

2 Write down both possible values of each square root. You will need to use your calculator for some of them.

a $\sqrt{225}$ **b** $\sqrt{289}$ **c** $\sqrt{441}$ **d** $\sqrt{625}$

e $\sqrt{1089}$ **f** $\sqrt{1369}$ **g** $\sqrt{3136}$ **h** $\sqrt{6084}$

i $\sqrt{40\,804}$ **j** $\sqrt{110\,889}$

3 Here is a number pattern using square roots and square numbers.

$\sqrt{1} = 1$

$\sqrt{1} + \sqrt{4} = 3$

$\sqrt{1} + \sqrt{4} + \sqrt{9} = 6$

a Write down the next three lines in the pattern.

b Describe any pattern you notice in the answers.

4 Write these in order, from smallest value to largest value.

2^2 $\sqrt{20}$ $\sqrt{10}$ 3^2

5 Between which two consecutive whole numbers does the square root of 40 lie?

6 A child has 125 square tiles that she is arranging into square patterns.
How many tiles will be in the biggest square she can make?

7 Alfie is tiling a square kitchen floor. Altogether he needs 121 square tiles.
How many tiles are there in each row?

5.7 Basic calculations on a calculator

Homework 5I

1 Subtract each number from 180.

a 87 **b** 171 **c** 214

2 Subtract 18 from 32 and divide by 2.

3 Work these out.

a $(4 - 2) \times 180 \div 4$

b $(8 - 2) \times 180 \div 8$

4 Work out the value of $\frac{5.4 - 6.3^2}{0.3}$.

5 Use your calculator to work out the value of $\frac{6.27 \times 4.52}{4.81 + 9.63}$.

a Write down all the figures on your calculator display.

b Write your answer to part **a** to an appropriate degree of accuracy.

6 Use your calculator to work out the value of $\frac{1}{2.73^2 - 3.86}$.

a Write down all the figures on your calculator display.

b Write your answer to part **a** to an appropriate degree of accuracy.

7 Use your calculator to work out $\frac{15.1 + 4.82}{6.2 - 3.7}$.

Write down all the figures on your calculator display.

8 Use your calculator to work out $\sqrt{\frac{6.32 - 2.8}{8.7 + 9.2}}$.

Write down all the figures on your calculator display.

9 Use your calculator to work out the value of $\frac{8.95 + \sqrt{7.84}}{2.03 \times 1.49}$.

a Write down all the figures on your calculator display.

b Write down your answer to part **a** correct to three significant figures.

10 Calculate the value of $\frac{\sqrt{9.83 - 1.67^2}}{23.8 - 4.47 \times 5.12}$.

6 Number: Approximations

6.1 Rounding whole numbers

Homework 6A

1. Round each number to the nearest 10.

   a 34 **b** 67 **c** 23 **d** 49 **e** 55
   f 11 **g** 95 **h** 123 **i** 109 **j** 125

2. Round each number to the nearest 100.

   a 231 **b** 389 **c** 410 **d** 777 **e** 850
   f 117 **g** 585 **h** 250 **i** 975 **j** 1245

3. Round each number to the nearest 1000.

   a 2176 **b** 3800 **c** 6760 **d** 4455 **e** 1204
   f 6782 **g** 5500 **h** 8808 **i** 1500 **j** 9999

4. The values of five houses in a village are shown below.

   Round the prices to the nearest thousand pounds.

5. Write these bus journey times to the nearest 5 minutes.

   a 16 minutes **b** 28 minutes **c** 34 minutes
   d 42 minutes **e** 23 minutes **f** 17 minutes

6. Margaret knows that, to the nearest £10, she has £2240 in her savings account.

   a What is the least amount in her account?
   b What is the greatest amount in her account?

7 The size of a crowd at a festival was reported to be 57 000 to the nearest thousand.

a What was the least number of people in the crowd?

b What was the greatest number of people in the crowd?

8 Rafiq and Ruba are playing a game with whole numbers.

a Rafiq is thinking of a number. Rounded to the nearest 10, it is 270.

What is the biggest number Rafiq could be thinking of?

b Ruba is thinking of a different number. Rounded to the nearest 100, it is 300.

If Ruba's number is less than 270, how many different numbers could she be thinking of?

9 The number of fish in a pond is 130, to the nearest 10. The number of frogs in the pond is 90, to the nearest 10.

What is the greatest possible number of fish and frogs in the pond?

6.2 Rounding decimals

Homework 6B

Hints and tips 5.852 will round to 5.85 to two decimal places.
7.156 will round to 7.16 to two decimal places.
0.284 will round to 0.3 to one decimal place.
15.3518 will round to 15.4 to one decimal place.

1 Round each number to one decimal place.

a 3.73 b 8.69 c 5.34 d 18.75 e 0.423

f 26.288 g 3.755 h 10.056 i 11.08 j 12.041

2 Round each number to two decimal places.

a 6.721 b 4.457 c 1.972 d 3.485 e 5.807

f 2.564 g 21.799 h 12.985 i 2.302 j 5.555

3 Round each number to the number of decimal places indicated.

a 4.572 (1 dp) b 0.085 (2 dp) c 5.7159 (3 dp) d 4.558 (2 dp)

e 2.099 (2 dp) f 0.7629 (3 dp) g 7.124 (1 dp) h 8.903 (2 dp)

i 23.7809 (3 dp) j 0.99 (1 dp)

4 Round each number to the nearest whole number.

a 6.7 b 9.3 c 2.8 d 7.5 e 8.38

f 2.82 g 2.18 h 1.55 i 5.252 j 3.999

5 Graham buys these car accessories.

Shampoo: £4.99 Wax: £7.29 Wheel cleaner: £4.81 Dusters: £1.08

By rounding each price to the nearest pound (£), work out an estimate for the total cost of the items.

6 Which of these numbers are correctly rounded values of 9.281?

9 9.2 9.28 9.3 9.30

7 When an answer is rounded to two decimal places, it is 6.14. Which of these could be the unrounded answer?

6.140 6.143 6.148 6.15

6.3 Approximating calculations

Homework 6C

1 Round each number to one significant figure.

a 46 313	**b** 57 123	**c** 30 569	**d** 94 558	**e** 85 299
f 54.26	**g** 85.18	**h** 27.09	**i** 96.432	**j** 167.77
k 0.5388	**l** 0.2823	**m** 0.005 84	**n** 0.047 85	**o** 0.000 876
p 9.9	**q** 89.5	**r** 90.78	**s** 199	**t** 999.99

2 Write down the smallest and greatest numbers of people that live in these villages.

a Hellaby population 900 (to 1 sf)

b Hook population 650 (to 2 sf)

c Hundleton population 1050 (to 3 sf)

3 A baker estimates that she has baked 100 loaves. This is correct to one significant figure.

She sells two loaves and now has 90 loaves, still correct to one significant figure.

How many loaves could she have had to start with? Work out all possible answers.

4 There are 500 cars in a car park, correct to one significant figure.

What is the smallest possible number of cars that could enter the car park, so that there are 700 cars in the car park, correct to one significant figure?

Homework 6D

1 Work out approximate answers to each of these.

a 4324×6.71	**b** 6170×7.311	**c** 72.35×3.142
d 4709×3.81	**e** $63.1 \times 4.18 \times 8.32$	**f** $320 \times 6.95 \times 0.98$
g $454 \div 89.3$	**h** $26.8 \div 2.97$	**i** $4964 \div 7.23$
j $316 \div 3.87$	**k** $2489 \div 48.58$	**l** $63.94 \div 8.302$

2 Work out the approximate monthly pay for each annual salary.

a £47 200 **b** £24 200 **c** £19 135

3 Work out each person's approximate annual pay.

a Trevor: £570 per week **b** Brian: £2728 per month

4 A farmer bought 350 kg of seed at a cost of £3.84 per kilogram. Estimate the total cost of this seed.

5 A greengrocer buys apples from a wholesaler at a cost of 9 pence per apple. He then sells a box of 250 apples for £47.

a Approximately how much did he sell each apple for?

b Approximately how much profit did he make per apple?

6 Keith runs about 15 km every day. Approximately how far does he run in:

a a week

b a month

c a year?

7 A litre of paint will cover an area of about 6.8 m^2. Approximately how many 1-litre cans will I need to paint a fence with a total surface area of 43 m^2?

8 A tour of London sets off at 10:13 am and costs £21. It returns at 12:08 pm.

What is the approximate cost of the tour per hour?

9 How many jars, each holding 119 cm^3 of water, can be filled from a 3-litre flask?

10 If I walk at an average speed of 62 metres per minute, how long will it take me to walk a distance of 4 km?

11 Helen earns £41 500 a year. She works 5 days a week for 45 weeks of the year.

How much does she earn per day?

12 If 10 g of gold costs £2.17, how much will 1 kg of gold cost?

13 The sides of a triangle are measured as 4 cm, 5 cm and 6 cm, to the nearest centimetre.

a Work out the error interval for each side.

b Work out the error interval of the perimeter.

7 Number: Decimals and fractions

7.1 Calculating with decimals

Homework 7A

1. Work these out.
   a 2.5×0.4
   b 0.14×0.5
   c 5.4×0.8
   d 5.81×0.4

2. Evaluate each of these.
   a $3.6 \div 0.9$
   b $6.4 \div 0.04$
   c $0.12 \div 4$
   d $16.9 \div 1.3$

3. For each calculation:
   i estimate by rounding each number to one significant figure
   ii calculate the exact answer
   iii work out the difference between your estimate and the exact answer.
   a 3.2×4.9
   b 9.5×8.7
   c 12.4×34.1
   d 42.9×65.5

4. **a** Use any method to work out 45×85.
   b Use your answer to part **a** to work out:
   i 4.5×8.5
   ii 0.45×0.85
   iii 450×0.85

5. You are given that $27.73 \div 0.47 = 59$.
   Use this information to work out the answers to these divisions.
   a $2.773 \div 0.47$
   b $277.3 \div 4.7$
   c $27.73 \div 47$

7.2 Fractions and reciprocals

Homework 7B

1. Write each of these proper fractions as a terminating or recurring decimal, as appropriate.

   a $\frac{3}{4}$ **b** $\frac{1}{15}$ **c** $\frac{1}{25}$ **d** $\frac{1}{11}$ **e** $\frac{1}{20}$

2. There are several patterns found in recurring decimals. For example:

   $\frac{1}{13} = 0.076\,923\,076\,923\,076\,923\ldots$

   $\frac{2}{13} = 0.153\,846\,153\,846\,153\,846\ldots$

   $\frac{3}{13} = 0.230\,769\,230\,769\,230\,769\ldots$

   and so on.

   a Write down the decimals for $\frac{4}{13}, \frac{5}{13}, \frac{6}{13}, \frac{7}{13}, \frac{8}{13}, \frac{9}{13}, \frac{10}{13}, \frac{11}{13}, \frac{12}{13}$ to 18 decimal places.

   b What do you notice about the numbers in the pattern?

3. Write each proper fraction as a decimal. Then write the proper fractions in order of size, smallest first.

   $\frac{2}{9}$ $\frac{1}{5}$ $\frac{23}{100}$ $\frac{2}{7}$ $\frac{3}{11}$

4. Convert these terminating decimals to proper fractions. Give each fraction in its simplest form.

   a 0.57 **b** 0.275 **c** 0.85 **d** 0.06 **e** 3.65

5. Work out the reciprocal of each number.

   a 4 **b** 8 **c** 32 **d** 40 **e** 100

6. Write down the reciprocal of each of fraction.

   a $\frac{2}{3}$ **b** $\frac{5}{8}$ **c** $\frac{9}{10}$ **d** $\frac{7}{12}$ **e** $\frac{17}{20}$

7.3 Fractions of quantities

Homework 7C

1. Write the first quantity as a fraction of the second. Give your answers as fractions in their simplest form.

   a 10 cm, 80 cm

   b 4 ml, 24 ml

   c £3, £7

   d 110 days, 140 days

   e 135 hours, 153 hours

   f 48 minutes, 52 minutes

   g 40 kg, 110 kg

   h 6p, 32p

2 A theme park makes £300 profit each day. $\frac{1}{3}$ of the profit comes from entry fees, $\frac{1}{6}$ comes from selling food and $\frac{2}{5}$ comes from selling gifts. How much daily profit is made from everything else?

3 LP builders charge £500 per day. $\frac{1}{5}$ of this charge covers their staff costs, $\frac{2}{7}$ covers their petrol costs and $\frac{2}{3}$ covers the cost for materials. The rest is profit. Calculate their daily profit.

4 Jim is 1 m 30 cm tall. His brother is $\frac{1}{10}$ taller . How tall is Jim's brother?

7.4 Adding and subtracting fractions

Homework 7D

1 Work these out.

a $\frac{1}{4}+\frac{3}{5}$ **b** $\frac{2}{3}+\frac{4}{9}$ **c** $\frac{3}{4}+\frac{7}{10}$

d $\frac{1}{8}+\frac{7}{25}$ **e** $\frac{9}{20}+\frac{5}{16}$ **f** $\frac{3}{8}+\frac{3}{16}+\frac{3}{4}$

g $\frac{17}{20}-\frac{5}{12}$ **h** $\frac{5}{8}-\frac{7}{24}$ **i** $\frac{9}{32}-\frac{1}{12}$

j $\frac{3}{5}+\frac{7}{16}-\frac{1}{3}$ **k** $\frac{9}{24}+\frac{5}{18}-\frac{1}{10}$ **l** $\frac{1}{4}+\frac{7}{9}-\frac{5}{13}$

2 Work these out.

a $5\frac{1}{4}+7\frac{3}{5}$ **b** $8\frac{2}{3}+1\frac{4}{9}$ **c** $6\frac{3}{4}+2\frac{7}{10}$ **d** $9\frac{1}{8}+3\frac{7}{25}$

e $7\frac{9}{20}+3\frac{5}{16}$ **f** $8\frac{3}{8}+1\frac{3}{16}+2\frac{3}{4}$ **g** $6\frac{17}{20}-5\frac{5}{12}$ **h** $2\frac{5}{8}-1\frac{7}{24}$

i $3\frac{9}{32}-1\frac{1}{12}$ **j** $4\frac{3}{5}+5\frac{7}{16}-8\frac{1}{3}$ **k** $1\frac{9}{24}+1\frac{5}{18}-1\frac{1}{10}$ **l** $5\frac{1}{4}+2\frac{7}{9}-6\frac{5}{13}$

3 A tank of water is two-thirds full. Anna then uses one-quarter of a tank and Ed pours in one-twelfth of a tank. How full is the tank?

4 Look at this road sign.

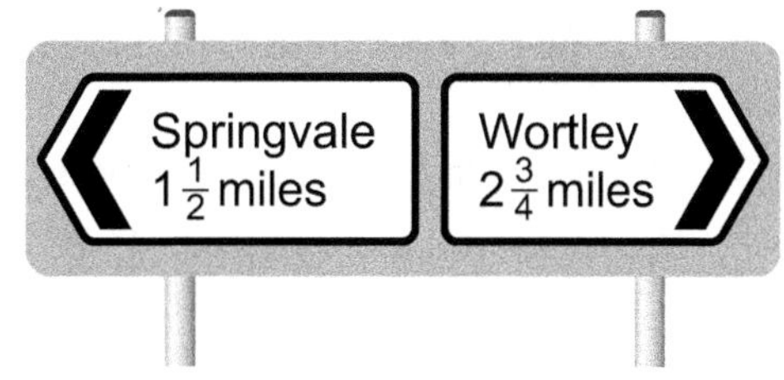

a What is the distance between Springvale and Wortley, using these roads?

b How much further is it to Wortley than to Springvale?

5 Write down what you would say to someone, in a telephone conversation, to explain how to work out the answer to this calculation using a calculator.

$\frac{1}{4}\times\frac{2}{3}$

6 A class had the same numbers of boys and girls. Three girls leave the class and are replaced by three boys. $\frac{3}{8}$ of the class are now girls. How many students are in the class?

7.5 Multiplying and dividing fractions

Homework 7E

1 Work these out. Give each answer as a mixed number or fraction in its simplest form.

a $\frac{1}{8} \times \frac{3}{4}$ **b** $\frac{5}{12} \times \frac{1}{4}$ **c** $\frac{3}{9} \times \frac{1}{3}$

d $\frac{2}{5} \times \frac{4}{10}$ **e** $\frac{3}{6} \times \frac{7}{8}$

2 Work these out. Give each answer as a mixed number or fraction in its simplest form.

a $2\frac{1}{6} \times 1\frac{2}{3}$ **b** $3\frac{2}{3} \times 3$ **c** $2\frac{2}{3} \times 3$

d $1\frac{1}{2} \times \frac{2}{3}$ **e** $1\frac{1}{4} \times \frac{2}{5}$

3 Work these out.

a $\frac{1}{4} \div \frac{1}{3}$ **b** $\frac{4}{5} \div \frac{2}{10}$ **c** $\frac{1}{2} \div \frac{2}{4}$

d $\frac{3}{5} \div \frac{6}{10}$ **e** $\frac{1}{4} \div \frac{4}{5}$

4 Work these out.

a $3\frac{1}{3} \div 2\frac{1}{2}$ **b** $4\frac{1}{3} \div 4\frac{1}{4}$ **c** $4\frac{4}{5} \div 2\frac{7}{10}$

d $4\frac{2}{5} \div 4\frac{3}{4}$ **e** $3\frac{3}{5} \div 2\frac{1}{2}$ **f** $3\frac{9}{10} \div 2\frac{2}{3}$

g $4\frac{1}{2} \div 4\frac{7}{10}$ **h** $4\frac{1}{5} \div 4\frac{4}{5}$ **i** $4\frac{1}{2} \div 4\frac{3}{4}$

J $3\frac{3}{5} \div 3\frac{3}{4}$

5 The formula for the area of a rectangle is area = length × width.
Work out the area of a rectangle of length $\frac{3}{4}$ m and width $\frac{1}{3}$ m.

6 Bricks are $\frac{1}{6}$ m long.
How many bricks would need to be placed end-to-end to form a line 2 m long?

7 A sphere with volume $19\frac{2}{5}$ cm^3 is cut into four equal pieces. Work out the volume of one piece.

8 The formula for average speed is average speed = distance ÷ time taken.
Work out the average speed of a car that travels $6\frac{3}{4}$ miles in a $\frac{1}{4}$ hour.

7.6 Fractions on a calculator

Homework 7F

In this exercise, try to key in each calculation as one continuous set of operations, without writing down any intermediate values.

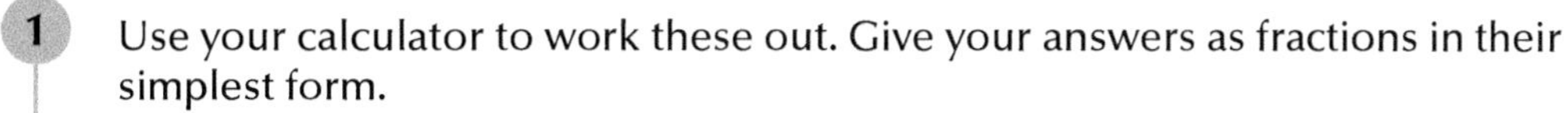

1 Use your calculator to work these out. Give your answers as fractions in their simplest form.

a $\frac{1}{4} \times \frac{3}{5}$ **b** $\frac{2}{3} \times \frac{4}{9}$ **c** $\frac{3}{4} \times \frac{7}{10}$

d $\frac{1}{8} \times \frac{7}{25}$ **e** $\frac{9}{20} \times \frac{5}{16}$ **f** $\frac{3}{8} \times \frac{3}{16} \times \frac{3}{4}$

g $\frac{17}{20} \div \frac{5}{12}$ **h** $\frac{5}{8} \div \frac{7}{24}$ **i** $\frac{9}{32} \div \frac{1}{12}$

j $\frac{3}{5} \times \frac{7}{16} \div \frac{1}{3}$ **k** $\frac{9}{24} \times \frac{5}{18} \div \frac{1}{10}$ **l** $\frac{1}{4} \times \frac{7}{9} \div \frac{5}{13}$

2 **a** Use your calculator to work out $\frac{2}{3} \times \frac{7}{11}$.

b Without using a calculator, write down the answer to $\frac{2}{11} \times \frac{7}{3}$.

3 **a** Use your calculator to work out $\frac{3}{4} \div \frac{7}{12}$.

b Use your calculator to work out $\frac{3}{4} \times \frac{12}{7}$.

c Use your calculator to work out $\frac{2}{9} \div \frac{2}{3}$.

d Use your calculator to work out $\frac{2}{9} \times \frac{3}{2}$.

4 Use your calculator to work these out. Give your answers as mixed numbers.

a $3\frac{1}{4} \times 2\frac{3}{5}$ **b** $6\frac{2}{3} \times 1\frac{4}{9}$ **c** $7\frac{3}{4} \times 2\frac{7}{10}$

d $5\frac{1}{8} \times 2\frac{7}{25}$ **e** $6\frac{9}{20} \times 4\frac{5}{16}$ **f** $1\frac{3}{8} \times 1\frac{3}{16} \times 1\frac{3}{4}$

g $4\frac{17}{20} \div 2\frac{5}{12}$ **h** $1\frac{5}{8} \div 1\frac{7}{24}$ **i** $2\frac{9}{32} \div 1\frac{1}{12}$

j $3\frac{3}{5} \times 2\frac{7}{16}$ **k** $2\frac{9}{24} \times 3\frac{5}{18} \div 1\frac{1}{10}$ **l** $4\frac{1}{4} \times 3\frac{7}{9} \times 2\frac{5}{13}$

5 Use your calculator to work these out.

a $\frac{4}{5} + \frac{3}{7}$ **b** $\frac{9}{11} - \frac{6}{13}$ **c** $\frac{5}{9} \times \frac{2}{5}$

d $3\frac{1}{3} \times 5\frac{1}{4}$ **e** $7\frac{5}{6} - 3\frac{2}{3}$ **f** $9\frac{7}{10} \div 5\frac{2}{3}$

g $6\frac{8}{11} + 2\frac{1}{2} \times 4\frac{1}{6}$ **h** $(7\frac{2}{7})^2 - 5\frac{1}{4} \div 3\frac{4}{5}$ **i** $5\frac{3}{4} \div (\frac{5}{9} \times 3\frac{3}{4}) + 2\frac{1}{3}$

6 **a** Use your calculator to work out $\frac{19}{23} - \frac{21}{25}$.

b Explain how your answer tells you that $\frac{19}{23}$ is less than $\frac{21}{25}$.

7 **a** Work out $\frac{10}{27} - \frac{3}{11}$ on your calculator.

b Work out $\frac{10}{27} - \frac{7}{16}$ on your calculator.

c Explain why your answers to parts **a** and **b** show that $\frac{10}{27}$ is a fraction in between $\frac{3}{11}$ and $\frac{7}{16}$.

8 Algebra: Linear graphs

8.1 Graphs and equations

Homework 8A

1 Write down the coordinates of points A to K.

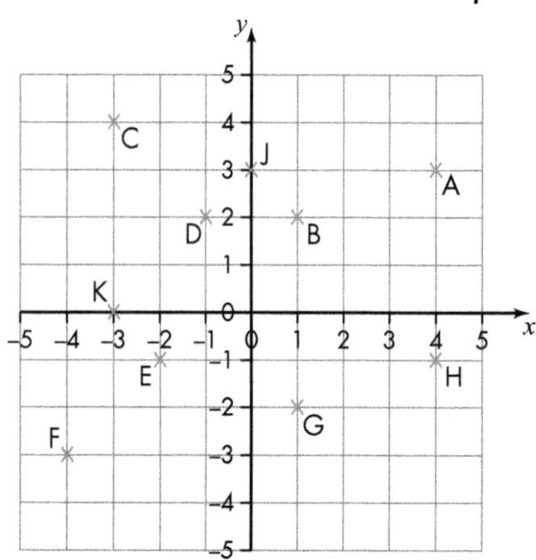

2 For each equation:

i copy and complete the table of values.

ii draw the graph on a pair of coordinate axes.

a $y = x - 3$

x	0	1	2	3	4	5
y						

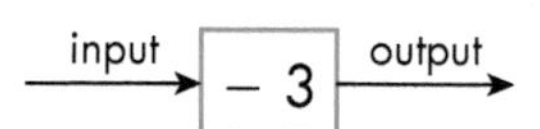

b $y = 2x + 1$

x	0	1	2	3	4	5
y						

c $y = 4x - 2$

x	0	1	2	3	4
y					

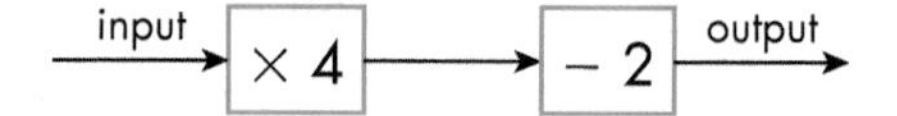

d $y = 5x$

x	0	1	2	3	4
y					

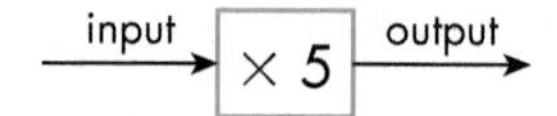

e $y = -3x - 1$

x	0	1	2	3	4	5
y						

3 Draw the graph of $y = 3x - 5$. Choose your own input and axes.

4 A teacher reads out this problem:

'I am thinking of a number. I multiply it by 8 and add 3.'

a Represent this problem using a flow diagram.

b If the input is x and the output is y, write down a relationship between x and y.

c Draw a graph for x-values from 0 to 5.

d The final answer is 27. How can you use the graph to work out the number the teacher is thinking of?

8.2 Drawing linear graphs by finding points

Homework 8B

Hints and tips Follow these steps when drawing graphs.

- Use the highest and lowest values of x given in the range.
- If the first part of the function is a division, pick x-values that divide exactly to avoid fractions.
- Always label your graph with its equation. This is particularly important when you are drawing two graphs on the same set of axes.
- Create a table of values.

1 Draw the graph of $y = 2x + 3$ for x-values from 0 to 5 ($0 \leqslant x \leqslant 5$).

2 Draw the graph of $y = 3x - 1$ for $0 \leqslant x \leqslant 5$.

3 Draw the graph of $y = \frac{x}{2} - 2$ for $0 \leqslant x \leqslant 12$.

4 Draw the graph of $y = 2x + 1$ for $-2 \leqslant x \leqslant 2$.

5 Draw the graph of $y = \frac{x}{2} + 5$ for $-6 \leqslant x \leqslant 6$.

6 **a** On the same set of axes, draw the graphs of $y = 3x - 1$ and $y = 2x + 3$ for $0 \leqslant x \leqslant 5$.

b At which point do the two lines intersect?

7 **a** On the same axes, draw the graphs of $y = 4x - 3$ and $y = 3x + 2$ for $0 \leqslant x \leqslant 6$.

b At which point do the two lines intersect?

8 **a** On the same axes, draw the graphs of $y = \frac{x}{2} + 1$ and $y = \frac{x}{3} + 2$ for $0 \leqslant x \leqslant 12$.

b At which point do the two lines intersect?

9 **a** On the same axes, draw the graphs of $y = 2x + 3$ and $y = 2x - 1$ for $0 \leqslant x \leqslant 4$.

b Do the graphs intersect? If not, give a reason.

10 **a** Copy and complete the table of values for the equation $x + y = 6$ and draw its graph.

x	0	1	2	3	4	5	6
y							

b Now draw the graph of $x + y = 3$ on the same axes.

11 CityCabs uses this formula to work out the cost of a journey of k kilometres:

$C = 2.5 + k$

TownCars uses this formula to work out the cost of a journey of k kilometres:

$C = 2 + 1.25k$

a On a copy of the grid, draw lines to show these formulae.

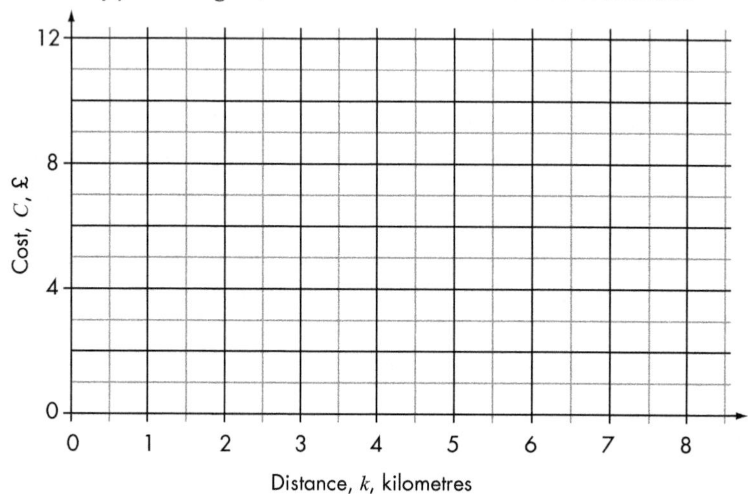

b At what length of journey do CityCabs and TownCars charge the same amount?

12 The diagram shows the graph of $x + y = 5$.

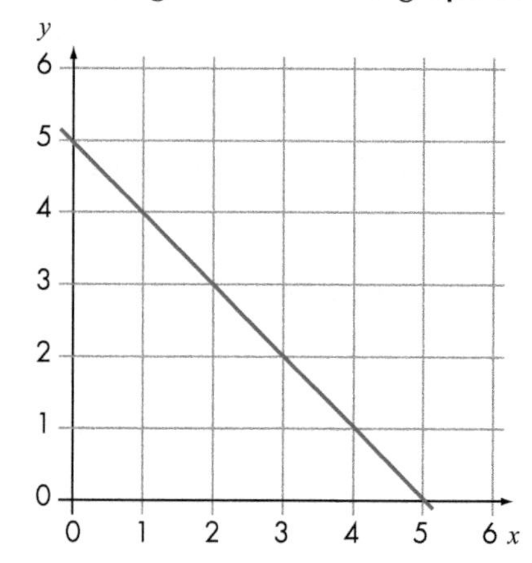

Draw a line of the form $x = a$ *and* a line of the form $y = b$ so that the area between the three lines is 4.5 square units.

8.3 Gradient of a line

Homework 8C

Hints and tips You can find the gradient using the formula $\frac{\text{change in } y}{\text{change in } x}$ or $\frac{y_2 - y_1}{x_2 - x_1}$.
For example, in Question 4, Line A has a gradient of $-\frac{2}{4} = -\frac{1}{2}$.
The origin is the point where the x and y axes cross.

1 Work out the gradient of lines A to J.

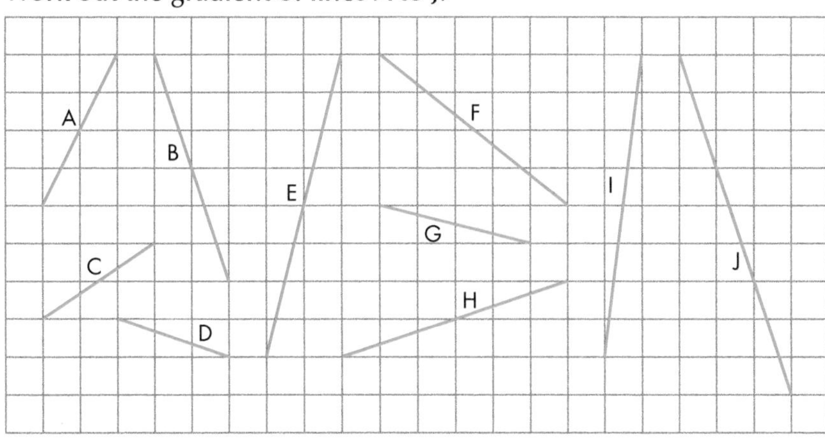

2 Draw lines with these gradients.

a 3 **b** $\frac{1}{2}$ **c** –1 **d** 8 **e** $\frac{3}{4}$ **f** $-\frac{1}{3}$

3 **a** Draw a pair of axes with x and y from –10 to 10.

b Draw one line with each of these gradients, starting from the origin each time. Remember to label each line.

i $\frac{1}{2}$ **ii** 1 **iii** 2 **iv** 4 **v** –4 **vi** –2 **vii** –1 **viii** $-\frac{1}{2}$

c Describe the symmetries of your diagram.

4 Work out the gradient of:

a line G **b** line J **c** line D **d** line E **e** a line parallel to B.

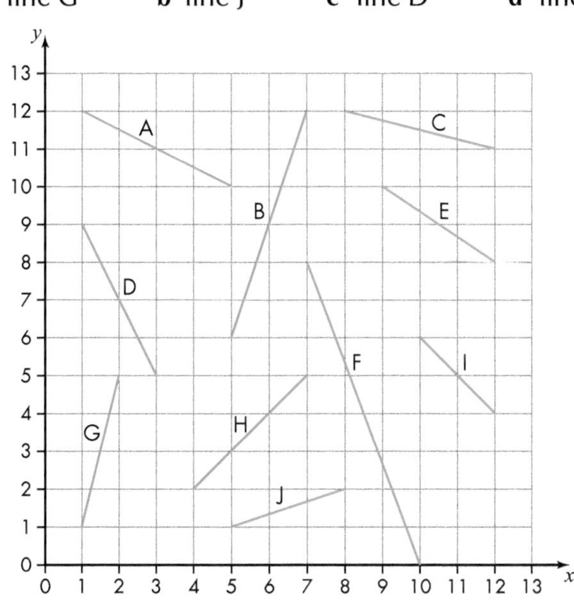

5 Work out the gradient of the line between each pair of points.

a (3, 5) and (4, 7) **b** (5, 9) and (7, 17) **c** (4, 6) and (5, 7)

d (1, 4) and (4, 19) **e** (0, 11) and (4, 23)

6 Work out the gradient of the line between each pair of points.

a (2, 5) and (3, –3) **b** (2, 8) and (3, 2) **c** (4, 8) and (8, 8)

d (8, 15) and (6, 33) **e** (7, 12) and (4, 42) **f** (4, 8) and (3, 14)

8.4 $y = mx + c$

Homework 8D

Hints and tips All straight-line graphs (linear graphs) have an equation of the form $y = mx + c$, for example, $y = 4x + 3$.
m is the gradient (slope) of the line and c is the y-intercept (where it crosses the y-axis).
For $y = 3x + 2$, the gradient is 3 and the y-intercept is 2.
For $y = x + 1$, both the gradient and the y-intercept are 1.

1 Write down **i** the gradient and **ii** the y-intercept of each line.

a $y = 4x + 3$ **b** $y = 3x - 2$ **c** $y = 2x + 1$

d $y = -3x + 3$ **e** $y = 5x$ **f** $y = -2x + 3$

g $y = x$ **h** $y = -\frac{1}{2}x + 3$ **i** $y = \frac{1}{4}x + 2$

2 Draw these lines using the gradient-intercept method. Use the same grid, taking x from –10 to 10 and y from –10 to 10.

a $y = 4x - 3$ **b** $y = 2x + 3$ **c** $y = -5x + 2$

d $y = 3x + 1$ **e** $y = x + 3$ **f** $y = -2x - 3$

Homework 8E

Draw these lines using the cover-up method. Use the same grid, taking x from –10 to 10 and y from –10 to 10.

1 $3y = 4x + 6$ **2** $2y = -5x - 4$

3 $8y = 4x - 12$ **4** $x - y = 0$

5 $y - 7 = 3x$ **6** $y - 2x - 4x = 0$

7 $6x + 4y = 12$ **8** $2x + 3y = 12$

9 $4x - 5y = 40$ **10** $x + y = 6$

11 $3x - 2y = 24$ **12** $x - y = -6$

8.5 Finding the equation of a line from its graph

Homework 8F

1 Write down the equation of each line. They all have positive gradients.

a

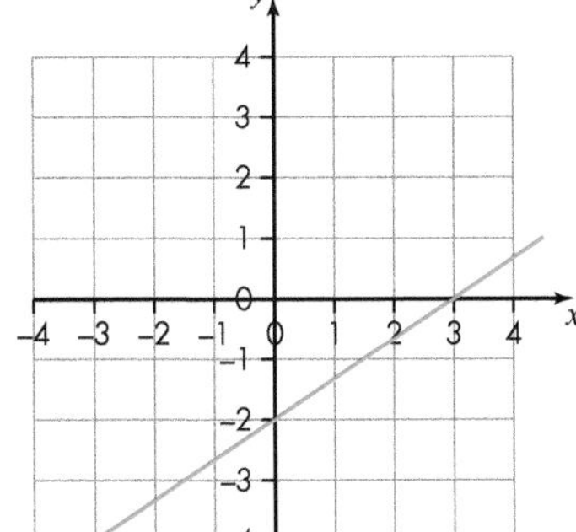

b

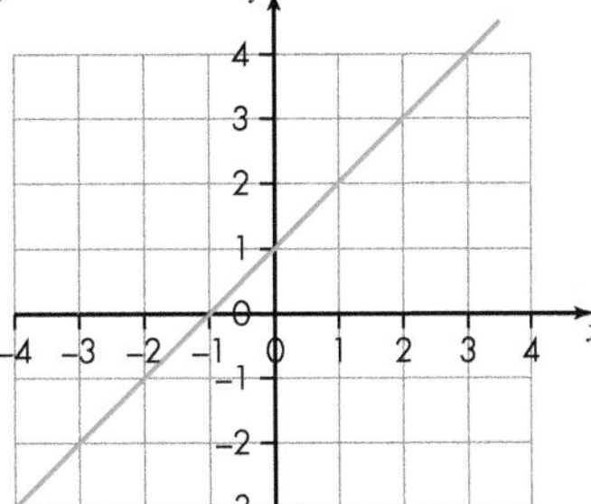

c

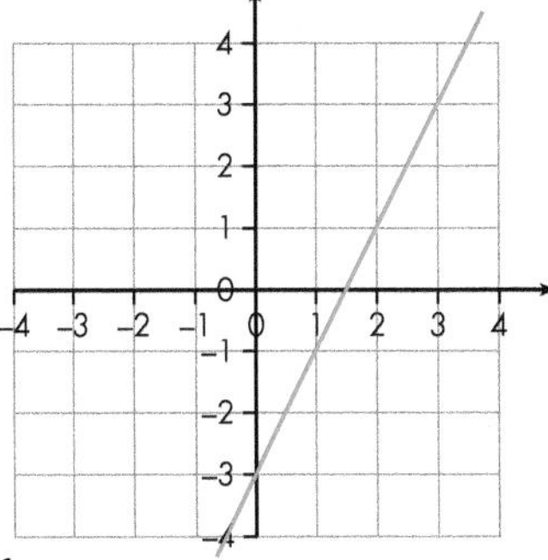

d

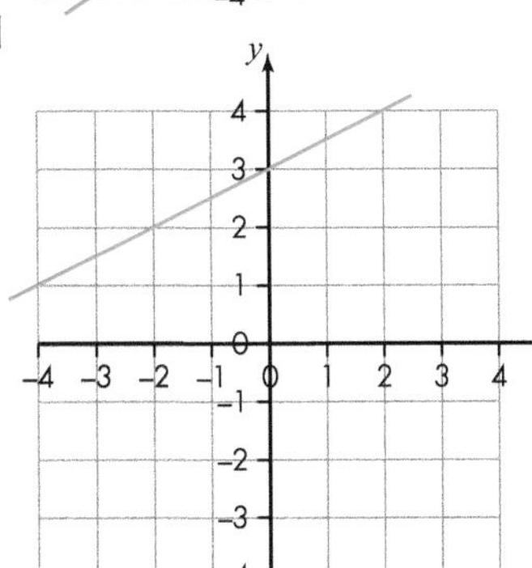

e

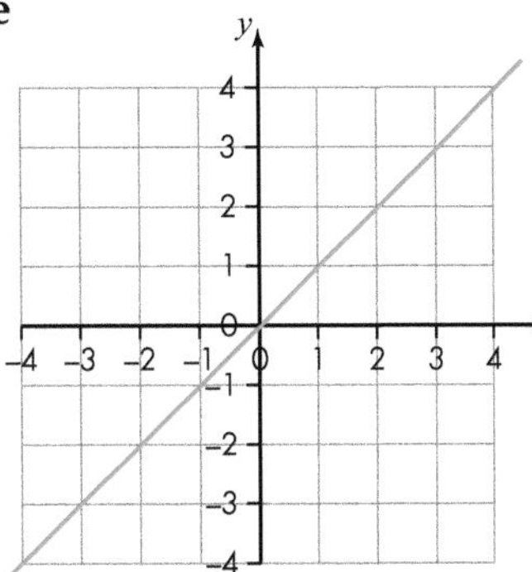

f

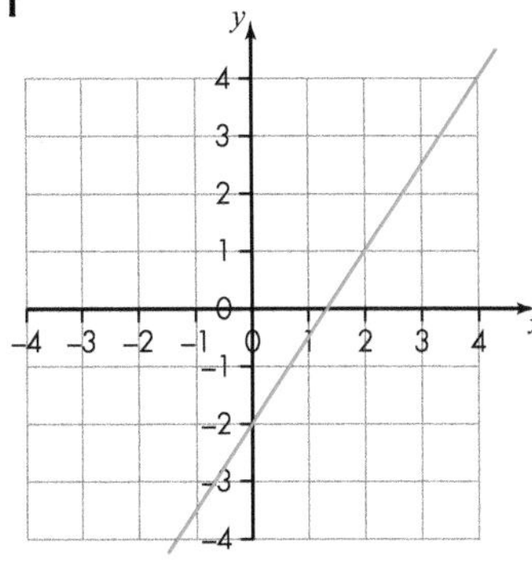

2 Write down the equation of each line. They all have negative gradients.

a

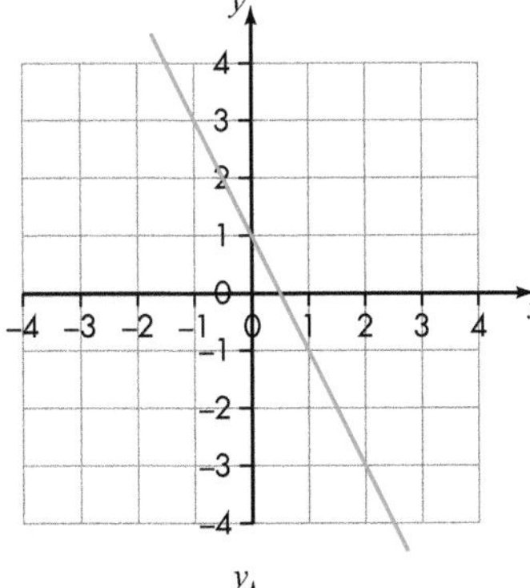

b

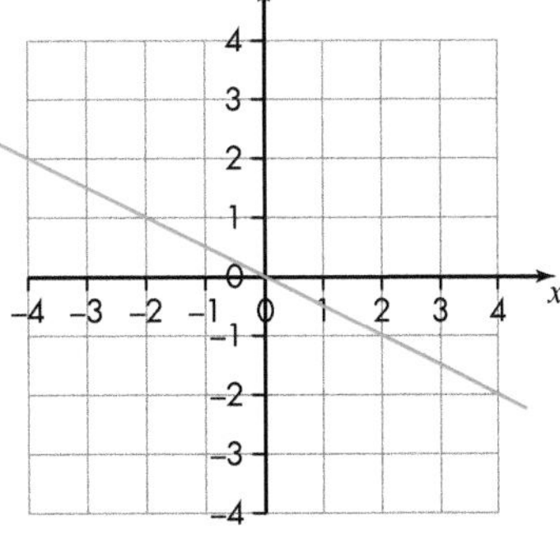

c

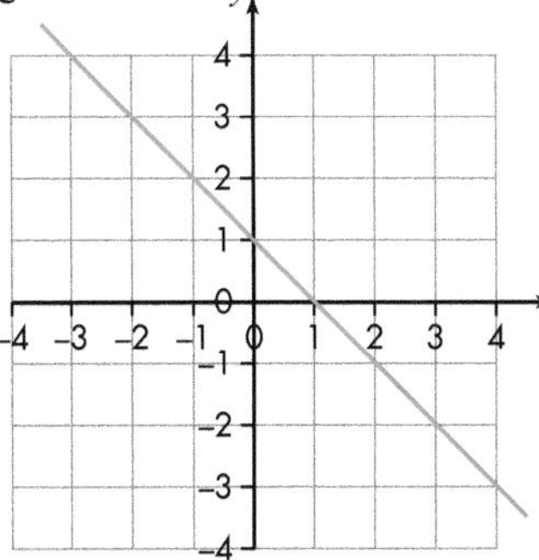

d

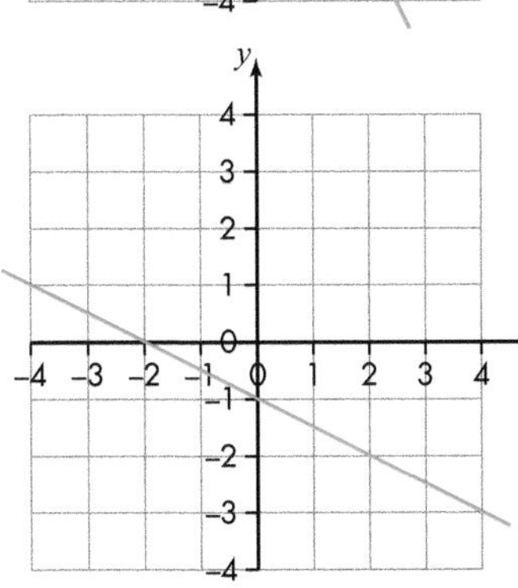

e

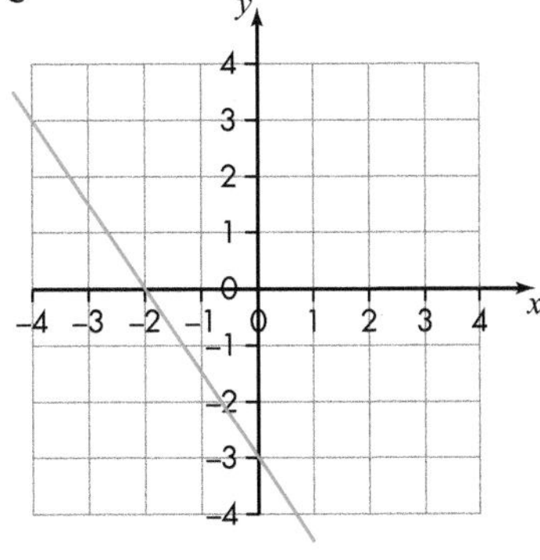

3 Use the facts to work out the equation of each line in the form $y = mx + c$.

a gradient is –4; y-intercept is 2

b gradient is 3; line passes through the point (6, 4)

c y-intercept is –5; line passes through the point (2, 11)

d line passes through the points (2, 18) and (5, 9)

e line passes through the points (–6, –7) and (18, 19)

4 Work out the equation of the line that passes through the points (2, 0) and (0, 7), giving the equation in the form $ax + by = c$.

8.6 The equation of a parallel line

Homework 8G

1 **a** On a copy of the grid below, draw the graph $y = x$.

b Draw three other lines that are parallel to $y = x$.

c Write down the equations of these lines.

d What do you notice about the equations?

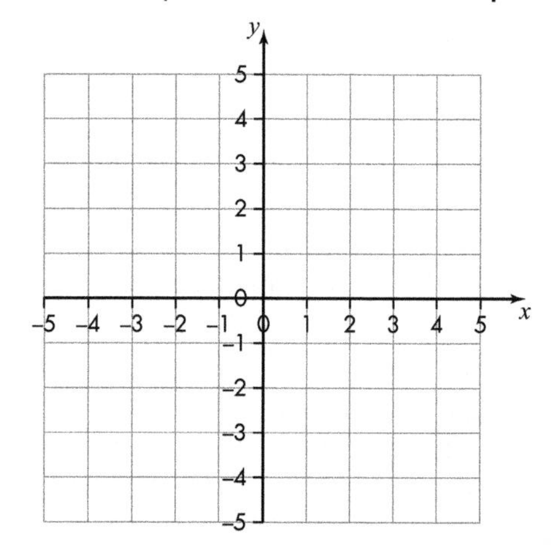

2 **a** On a copy of the grid above, draw the graph $y = -0.5x$.

b Draw three other lines that are parallel to $y = -0.5x$.

c Write down the equations of these lines.

d What do you notice about the equations?

3 Copy and complete this sentence.

If the gradient of a line is m, then the gradient of a line that is parallel is ___.

4 Work out the equation of a line that passes through (1, 10) and is parallel to the line $y = 2x + 7$.

5 Work out the equation of the line that passes through (0, –8) and is parallel to the line $y = -9x - 2$.

8.7 Real-life uses of graphs

Homework 8H

1 This is a conversion graph between yards and feet.

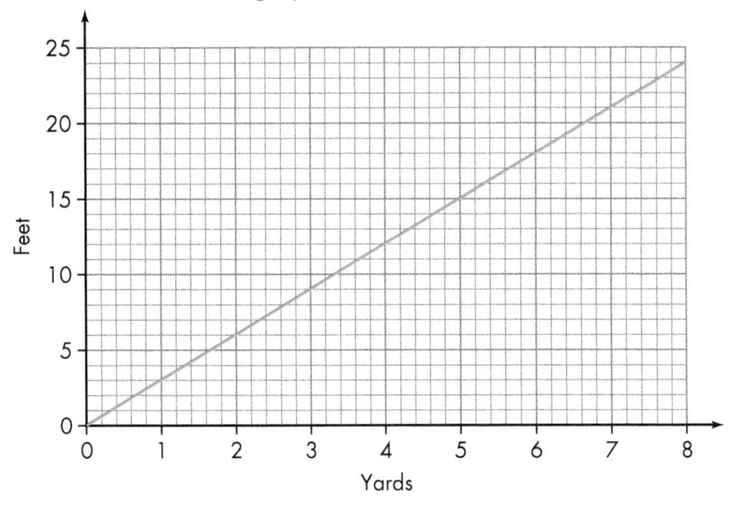

a Use the conversion graph to make an approximate conversion of:

i 5 yards to feet **ii** 24 feet to yards **iii** 10 yards to feet

b Ethan's living room is 20 feet long. Lydia's living room is 7 yards long. Ethan says his living room is longer. Explain why he is wrong.

2 The graph shows the charges for Jack's pay-as-you-go mobile phone.

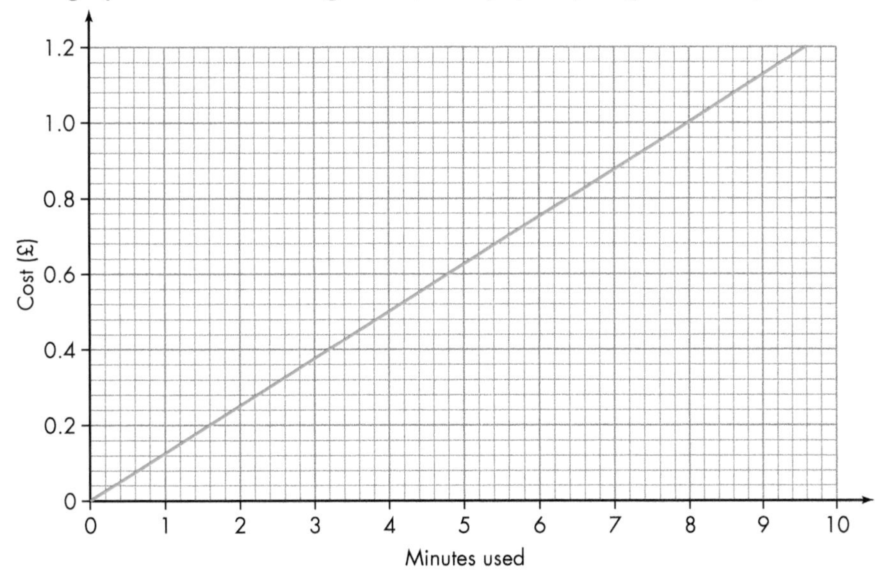

a How much will it cost Jack to make a 4-minute phone call?

b Jack only has £1 credit left on his phone. For how long can he speak before his credit runs out?

c Jack estimates he talks on his phone for 5 minutes each day. He sees a contract that offers 150 minutes a month for £20. Should he change to the contract?

Give reasons for your answer.

3. This is a conversion graph between miles and kilometres.

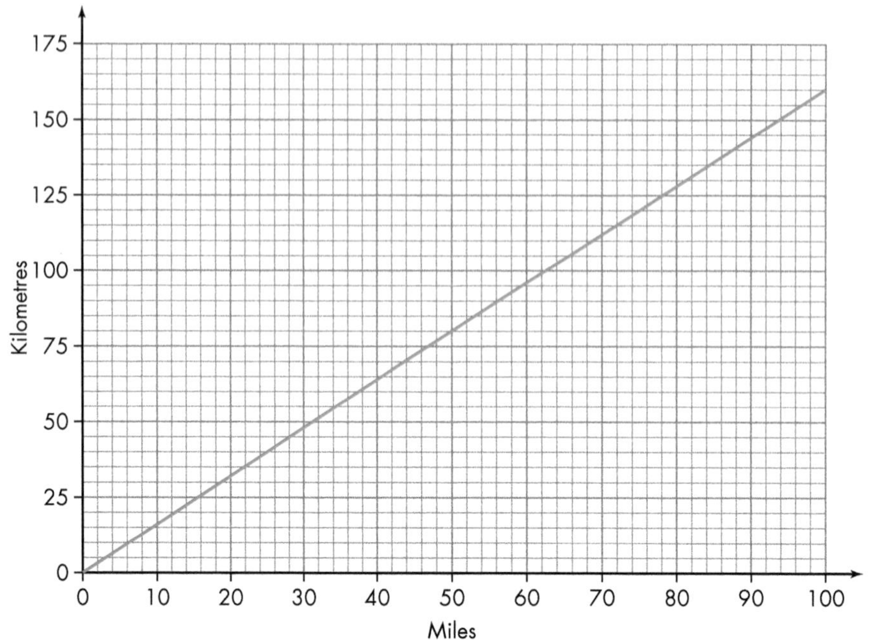

a How many kilometres is equivalent to 50 miles?

b How many miles is equivalent to 120 km?

c Joe drives 45 km in 30 minutes. Lia drives at 55 mph.
Who drives faster? Explain your answer.

8.8 Solving simultaneous equations using graphs

Homework 8I

In questions **1** to **8**, draw the graphs to work out the solution to each pair of simultaneous equations.

1. $y = 3x - 1$
   $y = 2x$

2. $y = 2x - 1$
   $y = x$

3. $y = 3x - 2$
   $y = x - 2$

4. $y = 3 - 2x$
   $y = x$

5. $-x + y = 5$
   $y = 2x - 1$

6. $2x + y = 6$
   $x + y = -6$

7. $x - y = 3$
   $x + y = 5$

8. $x + y = -5$
   $y = 4x$

9 Algebra: Expressions and formulae

9.1 Basic algebra

Homework 9A

1 Write down the algebraic expression for each of these.

a 4 more than x
b 7 less than x
c k more than 3
d t less than 8
e x added to y
f x multiplied by 4
g 5 multiplied by t
h a multiplied by b
i m divided by 2
j p divided by q

2 Val is x years old. Dave is four years older than Val. Ella is five years younger than Val.

a How old is Dave?
b How old is Ella?

3 **a** How many days are there in three weeks?

b How many days are there in z weeks?

4 **a** Jenson has £20 and spends £16. How much does he have left?

b Mia has £10 and spends £a. How much does she have left?

c Steve has £b and spends £c. How much does he have left?

5 **a** Nanny Margaret divides £50 equally between her five grandchildren. How much does each receive?

b Granny Mag divides £r equally between her two grandchildren. How much does each receive?

c Nanny Jess divides £p equally between her q grandchildren. How much does each receive?

6 My brother is five years younger than me. The sum of our ages is 27. How old am I?

7 Sue has p pets. Frank has two more pets than Sue. Chloe has three fewer pets than Sue. Lizzie has twice as many pets as Sue.

How many pets does each person have?

8 Bill has 78p and Ben has 62p. How much should Bill give Ben so they both have the same amount?

9.2 Substitution

Homework 9B

Example

When $x = 1$, the expression $3x + 2$ has the value 5.

When $x = 4$, the same expression has the value 14.

1 Work out the value of $2x + 3$ for each value of x.

a $x = 2$ **b** $x = 5$ **c** $x = 10$

2 Work out value of $3k - 4$ for each value of k.

a $k = 2$ **b** $k = 6$ **c** $k = 12$

3 Work out the value of $4 + t$ for each value of t.

a $t = 4$ **b** $t = 20$ **c** $t = \frac{1}{2}$

4 Evaluate $10 - 2x$ for each value of x.

a $x = 3$ **b** $x = 5$ **c** $x = 6$

5 Evaluate $5y + 10$ for each value of y.

a $y = 5$ **b** $y = 10$ **c** $y = 15$

6 Evaluate $6d - 2$ for each value of d.

a $d = 2$ **b** $d = 5$ **c** $d = \frac{1}{2}$

7 Two of the first recorded units of measurement were the 'cubit' and the 'palm'.

The cubit is the distance from the fingertip to the elbow and the palm is the distance across the hand.

A cubit is four and a half palms.

The actual length of a cubit varied throughout history, but it is now accepted to be 54 cm.

a How many centimetres is a palm?

b If a boat was recorded as being 300 cubits long by 50 cubits wide by 30 cubits high, what are the dimensions of the boat in metres?

8 Work out the value of $\frac{x + 2}{4}$ for each value of x.

a $x = 6$ **b** $x = 10$ **c** $x = 18$

9 Work out the value of $\frac{3x - 1}{2}$ for each value of x.

a $x = 1$ **b** $x = 3$ **c** $x = 4$

10 Evaluate $\frac{20}{p}$ for each value of p.

a $p = 1$ **b** $p = 3$ **c** $p = 4$

11 Work out the value of $3(2y + 5)$ for each value of y.

a $y = 1$ **b** $y = 3$ **c** $y = 5$

12 The rule for converting degrees Fahrenheit (F) into degrees Celsius (C) is:

$$C = \frac{5}{9}(F - 32)$$

a Use this rule to convert 68 °F into degrees Celsius.

b Show, with a suitable substitution, that degrees Fahrenheit and degrees Celsius have the same value at –40°.

9.3 Expanding brackets

Homework 9C

1 Copy the diagram and match the equivalent algebraic expressions. One has been done for you.

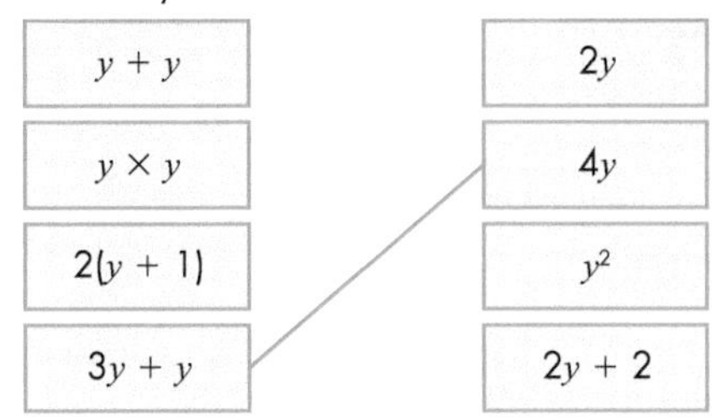

2 Expand these expressions.

a $3(4 + m)$ **b** $6(3 + p)$ **c** $4(4 - y)$ **d** $3(6 + 7k)$

e $4(3 - 5f)$ **f** $2(4 - 23w)$ **g** $7(g + h)$ **h** $4(2k + 4m)$

i $6(2d - n)$ **j** $t(t + 5)$ **k** $m(m + 4)$ **l** $k(k - 2)$

m $g(4g + 1)$ **n** $y(3y - 21)$ **o** $p(7 - 8p)$ **p** $2m(m + 5)$

q $3t(t - 2)$ **r** $3k(5 - k)$ **s** $2g(4g + 3)$ **t** $4h(2h - 3)$

3 An approximate rule for converting degrees Fahrenheit into degrees Celsius is:

$C = 0.5(F - 30)$

Expand these expressions to work out which of the following is an approximate rule for converting degrees Celsius into degrees Fahrenheit.

$F = 2(C + 30)$ $F = 0.5(C + 30)$ $F = 2(C + 15)$ $F = 2(C - 15)$

Homework 9D

1 Simplify these expressions.

a $5t + 4t$ **b** $4m + 3m$ **c** $6y + y$ **d** $2d + 3d + 5d$

e $7e - 5e$ **f** $6g - 3g$ **g** $3p - p$ **h** $5t - t$

i $t^2 + 4t^2$ **j** $5y^2 - 2y^2$ **k** $4ab + 3ab$ **l** $5a^2d - 4a^2d$

2 Expand and simplify.

a $3(2 + t) + 4(3 + t)$ **b** $6(2 + 3k) + 2(5 + 3k)$

c $5(2 + 4m) + 3(1 + 4m)$ **d** $3(4 + y) + 5(1 + 2y)$

e $5(2 + 3f) + 3(6 - f)$ **f** $7(2 + 5g) + 2(3 - g)$

3 Expand and simplify.

a $2(3 + h) - 3(5 + 3h)$ **b** $3(2g + 1) - 2(g + 5)$

c $2(3y + 2) - 3(3y + 1)$ **d** $4(2t + 1) - 3(3t + 1)$

e $2(5k + 3) - 3(2k - 1)$ **f** $4(2e + 3) - 3(3e + 2)$

4 Expand and simplify.

a $m(5 + p) + p(2 + m)$ b $k(4 + h) + h(5 + 2k)$ c $t(1 + 2n) + n(3 + 5t)$
d $p(5q + 1) + q(3p + 5)$ e $2h(3 + 4j) + 3j(h + 4)$ f $3y(4t + 5) + 2t(1 + 4y)$

5 Don wrote the following.

$2(3x - 1) + 5(2x + 3) = 5x - 2 + 10x + 15 = 15x - 13$

Don has made two mistakes in his working.

Explain the mistakes that Don has made.

9.4 Factorisation

Homework 9E

1 Factorise each expression.

a $9m + 12t$ b $9t + 6p$ c $4m + 12k$
d $4r + 6t$ e $4w - 8t$ f $10p - 6k$
g $12h - 10k$ h $2mn + 3m$ i $4g^2 + 3g$
j $4mp + 2mk$ k $4bc + 6bk$ l $8ab + 4ac$

2 Factorise each expression.

a $3y^2 + 4y$ b $5t^2 - 3t$ c $3d^2 - 2d$
d $6m^2 - 3mp$ e $3p^2 + 9pt$ f $8pt + 12mp$
g $8ab - 6bc$ h $4a^2 - 8ab$ i $8mt - 6pt$
j $20at^2 + 12at$ k $4b^2c - 10bc$ l $4abc + 6bed$
m $6a^2 + 4a + 10$ n $12ab + 6bc + 9bd$ o $6t^2 + 3t + at$
p $96mt^2 - 3mt + 69m^2t$ q $6ab^2 + 2ab - 4a^2b$ r $5pt^2 + 15pt + 5p^2t$

3 Factorise these expressions where possible. List those that do not factorise.

a $5m - 6t$ b $3m + 2mp$ c $t^2 - 5t$ d $6pt + 5ab$
e $8m^2 - 6mp$ f $a^2 + c$ g $3a^2 - 7ab$ h $4ab + 5cd$
i $7ab - 4b^2c$ j $3p^2 - 4t^2$ k $6m^2t + 9t^2m$ l $5mt + 3pn$

4 An ink cartridge is priced at £9.99.

The shop has a special offer of 20% off if you buy five or more.

20% of £9.99 is £1.99.

Tom wants six cartridges. Tess wants eight cartridges.

Tom writes down the calculation $6 \times 9.99 - 6 \times 1.99$ to work out how much he must pay.

Tess writes down the calculation $8 \times (9.99 - 1.99)$ to work out how much she must pay.

Both calculations are correct.

a Who has the easier calculation and why?

b How much will each of them pay for their cartridges?

5 a Factorise these expressions.

i $4x + 3 + 5x - 7 - 7x - 4 = 2x - 8$ ii $3x - 12$ iii $x^2 - 4x$

b What do all the answers in part **a** have in common?

6 A class of students were asked to add up all the numbers from 1 to 100 (i.e. $1 + 2 + 3 + 4 + ... + 98 + 99 + 100$).

Two minutes later, a student said she had the correct answer.

The teacher asked the student to show the class her method.

The student wrote:

$(1 + 100) + (2 + 99) + (3 + 98) + ... (50 + 51) = 50 \times 101$

a Explain why this gives the correct answer.

b What is the sum of all the numbers from 1 to 100?

9.5 Quadratic expansion

Homework 9F

1 Use the expansion method to expand these expressions. Then simplify your answers.

a $(x + 2)(x + 4)$ **b** $(x - 3)(x + 1)$ **c** $(x + 4)(x - 1)$ **d** $(x - 5)(x - 2)$

e $(x + 3)(x - 3)$ **f** $(x - 3)(x - 3)$ **g** $(x + 6)(x + 1)$ **h** $(x - 6)(x - 1)$

2 Write down the mistake that has been made in each expansion.

a $(x + 3)(x + 2) = x^2 + 5x + 5$ **b** $(x + 11)(x - 7) = x^2 + 18x + 77$

c $(x - 2)(x + 9) = x^2 - 7x + 18$ **d** $(x - 2)(x - 12) = x^2 - 10x + 24$

Homework 9G

1 Use the FOIL method to expand these expressions. Then simplify your answers.

a $(x + 11)(x + 4)$ **b** $(x - 4)(x + 1)$ **c** $(x + 4)(x - 5)$ **d** $(x - 5)(x - 11)$

e $(x + 3)(x - 2)$ **f** $(x - 7)(x - 3)$ **g** $(x + 2)(x + 8)$ **h** $(x - 7)(x - 1)$

Homework 9H

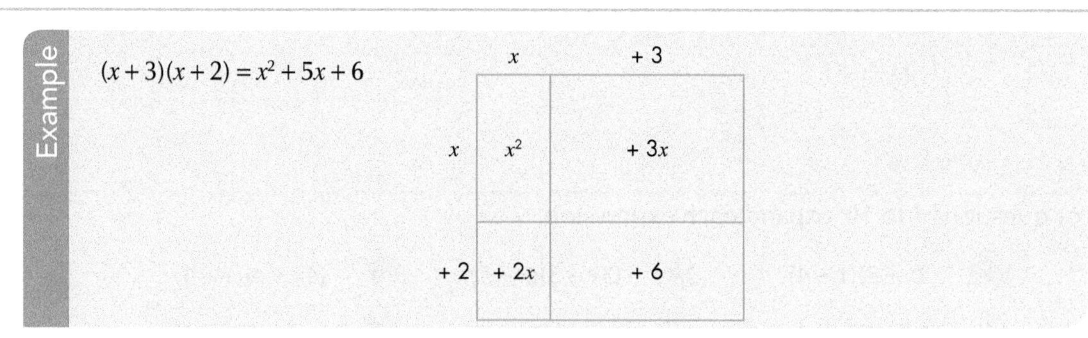

Copy and complete the boxes to expand each expression.

1 $(x + 2)(x + 6) =$

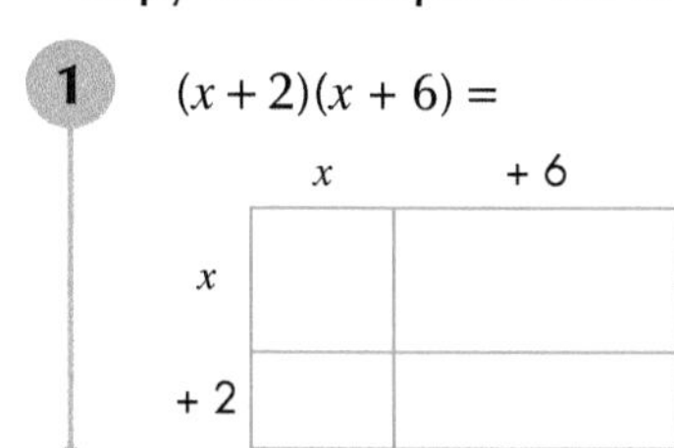

2 $(x + 1)(x + 5) =$

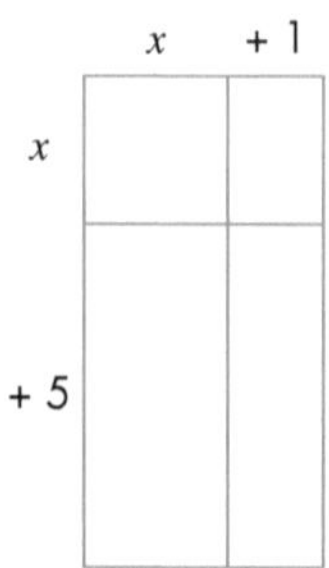

3 $(x - 9)(x + 4) =$

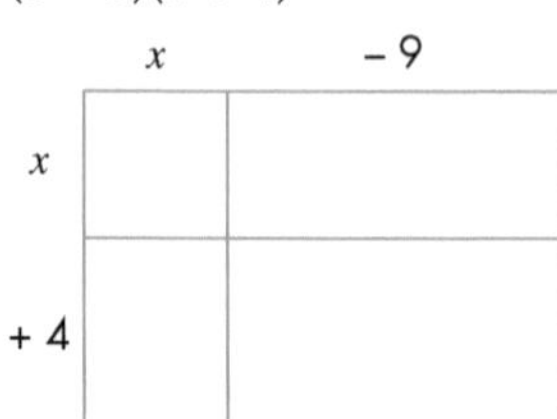

4 $(x + 3)(x + 3) =$

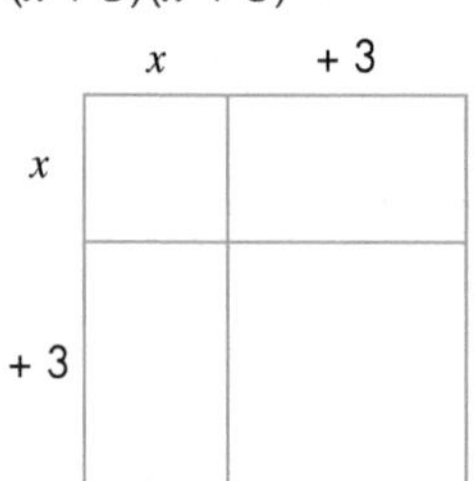

5 Use the box method to expand each expression.

a $(x + 2)(x - 5)$

b $(x - 8)(x + 3)$

c $(x - 4)(x - 4)$

d $(x + a)(x + b)$

e $(x + a)(x + a)$

f $(x + 2)(x - 2)$

Homework 9I

For questions **1** to **10**, expand each expression.

1 $(2x + 2)(x + 4)$

2 $(3x - 3)(x + 1)$

3 $(4x + 4)(x - 1)$

4 $(5x - 5)(x - 2)$

5 $(x + 3)(3x - 3)$

6 $(x - 3)(2x - 3)$

7 $(3x + 6)(2x + 1)$

8 $(4x - 6)(5x - 1)$

9 $(3x + 1)(2x + 5)$

10 $(2y - 2)(2y + 4)$

11 A sports ground is $(2x + 13)$ m long and $(2x - 7)$ m wide. Write an expression for the area of the sports ground.

Homework 9J

1 Expand the squares and simplify.

a $(x+1)^2$ **b** $(x-2)^2$ **c** $(x-9)^2$ **d** $(x+3)^2$ **e** $(x+5)^2$

2 Expand the squares and simplify.

a $(2x-9)^2$ **b** $(a+b)^2$ **c** $(a-b)^2$ **d** $(m-2n)^2$ **e** $(x+y)^2$

f $(2a+3b)^2$ **g** $(3a-6b)^2$

9.6 Quadratic factorisation

Homework 9K

1 Complete the boxes to factorise these expressions. The first one has been done for you.

a $(x+6)(x+1) = x^2 + 7x + 6$

	x	$+1$
x	x^2	x
$+6$	$6x$	$+6$

b (............)(............) $= x^2 + 5x + 6$

c (............)(............) $= x^2 + 14x + 49$

d (............)(............) $= x^2 - 2x - 15$

Factorise the expressions in questions **2** to **5**.

2 **a** $x^2 + 3x + 2$ **b** $x^2 + 9x + 14$ **c** $x^2 - 11x + 28$

3 **a** $x^2 + 7x - 30$ **b** $x^2 - x - 56$ **c** $x^2 + 4x - 21$

4 **a** $x^2 + 10x + 21$ **b** $x^2 + 13x + 40$ **c** $x^2 - 9x + 14$

5 **a** $x^2 + 13x + 36$ **b** $x^2 - 7x - 8$ **c** $x^2 + 3x - 28$

Homework 9L

Each of the expressions in questions **1** to **8** is the difference of two squares. Factorise them.

1 $x^2 - 1$ **2** $x^2 - 121$ **3** $x^2 - 169$ **4** $x^2 - 100$

5 $9 - x^2$ **6** $16 - x^2$ **7** $225 - x^2$ **8** $196 - x^2$

9.7 Changing the subject of a formula

Homework 9M

1 $y = 2x + 3$ Make x the subject.

2 $v = u - 10$ Make u the subject.

3 $T = 2 + 3y$ Make y the subject.

4 $p = q^2$ Make q the subject.

5 $p = \frac{q}{L}$ Make q the subject.

6 $2a = 5b + 1$ Make b the subject.

7 A rocket is fired vertically upwards with an initial velocity of u metres per second. After t seconds the rocket's velocity, v metres per second, is given by the formula $v = u + 10t$.

a Calculate v when $u = 120$ and $t = 6$.

b Rearrange the formula to make t the subject.

c Calculate t when $u = 20$ and $v = 100$.

8 A restaurant has a large oven that can cook up to 10 chickens at a time.

The chef uses the formula:

$T = 10n + 55$

to calculate the length of time, T, it takes to cook n chickens

A large group is booked for a chicken dinner at 7 pm. They will need a total of eight chickens.

a It takes 15 minutes to get the chickens out of the oven and prepare them for serving.

At what time should the chef put the eight chickens into the oven?

b Rearrange the formula to make n the subject

c Another large group is booked for 8 pm the following day. The chef calculates she will need to put the chickens in the oven at 5:50 pm.

How many chickens is the chef cooking for this party?

9 Fern notices that the price of six coffees is 90 pence less than the price of nine teas.

Let the price of a coffee be x pence and the price of a tea be y pence.

a Express the cost of a tea, y, in terms of the price of a coffee, x.

b If the price of a coffee is £1.20, how much is a tea?

10 Distance, speed and time are connected by the formula:

distance = speed × time

A delivery driver drove 90 miles at an average speed of 60 miles per hour.

On the return journey, he was held up at some road works for 30 minutes.

What was his average speed on the return journey?

10 Ratio, proportion and rates of change: Ratio, speed and proportion

10.1 Ratio

Homework 10A

Example

Simplify 5 : 20.

5 : 20 = 1 : 4 (Divide both sides of the ratio by 5.)

Example

Simplify 20p : £2.

20p : 200p = 1 : 10 (Change to a common unit, then divide both sides by 20.)

Example

A garden is divided into lawn and shrubs in the ratio 3 : 2. What fraction of the garden does each cover?

The lawn covers $\frac{3}{5}$ of the garden and the shrubs cover $\frac{2}{5}$ of the garden.

1 Express each ratio in its simplest form.

a 3 : 9 **b** 5 : 25 **c** 4 : 24 **d** 10 : 30 **e** 6 : 9

f 12 : 20 **g** 25 : 40 **h** 30 : 4 **i** 14 : 35 **j** 125 : 50

2 Write each ratio of quantities in its simplest form. (Remember to change to a common unit first, where necessary.)

a £2 to £8 **b** £12 to £16 **c** 25 g to 200 g

d 6 miles : 15 miles **e** 20 cm : 50 cm **f** 80p : £1.50

g 1 kg : 300g **h** 40 seconds : 2 minutes **i** 9 hours : 1 day

j 4 mm : 2 cm

3 Bob and Kathryn share £20 in the ratio 1 : 3.

a What fraction of the £20 does Bob receive?

b What fraction of the £20 does Kathryn receive?

4 In a class of students, the ratio of boys to girls is 2 : 3.

a What fraction of the class are boys?

b What fraction of the class are girls?

5 Pewter is an alloy containing lead and tin in the ratio 1 : 9.

a What fraction of pewter is lead?

b What fraction of pewter is tin?

6 Roy wins $\frac{2}{3}$ of his snooker matches. He loses the rest.

What is his ratio of wins to losses?

7 In the 2013 Ashes cricket series, the numbers of wickets taken by Chris Broad and Graham Swann were in the ratio 5 : 1.

The ratio of the number of wickets taken by James Anderson to those taken by Chris Broad was 2 : 1.

What fraction of the wickets taken by these three bowlers was taken by Graham Swann?

Homework 10B

1 Divide each amount in the given ratio.

a £10 in the ratio 1 : 4 **b** £12 in the ratio 1 : 2 **c** £40 in the ratio 1 : 3

d 60 g in the ratio 1 : 5 **e** 10 hours in the ratio 1 : 9

2 The ratio of female to male members at a sports centre is 3 : 1. The total number of members of the centre is 400.

a How many members are female?

b What percentage of the members are male?

3 A 20-metre length of cloth is cut into two pieces in the ratio 1 : 9. How long is each piece?

4 Divide each amount in the given ratio.

a 25 kg in the ratio 2 : 3 **b** 30 days in the ratio 3 : 2

c 70 m in the ratio 3 : 4 **d** £5 in the ratio 3 : 7

e 1 day in the ratio 5 : 3

5 James collects coasters. The ratio of British coasters to foreign coasters in his collection is 5 : 2. He has 1400 coasters. How many foreign coasters does he have?

6 Patrick and Jane share a box of sweets in the ratio of their ages. Patrick is 9 years old and Jane is 11 years old. If there are 100 sweets in the box, how many does Patrick get?

7 Emily is given £30 for her birthday. She decides to spend four times as much as she saves. How much does she save?

8 You can simplify a ratio by changing it into the form $1 : n$. For example, you can rewrite 5 : 7 as 1 : 1.4 by dividing each side of the ratio by 5.

Rewrite each ratio in the form $1 : n$.

a 2 : 3 **b** 2 : 5 **c** 4 : 5 **d** 5 : 8 **e** 10 : 21

9 The amount of petrol and diesel sold at a garage is in the ratio 2 : 1. One-tenth of the diesel sold is bio-diesel.

What fraction of all the fuel sold is bio-diesel?

Homework 10C

Example

Two business partners, John and Ben, divided their total profit in the ratio 3 : 5.

John received £2100. How much did Ben get?

John's £2100 was $\frac{3}{8}$ of the total profit.

So, $\frac{1}{8}$ of the total profit = £2100 ÷ 3 = £700.

Therefore, Ben's share, which was $\frac{5}{8}$, came to £700 × 5 = £3500.

1 Peter and Margaret's ages are in the ratio 4 : 5. If Peter is 16 years old, how old is Margaret?

2 Cans of lemonade and packets of crisps were bought for the school disco in the ratio 3 : 2. The organiser bought 120 cans of lemonade. How many packets of crisps did she buy?

3 Manuel is making fruit punch from fruit juice and iced soda water in the ratio 2 : 3. Manuel uses 10 litres of fruit juice.

a How many litres of soda water does he use?

b How many litres of fruit punch does he make?

4 Cupro-nickel coins are minted by mixing copper and nickel in the ratio 4 : 1.

a How much copper is needed to mix with 20 kg of nickel?

b How much nickel is needed to mix with 20 kg of copper?

5 The ratio of male to female spectators at a school inter-form football match is 2 : 1.

If 60 males watched the game, how many spectators were there in total?

6 Marmalade is made from sugar and oranges in the ratio 3 : 5. A jar of 'Savilles' marmalade contains 120 g of sugar.

a What is the mass of oranges in the jar?

b What is the total mass of the marmalade in the jar?

7 Each year Abbey School holds a sponsored walk for charity. The money raised is shared between a local charity and a national charity in the ratio 1 : 2. Last year the school gave £2000 to the local charity.

a How much did the school give to the national charity?

b How much did the school raise in total?

8 Fred's blackcurrant juice is made from 4 parts blackcurrant and 1 part water.

Jodie's blackcurrant juice is made from blackcurrant and water in the ratio 7 : 2.

Which juice contains the greater proportion of blackcurrant?

Show how you work out your answer.

10.2 Speed, distance and time

Homework 10D

Hints and tips

The relationship between speed, time and distance can be expressed in three ways.

$$\text{distance} = \text{speed} \times \text{time} \qquad \text{speed} = \frac{\text{distance}}{\text{time}} \qquad \text{time} = \frac{\text{distance}}{\text{speed}}$$

Remember, when you calculate a time and get a decimal answer, do not mistake the decimal part for minutes. You must either:

- leave the time as a decimal number and give the unit as hours, or
- change the decimal part to minutes by multiplying it by 60 (1 hour = 60 minutes) and give the answer in hours and minutes.

1 A cyclist travels a distance of 60 miles in 4 hours. What was his average speed?

2 How far along a motorway would you travel if you drove at an average speed of 60 mph for 3 hours?

3 Mr Baylis drives from Manchester to London in $4\frac{1}{2}$ hours. The distance is 207 miles. What is his average speed?

4 The distance from Leeds to Birmingham is 125 miles. The train I catch travels at an average speed of 50 mph. If I catch the 11:30 am train from Leeds, at what time should I expect to arrive in Birmingham?

5 Copy and complete this table.

	Distance travelled	Time taken	Average speed
a	240 miles	8 hours	
b	150 km	3 hours	
c		4 hours	5 mph
d		$2\frac{1}{2}$ hours	20 km/h
e	1300 miles		400 mph
f	90 km		25 km/h

6 A coach travels at an average speed of 60 km/h for 2 hours on a motorway, then slows down to do the last 30 minutes of its journey at an average speed of 20 km/h.

a What is the total distance of this journey?

b What is the average speed of the coach over the whole journey?

7 Hilary cycles 6 miles to work each day. She cycles the first 5 miles at an average speed of 15 mph and then the last mile in 10 minutes.

a How long does it take Hilary to get to work?

b What is her average speed for the whole journey?

8 Martha drives home from work in 1 hour 15 minutes. She drives home at an average speed of 36 mph.

a Change 1 hour 15 minutes to decimal time in hours.

b How far is it from Martha's work to her home?

9 A tram from A to B takes 15 minutes at an average speed of 16 mph.

The route from A to B by car is 2 miles longer.

How fast would a car need to travel to get from A to B in the same time as the tram?

10.3 Direct proportion problems

Homework 10E

Example

If eight pens cost £2.64, what is the cost of five pens?

First find the cost of one pen. This is £2.64 ÷ 8 = £0.33. The cost of five pens is £0.33 × 5 = £1.65.

1 If four DVDs cost £3.20, what would 10 DVDs cost?

2 Five oranges cost 90p. Work out the cost of 12 oranges.

3 Dylan earns £18.60 in 3 hours. How much will he earn in 8 hours?

4 Barbara bought 12 postcards for 3 euros when she was on holiday in Tenerife.

a How many euros would she have paid for 9 postcards?

b How many postcards could she have bought for 5 euros?

5 Five 'Day-Rover' bus tickets cost £8.50.

a How much will 16 tickets cost?

b Pat has £20. She wants to buy 12 'Day-Rover' bus tickets.

Does she have enough money?

Show your working.

6 A car uses 8 litres of petrol on a trip of 72 miles.

a How much would the same car use on a trip of 54 miles?

b What distance would the car travel on a full tank of 45 litres of petrol?

7 It takes a photocopier 18 seconds to produce 12 copies. How long will it take to produce 32 copies?

8 This is Val's recipe for making 12 flapjacks.

100 g margarine

100 g golden syrup

80 g granulated sugar

200 g rolled oats

a What quantities are needed for:

i 6 flapjacks **ii** 24 flapjacks **iii** 30 flapjacks?

b What is the maximum number of flapjacks she can make if she has 1 kg of each ingredient?

9 Greg the baker sells bread rolls in packs of 6 for £1.

Dom the baker sells bread rolls in packs of 24 for £3.19.

I have £5 to spend on bread rolls.

How many more rolls can I buy from Greg than from Dom?

10.4 Best buys

Homework 10F

1 Compare the prices of the products in each pair. State which, if either, is the better buy.

a

Mouthwash:

£1.50 for a bottle
£2.50 for a twin-pack

b

Deodorant:

£2.20 for 1
£4.45 for 2

c

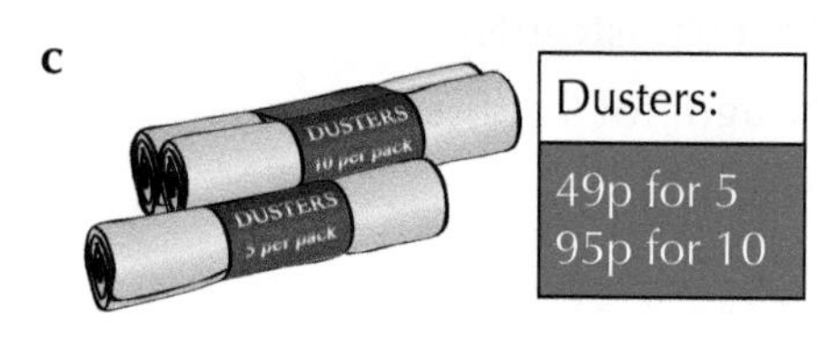

d

2 Compare the products in each pair. State which is the better buy. Explain your choice.

a Tomato ketchup: a medium bottle (200 g) for 55p, a large bottle (350 g) for 87p

b Milk chocolate: a small bar (125 g) for 77p, a large bar (200 g) for 92p

c Coffee: a large tin (750 g) for £11.95, a small tin (500 g) for £7.85

d Honey: a large jar (900 g) for £2.35, a small jar (225 g) for 65p

3 Bottles of cola are sold in different sizes.

a Copy and complete the table.

Size of bottle	Price	Cost per litre
$\frac{1}{2}$ litre	36p	
$1\frac{1}{2}$ litres	99p	
2 litres	£1.40	
3 litres	£1.95	

b Which size of bottle gives the best value for money?

4 Boxes of 'Wetherels' teabags are sold in three different sizes.

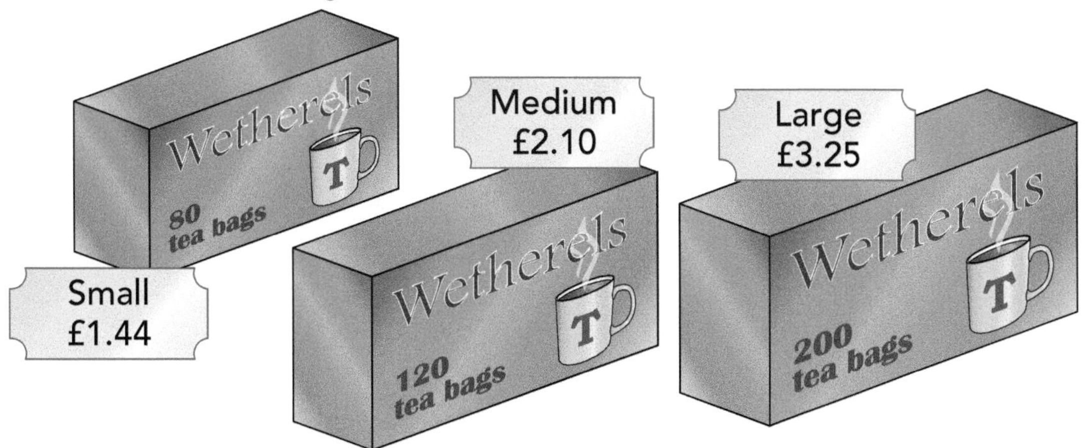

Which size of box gives the best value for money?

5 These product were being promoted by a supermarket.

The large box costs £1.99 and the small box is on offer for 3 for 2. Which is best value? Explain why.

6 Hannah scored 17 out of 20 in a test. John scored 40 out of 50 in a test of the same standard.

Who got the better mark?

11 Geometry and measures: Perimeter and area

11.1 Rectangles

Homework 11A

1 Work out the perimeter of each shape.

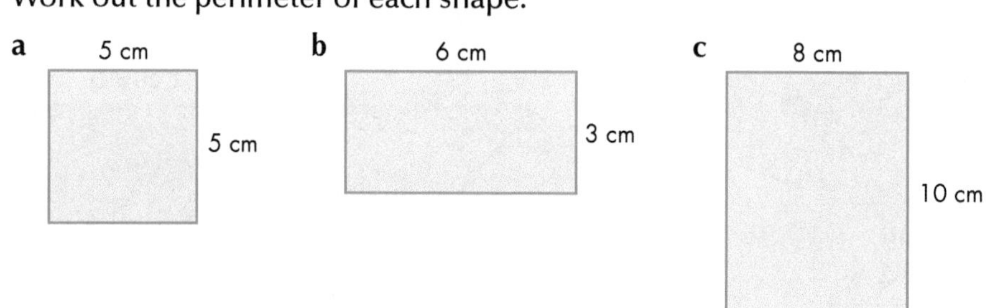

2 Draw as many different rectangles as possible with a perimeter of 14 cm.

3 For each rectangle, calculate **i** the area and **ii** the perimeter.

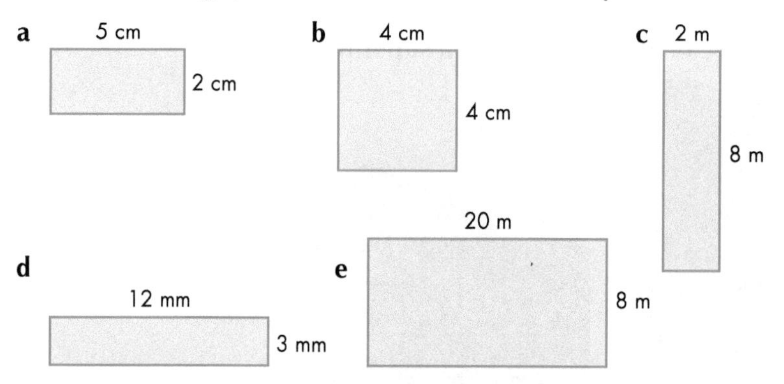

4 Is it possible to draw a rectangle with a perimeter of 9 cm? Explain your answer.

5 Which shape is the odd one out? Give a reason for your answer.

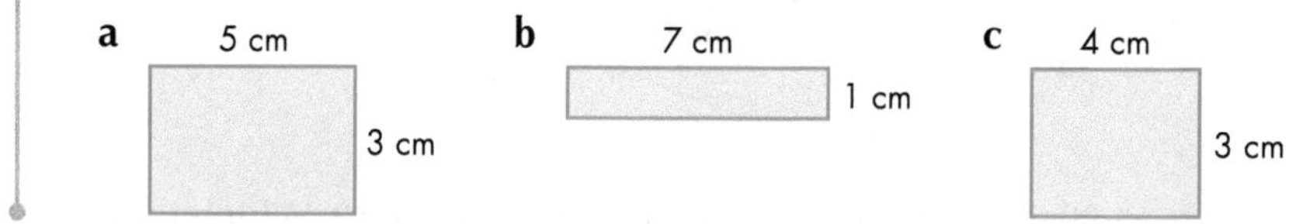

6 Tom wants to put a fence around three sides of this lawn. How much fencing does he need?

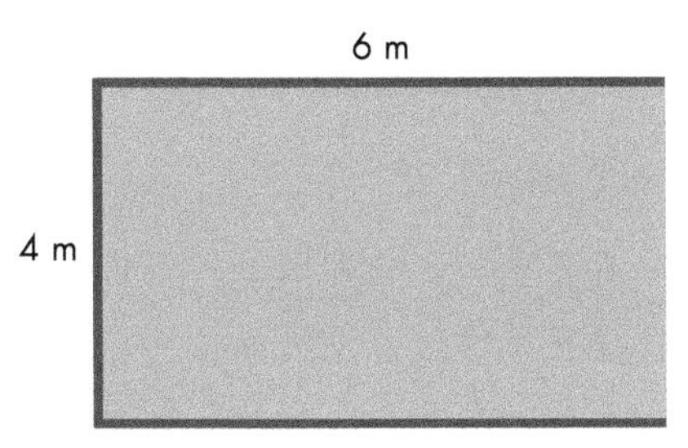

7 Copy and complete the table for rectangles **a** to **e**.

	Length	Width	Perimeter	Area
a	4 cm	2 cm		
b	7 cm	4 cm		
c	6 cm		22 cm	
d		3 cm		15 cm^2
e			30 cm	50 cm^2

8 A square has a perimeter of 24 cm. What is its area?

9 This shape is made from four rectangles that are all the same size.

Work out the area of one of the rectangles.

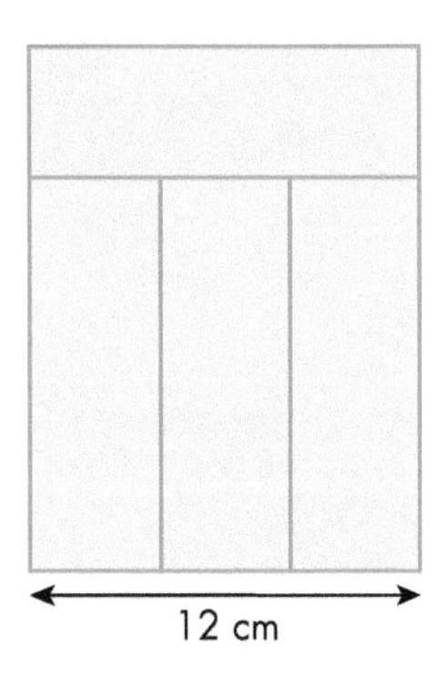

10 The diagrams show the dimensions of Mag's kitchen wall and the size of the square tiles she wants to use to tile the wall. They are not drawn to scale.

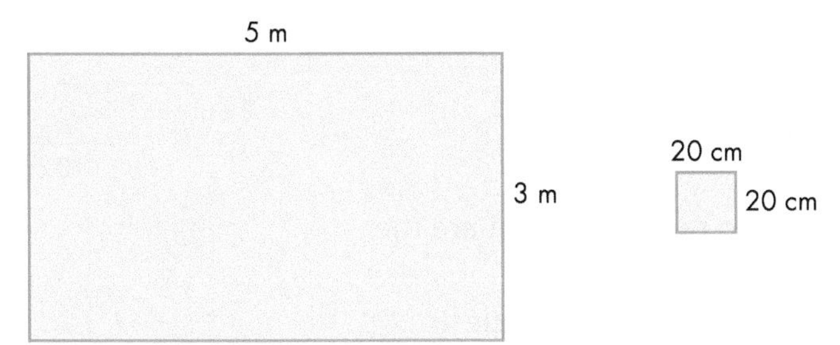

What is the minimum number of tiles Mag will need to cover the wall?

Hints and tips Remember to change the measurements of the wall into centimetres first.

11.2 Compound shapes

Homework 11B

1 Calculate **i** the perimeter and **ii** the area of each shape.

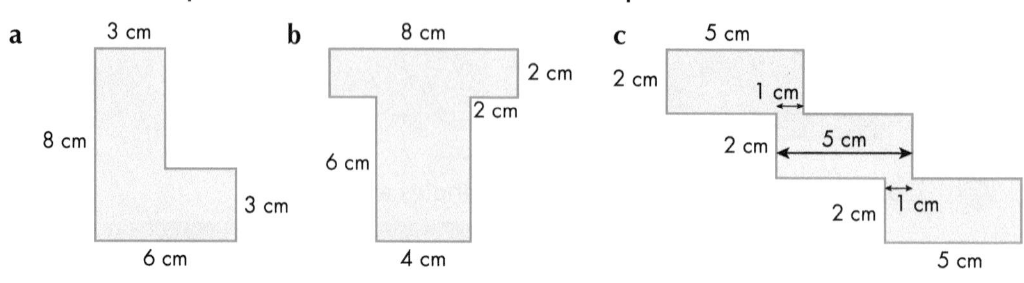

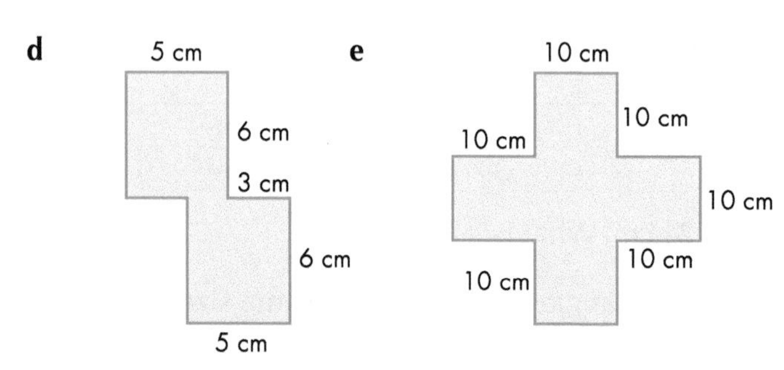

2 Mr Jackson wants to fix laminate onto his kitchen worktop.

The diagram shows the plan view of the worktop.

The laminate comes in rolls that are 5 metres long and 0.5 metres wide.

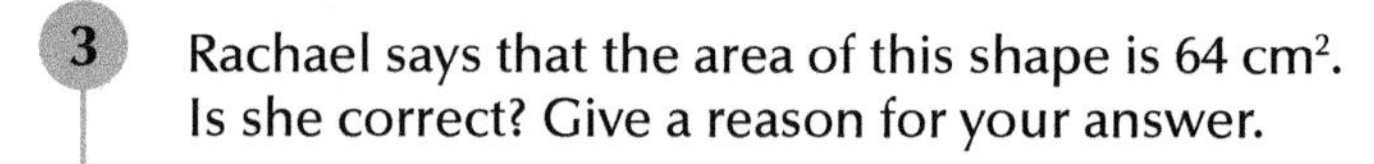

a Work out the area of the worktop.

b Is one roll of laminate enough to cover the worktop?

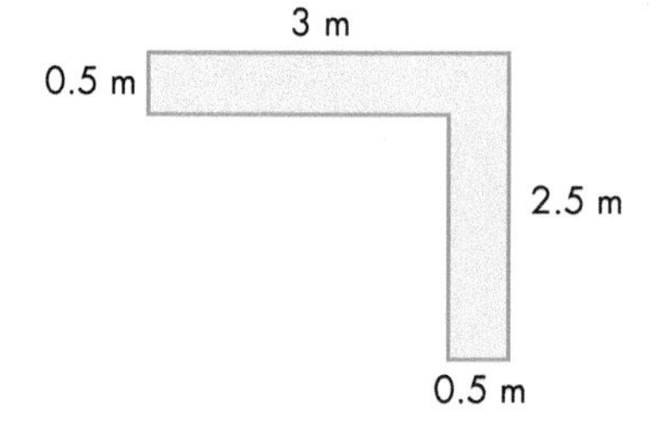

3 Rachael says that the area of this shape is 64 cm².
Is she correct? Give a reason for your answer.

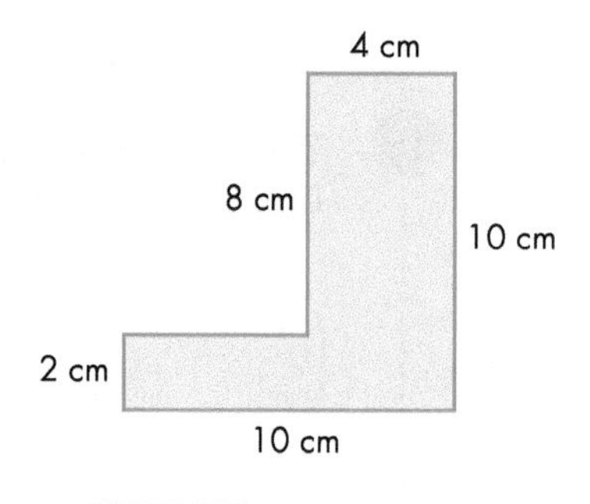

4 This L-shape is made from two rectangles that are the same size. It has an area of 48 cm².

Work out the length and width of the rectangles.

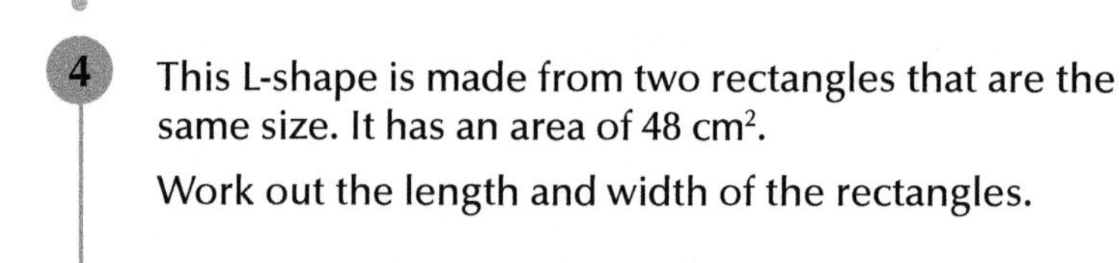

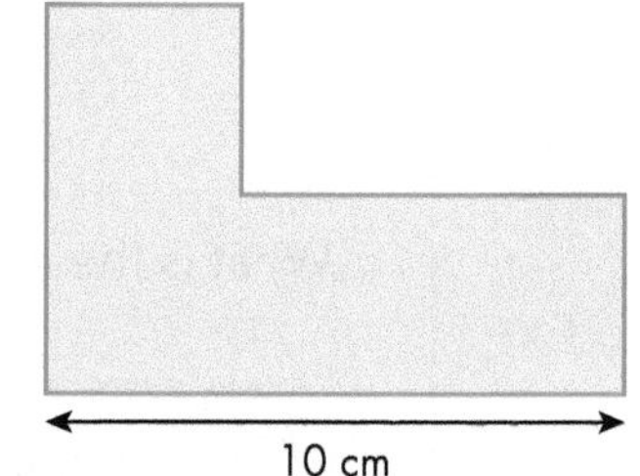

11.3 Area of a triangle

Homework 11C

Example

Work out the area of this triangle.

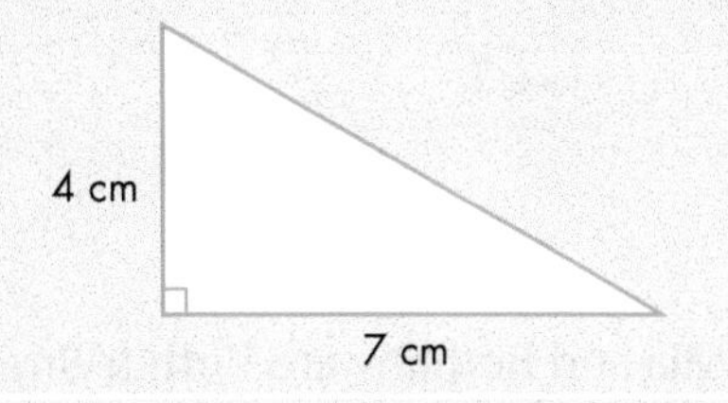

$$\text{Area} = \frac{1}{2} \times 7 \times 4$$
$$= \frac{1}{2} \times 28$$
$$= 14 \text{ cm}^2$$

1 Work out the perimeter and area of each triangle.

a

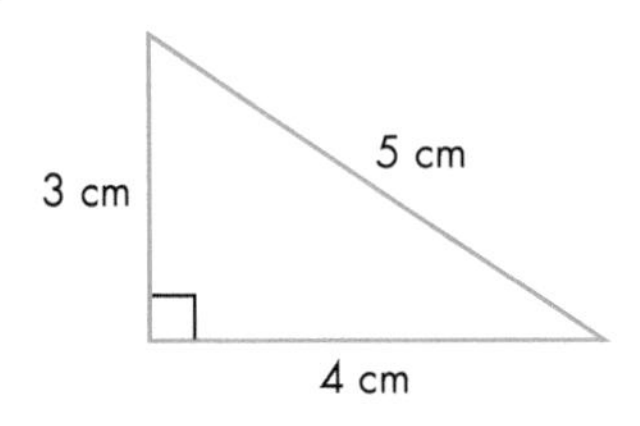

b

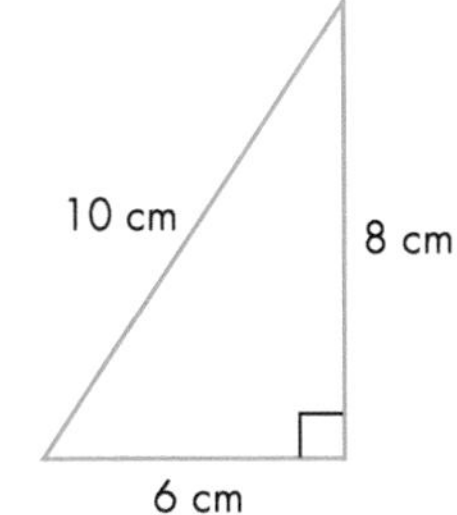

c

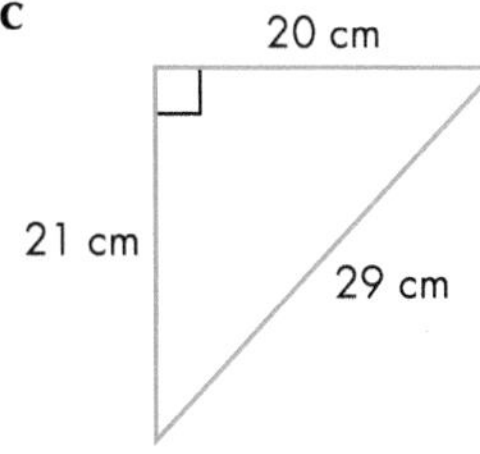

2 These compound shapes are made from rectangles and right-angled triangles. Work out the area of each shape.

a

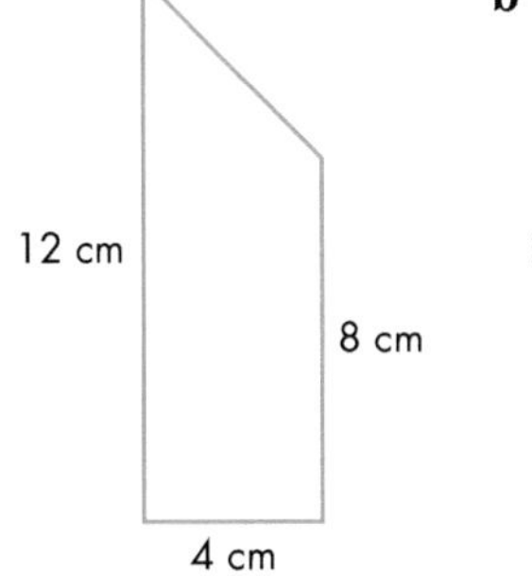

b

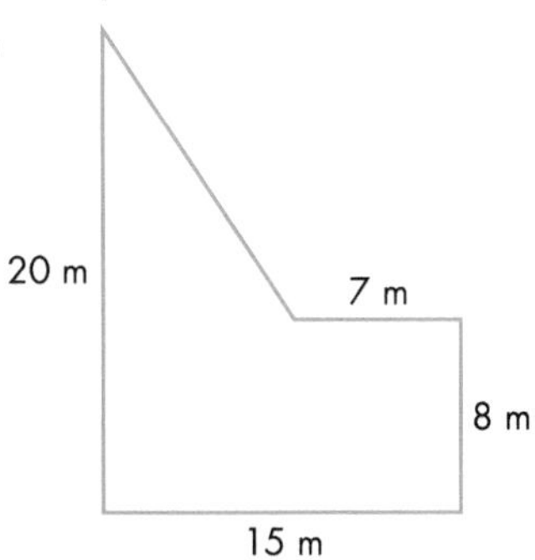

c

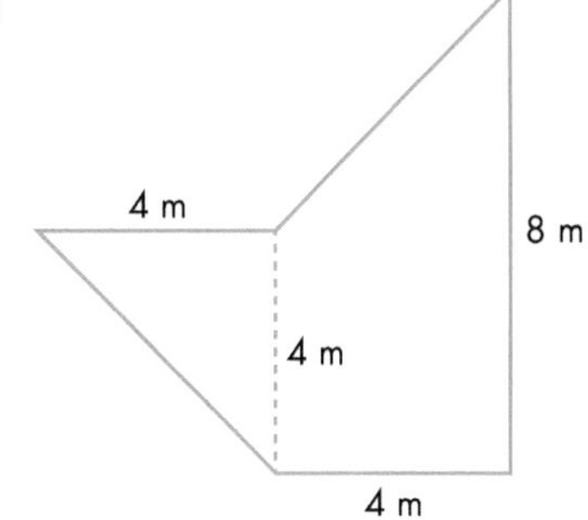

3 Work out the area of the wood on this blackboard set square.

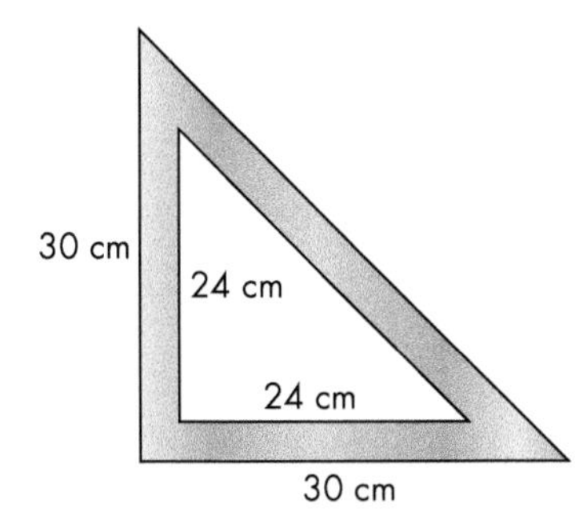

4 Which of these three triangles has the smallest area?

A

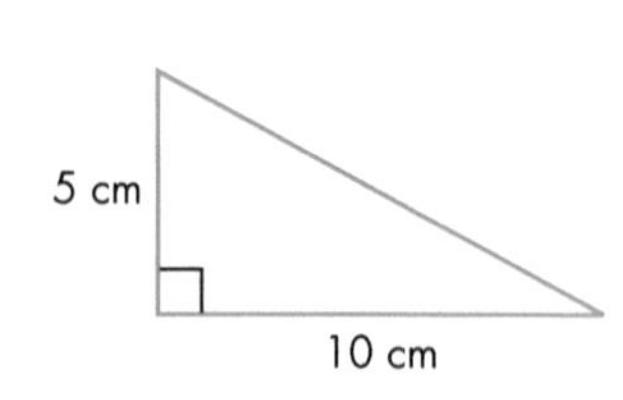

B

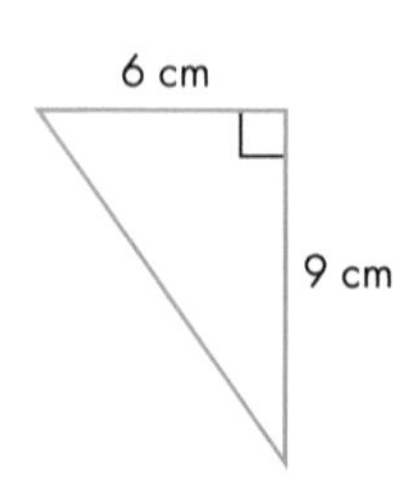

C

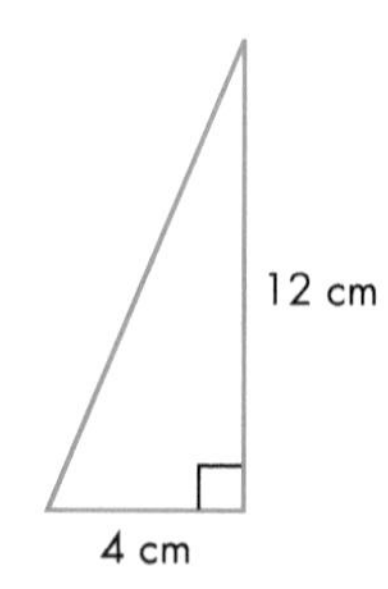

5 Mia and Bethany are both trying to work out the area of the right-angled triangle below.

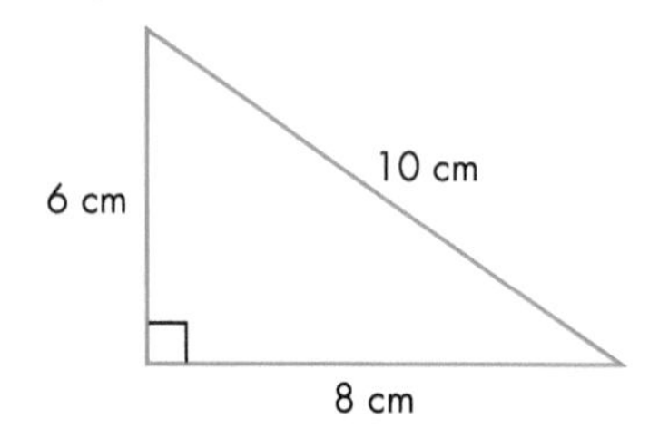

Mia's working	Bethany's working
$A = \frac{1}{2} \times 8 \times 6$	$A = \frac{1}{2} \times 8 \times 10$
$= 4 \times 6$	$= 4 \times 10$
$= 24\ cm^2$	$= 40\ cm^2$

Who is correct?

Give a reason for your answer.

6 Work out the area of this rhombus.

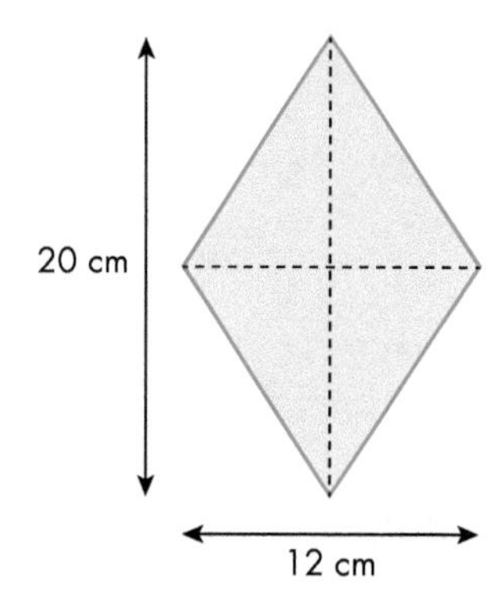

Hints and tips The diagonals of the rhombus intersect at right angles.

Example

Work out the area of this triangle.

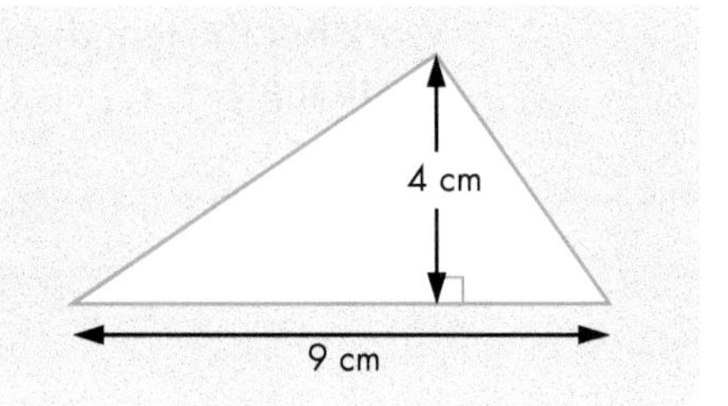

$\text{Area} = \frac{1}{2} \times 9 \times 4$

$= \frac{1}{2} \times 36$

$= 18 \text{ cm}^2$

1 Calculate the area of each triangle.

a

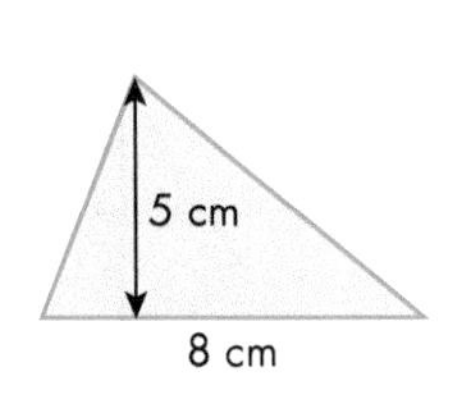

b

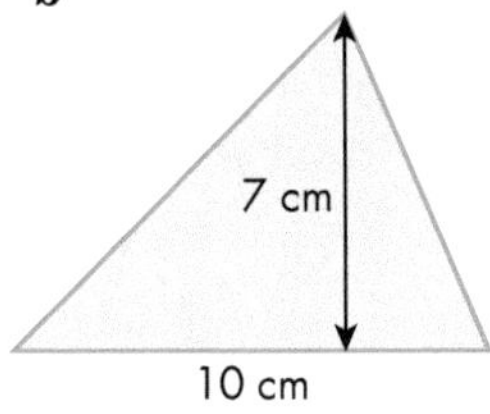

c

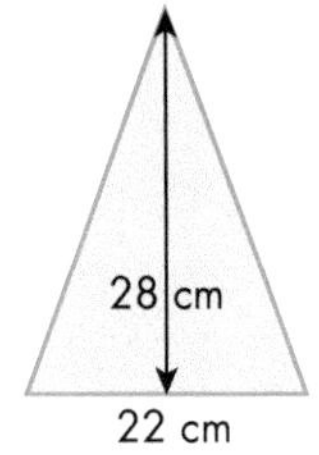

d

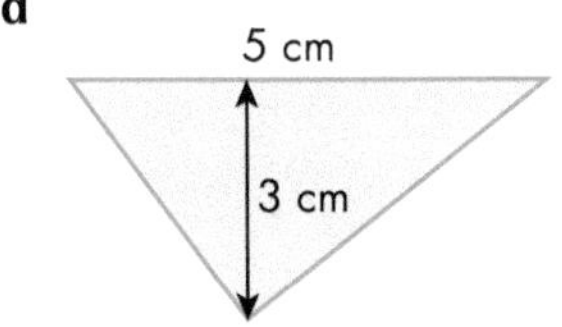

e

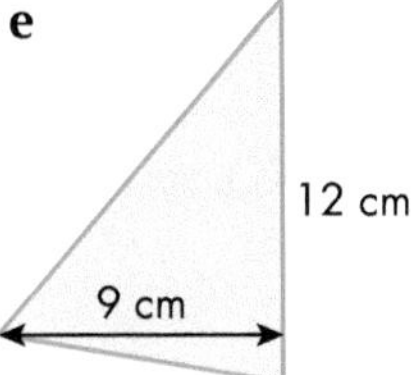

f

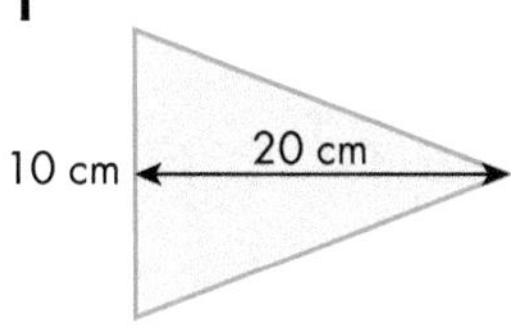

2 Copy and complete the following table for triangles **a** to **e**.

	Base	Vertical height	Area
a	6 cm	8 cm	
b	10 cm	7 cm	
c	5 cm	5 cm	
d	4 cm		12 cm²
e		20 cm	50 cm²

3 Work out the shaded area of each shape.

a

10 cm

40 cm

80 cm

b

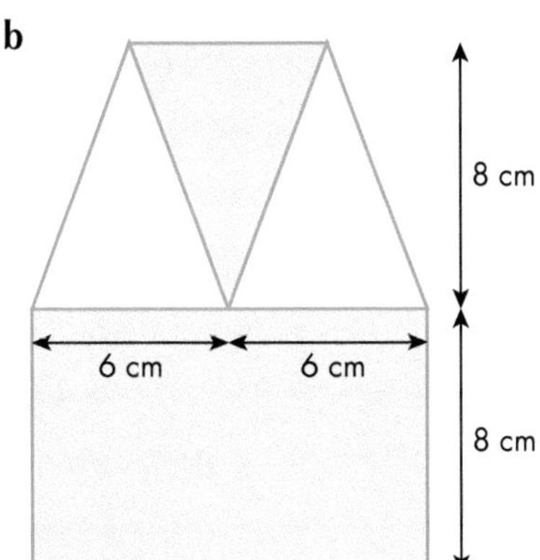

c

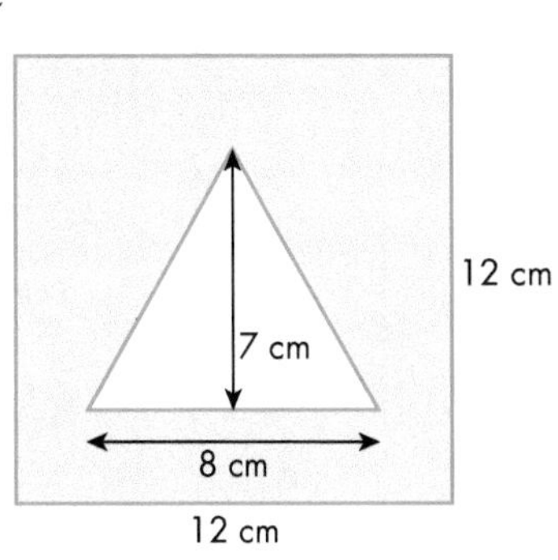

4 Draw diagrams to show triangles with different dimensions that have an area of 40 cm².

5 The rectangle and triangle on the right have the same area.

Work out the length of the base of the triangle.

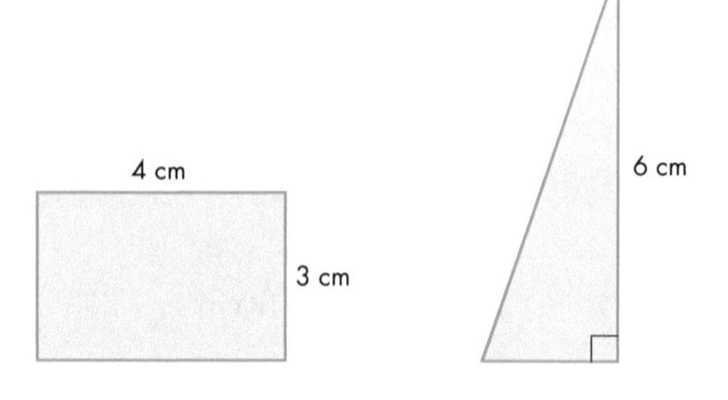

6 Compare these triangles.

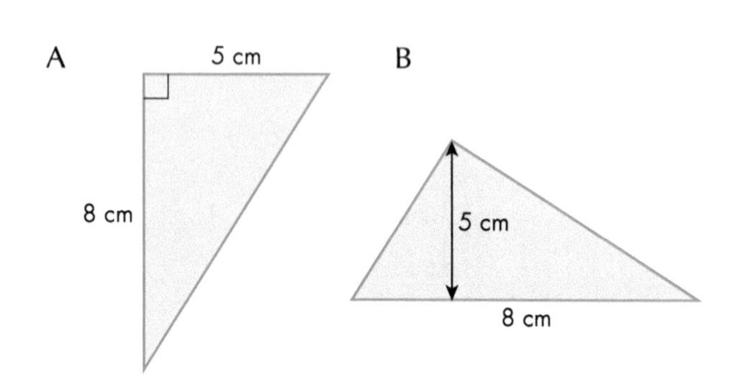

11.4 Area of a parallelogram

Homework 11E

Example

Work out the area of this parallelogram.

6 cm

8 cm

Area $= 8 \times 6$

$= 48 \text{ cm}^2$

1 Calculate the area of each parallelogram.

a 3 cm, 5 cm

b 5 cm, 8 cm

c 4 cm, 4 cm

d 10 cm, 24 cm

2 Work out the area of the shaded part of this shape.

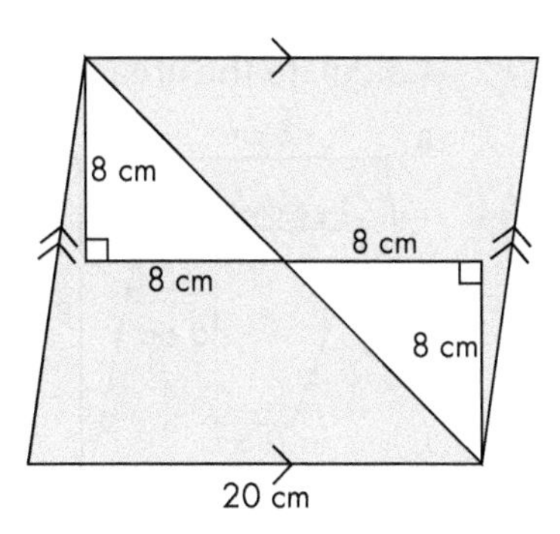

3 Which two of these shapes have the same area? Show your working.

a

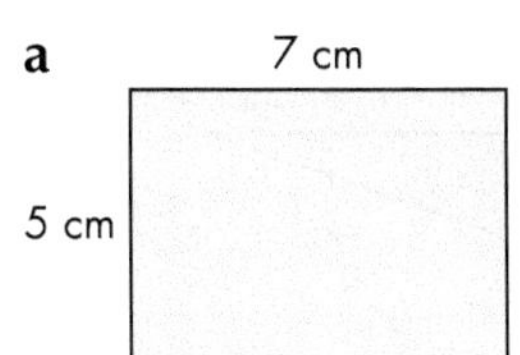

b

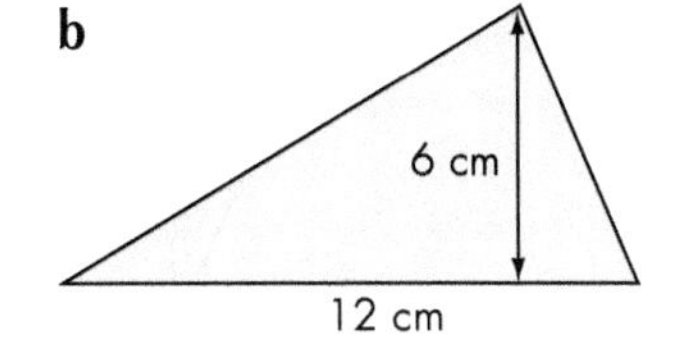

c

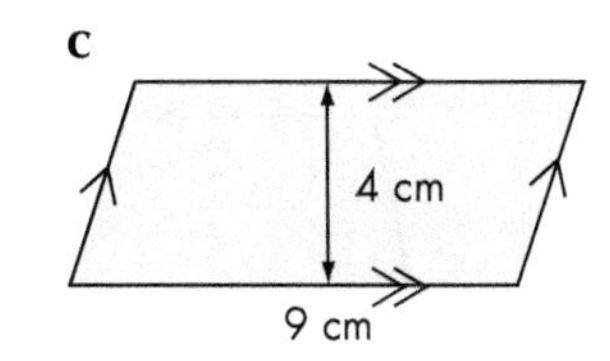

4 A square has the same area as this parallelogram.
What is the perimeter of the square?

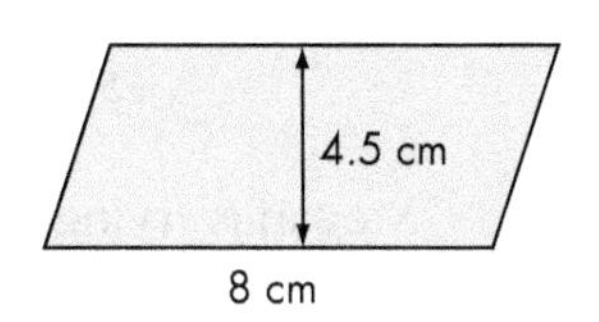

11.5 Area of a trapezium

Homework 11F

1 Calculate **i** the perimeter and **ii** the area of each trapezium.

a

5 cm

5 cm

4 cm

4.1 cm

9 cm

b

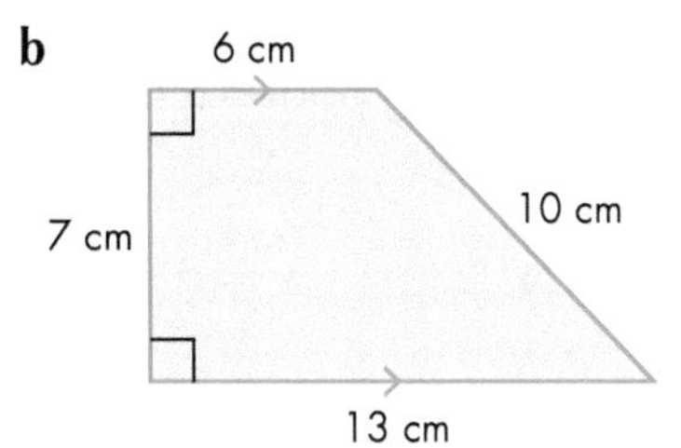

Hints and tips Be careful not to use the slanting side as the height.

2 Calculate the area of each of these compound shapes.

a

7 m

4 m

3 m

3 m

15 m

b

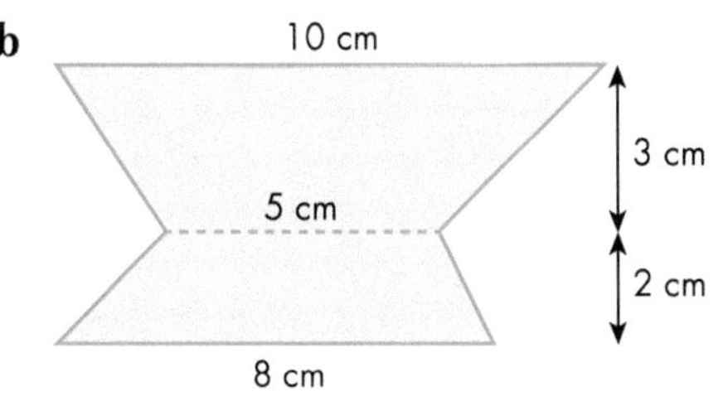

3 Calculate the area of the shaded part in each diagram.

a

b

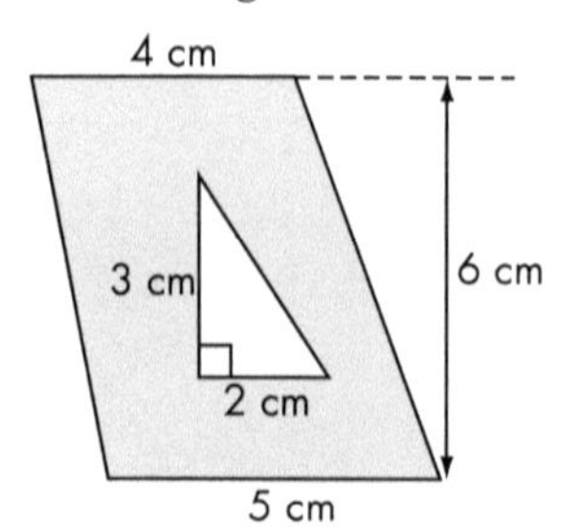

4 Which of these shapes has a larger area?

a

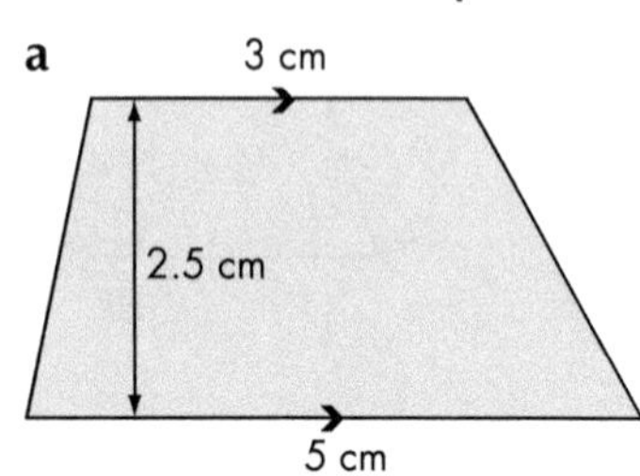

b

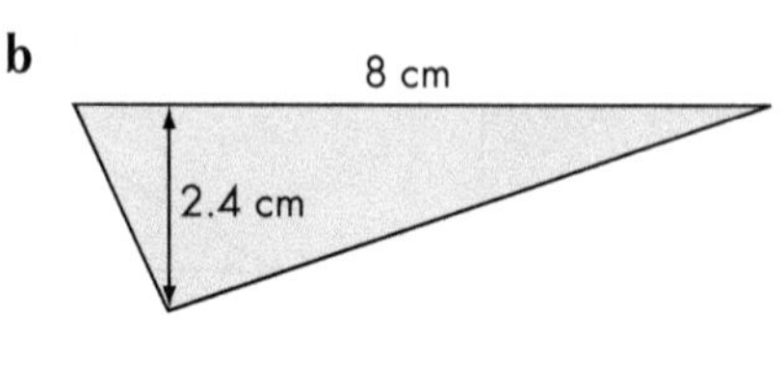

5 Megan is trying to work out the area of this trapezium.

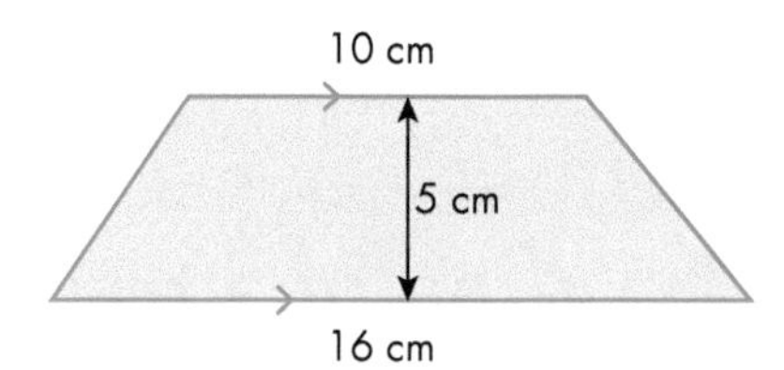

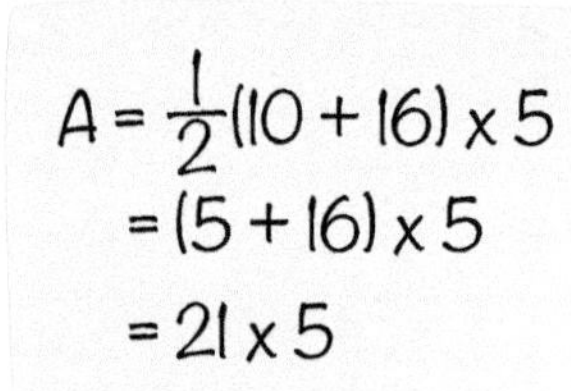

She has made three mistakes. Write out a correct solution to the question.

6 The side of a swimming pool is a trapezium, as shown in the diagram. Calculate its area in square metres.

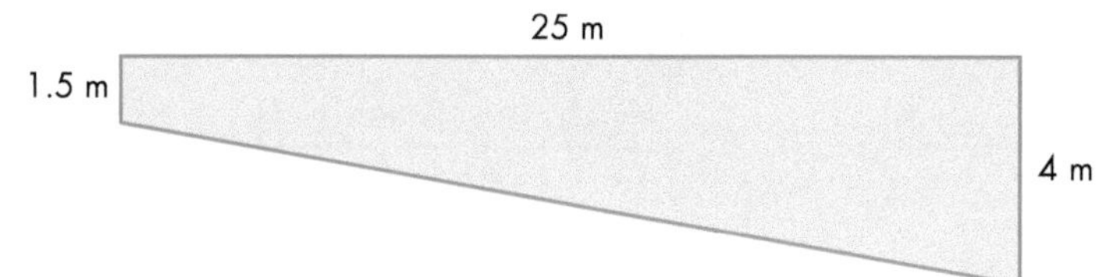

7 The area of this trapezium is 40 cm². Work out possible values for *a* and *b*.

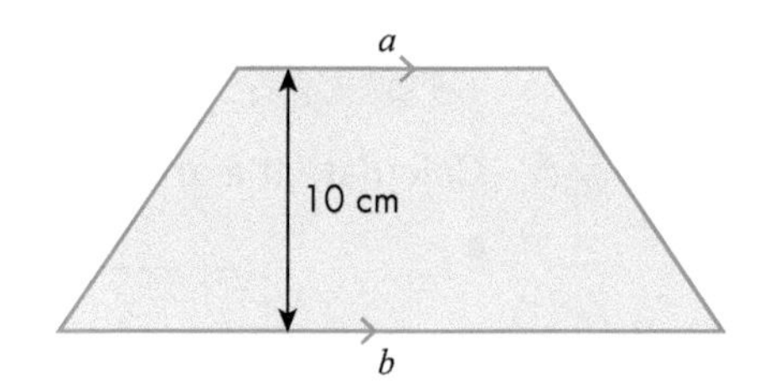

11.6 Circles

Homework 11G

1 Copy the diagram and label it using the circle terms below.

Arc Radius Tangent Diameter Sector Segment
Chord Circumference

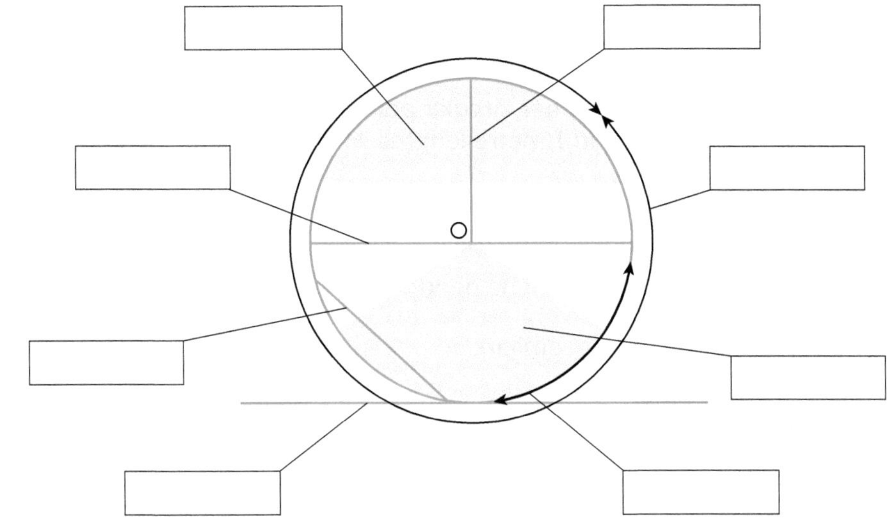

Homework 11H

Example

Work out the circumference of a circle with a diameter of 4 cm.

Remember, $C = \pi d$

So, $C = \pi \times 4$

$= 12.6$ cm (1 dp)

1 Calculate the circumference of each circle. Give your answers to one decimal place.

a 3 cm

b 9 cm

c 10 cm

d

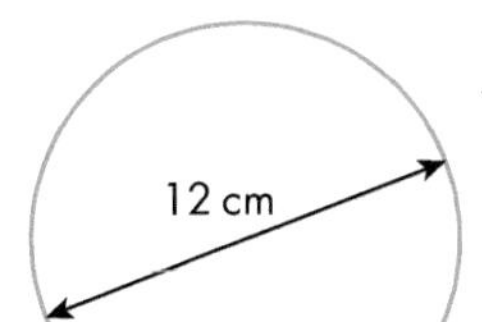

e

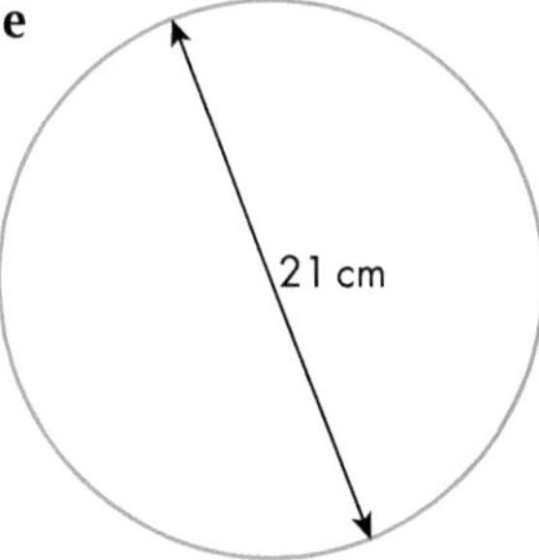

2 Calculate the circumference of each circle. Give your answers to one decimal place.

a 2 cm **b** 3.5 cm **c** 7 cm **d** **e**

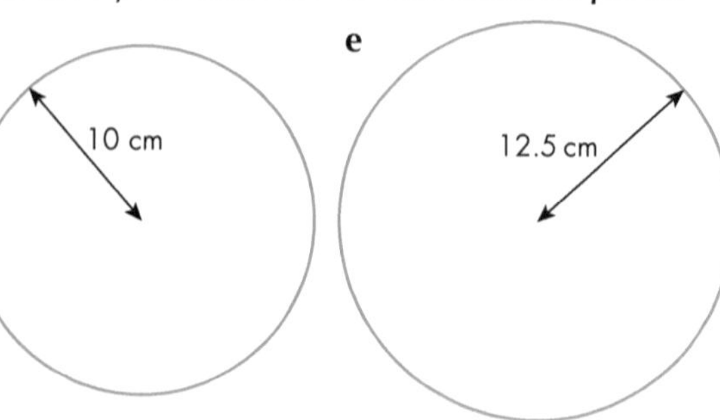

3 Pat wants to put a fence around her circular pond. The pond has a diameter of 15 m. She plans to buy the fencing in 1-metre lengths. How much fencing does she need?

4 Roger trains by running around a circular track that has a radius of 50 m.

a Calculate the circumference of the track. Give your answer to one decimal place.

b How many complete circuits will he need to run to be sure of running 5000 m?

5 Calculate the perimeter of this semicircle.

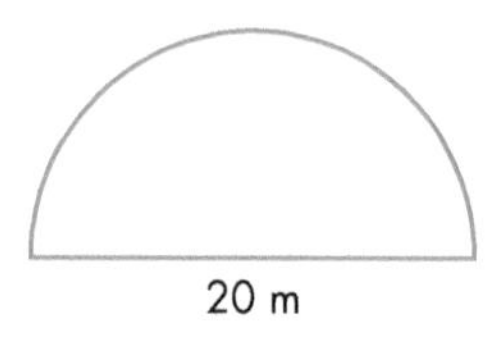

6 What is the diameter of a circle with a circumference of 40 cm? Give your answer to one decimal place.

7 A trundle wheel is used by surveyors to measure distances. One complete turn of the wheel is 1 m. What is the radius of a trundle wheel?

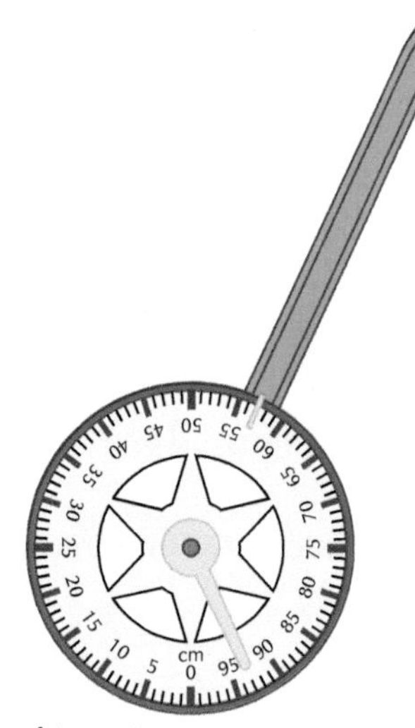

8 A circle has a radius of r cm. Another circle has a radius of $(r + 1)$ cm.

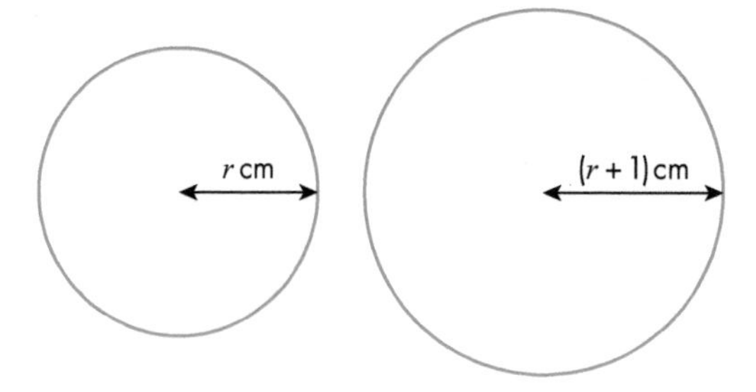

Show that the difference between the two circumferences is 2π cm.

9 The diameter of a cotton reel is 3 cm.

Cotton is wound onto the reel by rotating it on a machine.

A manufacturer wants to put on 80 m of cotton. How many rotations is this?

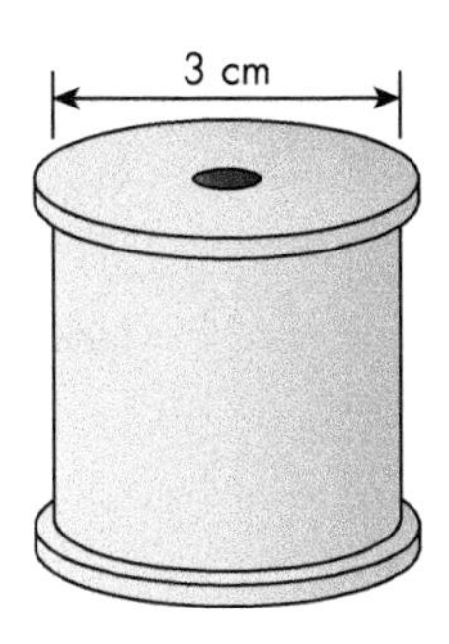

11.7 The area of a circle

Homework 11I 

Example

Calculate the area of a circle with a radius of 7 cm.

$A = \pi r^2$

So, $A = \pi \times 7^2$

$= \pi \times 49$

$= 153.9 \text{ cm}^2$ (1 dp)

1 Calculate the area of each circle. Give your answers to one decimal place.

a 2 cm **b** 6 cm **c** 8 cm **d** 10 cm **e**

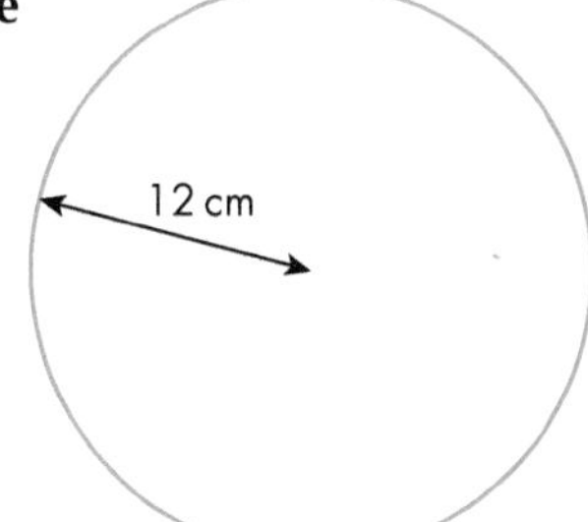

2 Calculate the area of each circle. Give your answers to one decimal place.

a **b** **c** **d** **e**

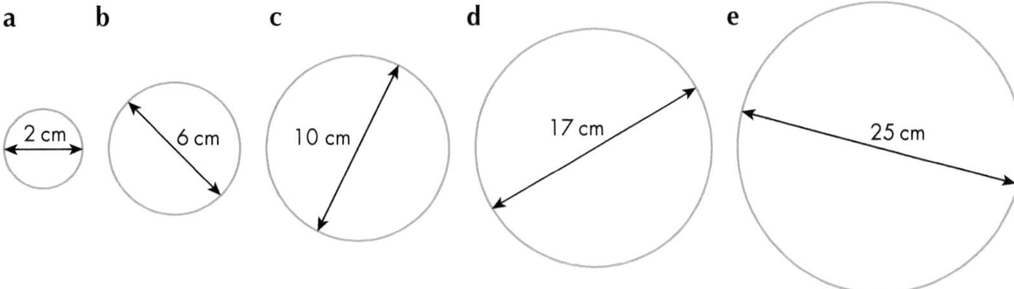

3 Helen is serving a meal for six people. Her circular table has a diameter of 80 cm.

a To sit in comfort around the table, each person needs at least 40 cm, plus 30 cm 'elbow room'. Is the table big enough for six people to sit comfortably? Give a reason for your answer.

b Helen wants her tablecloth to overhang the table by 10 cm. What size of circular tablecloth should she use?

4 The diagram shows the dimensions of Sasha's pond.

She wants to buy water lilies for the pond and would like six plants per square metre.

How many plants should she buy?

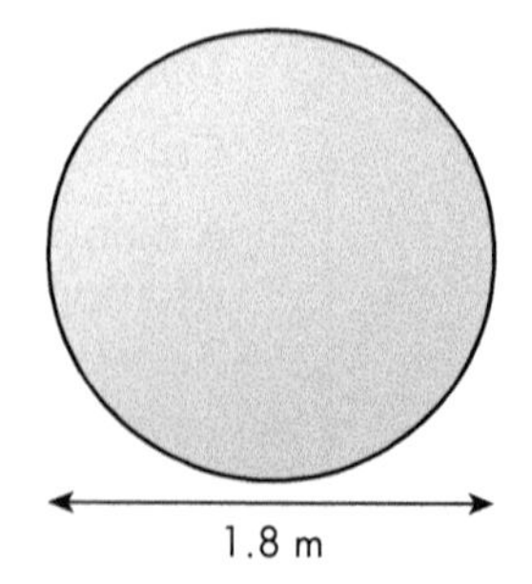

5 The diagram shows a circular path around a flowerbed. The radius of the flowerbed is 6 m and the width of the path is 1 m.

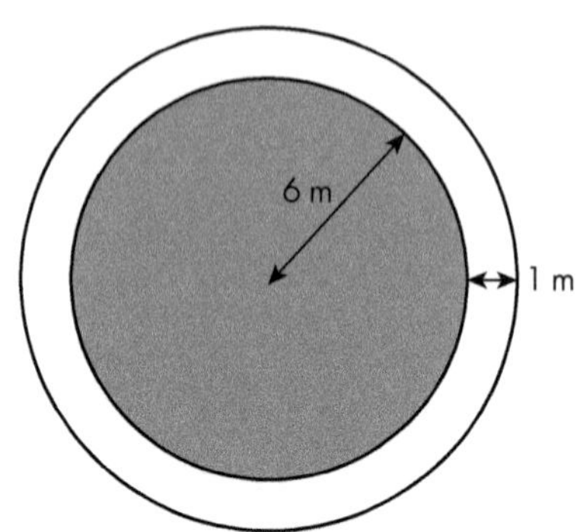

a Calculate the area of the flowerbed.

b Write down the radius of the large circle.

c Calculate the area of the large circle.

d Calculate the area of the path.

e Concrete costs £12 per square metre. The budget available is £300. Is this enough for a concrete path?

6 The diagram shows a running track.

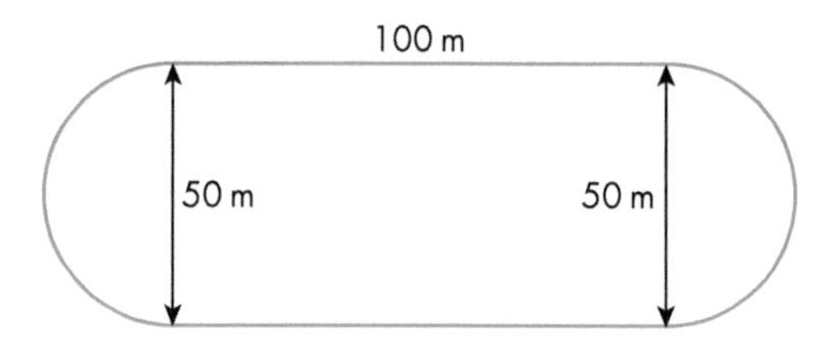

a Calculate the perimeter of the track. Give your answer to the nearest whole number.

b Calculate the total area inside the track. Give your answer to the nearest whole number.

7 A circle has a circumference of 50 cm.

a Calculate the diameter of the circle to one decimal place.

b What is the radius of the circle to one decimal place?

c Calculate the area of the circle to one decimal place.

8 The diagram shows a metal ring.

Calculate the area of the ring. Give your answer to one decimal place.

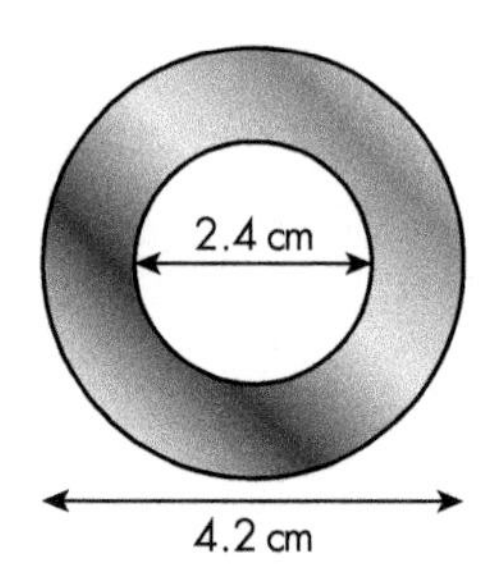

9 The diameter of a circle is d.

Because the diameter (d) is easier to measure than the radius (r), engineers often use the formula $A = \frac{\pi d^2}{4}$ to calculate the area of a circle. Show that this formula is equivalent to $A = \pi r^2$.

10 The diagram shows a shape made from a semicircle and a rectangle.

Calculate the area of this shape.

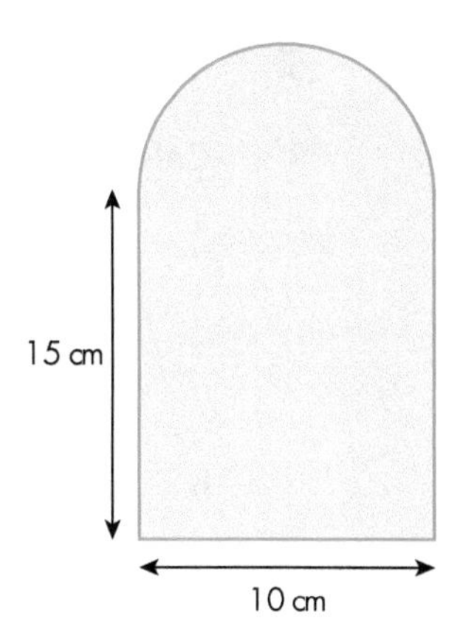

11.8 Answers in term of π

Homework 11J

Example

Write down the circumference (C) and area (A) of a circle with a radius of 5 cm. Give your answers in terms of π.

$C = \pi d$

$= \pi \times 10$

$= 10\pi$ cm

$A = \pi r^2$

$= \pi \times 25$

$= 25\pi$ cm^2

In this exercise, give all your answers in terms of π.

1 State the circumference of circles with each of these measures.

a diameter 7 cm
b radius 5 cm
c diameter 19 cm
d radius 3 cm

2 State the area of circles with each of these measures.

a radius 8 cm
b diameter 7 cm
c diameter 18 cm
d radius 9 cm

3 Sean is trying to calculate the area of a circle that has a radius of 8 cm. He has been asked to write down his answer in terms of π.

Sean writes down: $A = 16\pi$ cm^2

Explain why this is incorrect.

4 State the diameter of a circle with a circumference of 4π cm.

5 State the radius of a circle with an area of 36π cm^2.

6 State the diameter of a circle with a circumference of 20 cm.

7 State the radius of a circle with an area of 20 cm^2.

8 Calculate **i** the perimeter and **ii** the area of each shape.

a

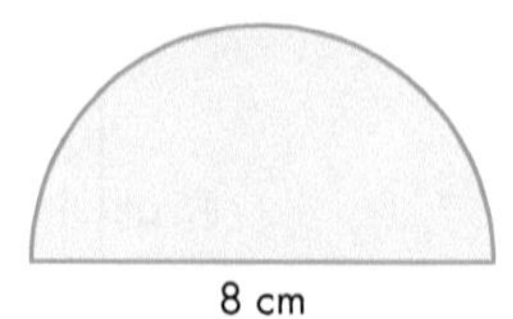

b

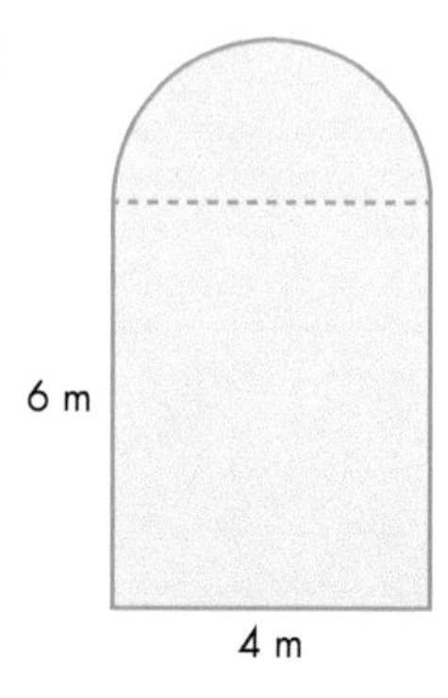

9 A star shape is made by cutting four quadrants from a square with side length $2a$.

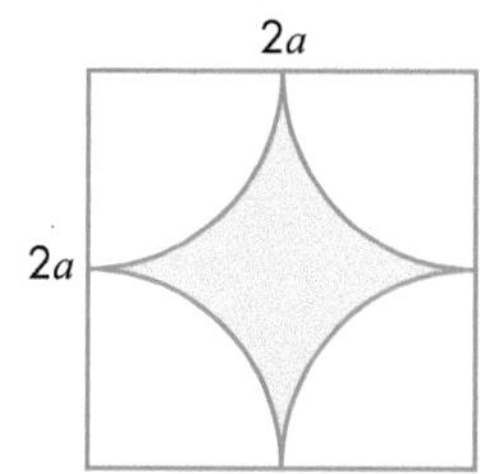

Write down the formula for the area of the star in terms of π and a.

12 Geometry and measures: Transformations

12.1 Rotational symmetry

Homework 12A

You can use tracing paper to help with this exercise.

1 Copy these shapes and write its order of rotational symmetry below each one.

a **b** **c**

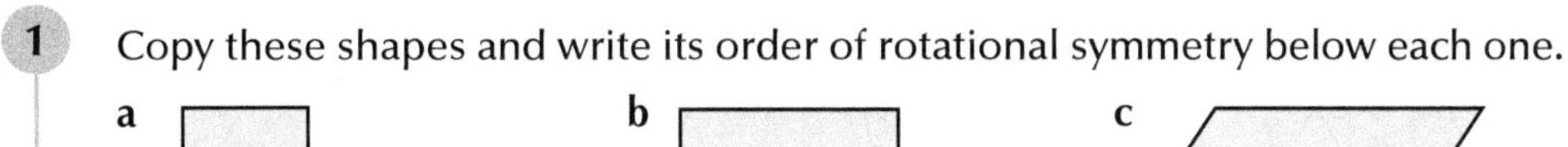

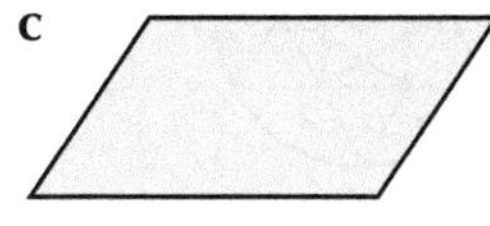

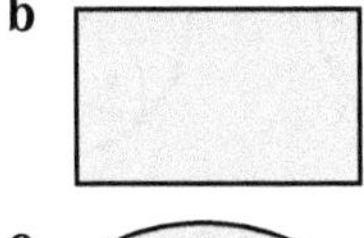

d **e**

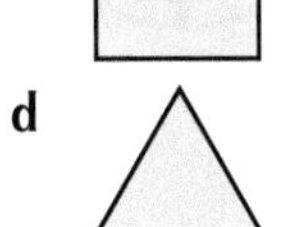

2 Write down the order of rotational symmetry for each shape.

a 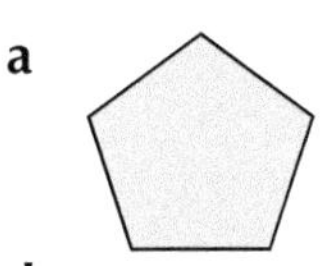

b 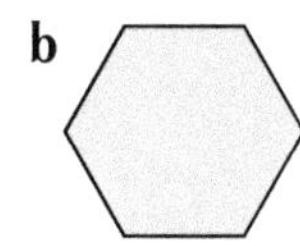

c 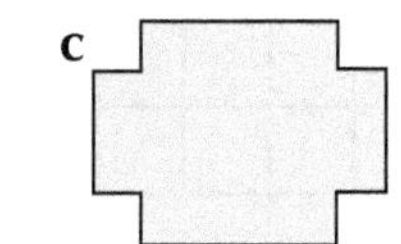

d 

e 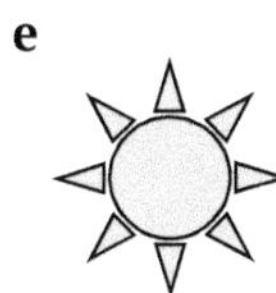

3 Write down the order of rotational symmetry for each symbol.

a 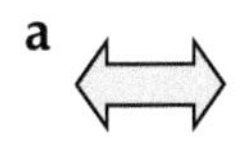

b 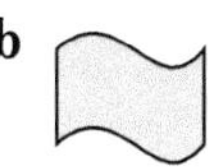

c 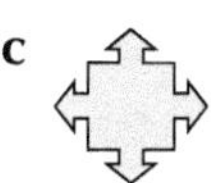

d 

e 

4 The upright capital letter A fits exactly onto itself only once. So, its order of rotational symmetry is 1. This means that it has no rotational symmetry. Write down the order of rotational symmetry for each of these letters.

a E **b** H **c** I **d** L **e** N

f Q **g** S **h** Z

5 Draw two copies of the diagram below.

a On the first copy, shade in two more squares so that the diagram has rotational symmetry of order 2 and no lines of symmetry.

b On the second copy, shade in two more squares so that the diagram has rotational symmetry of order 1 and exactly 1 line of symmetry.

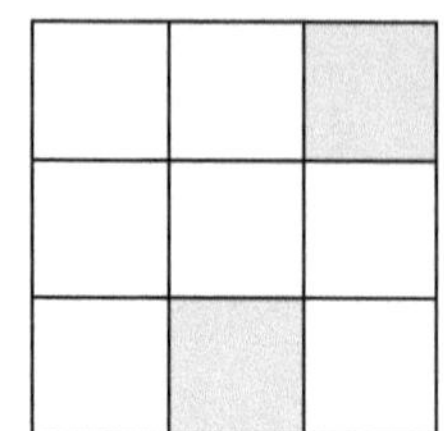

6 These patterns are taken from old Turkish coins.

What is the order of rotational symmetry for each one?

a  **b**  **c**  **d** 

7 On a copy of this shape, shade in four more squares so that the shape has rotational symmetry of order 2.

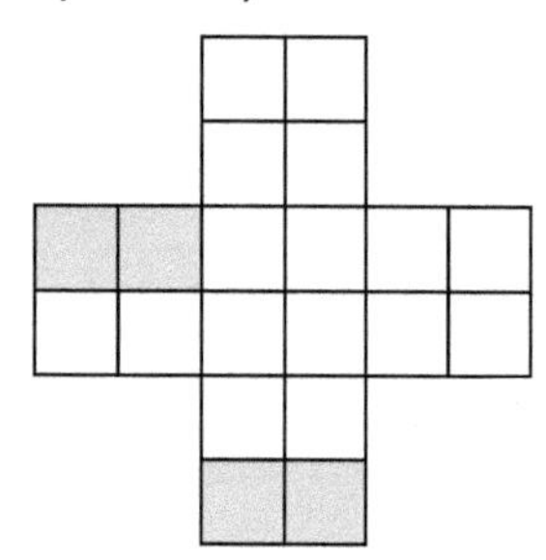

8 Lizzie is drawing shapes that have rotational symmetry of order 3.

Here are some of her examples.

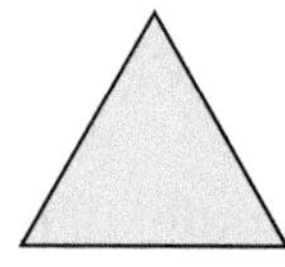 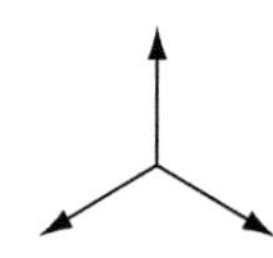 

She says that all shapes that have rotational symmetry of order 3 must have three lines of symmetry.

Draw an example to show that she is wrong.

12.2 Translations

Homework 12B

Copy each shape onto squared paper and draw its image after the given translation.

a 4 squares right

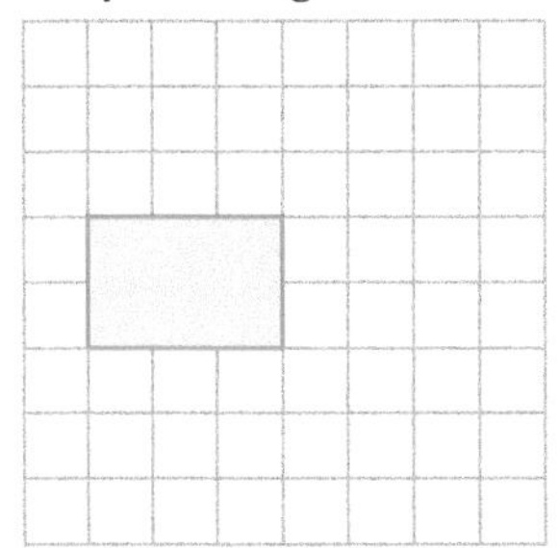

b 4 squares up

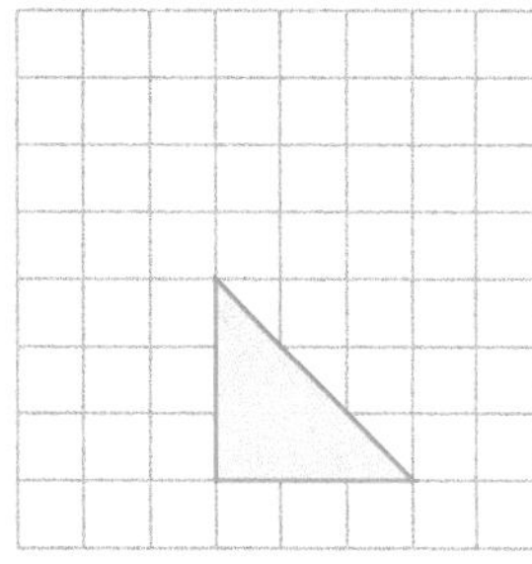

c 4 squares down

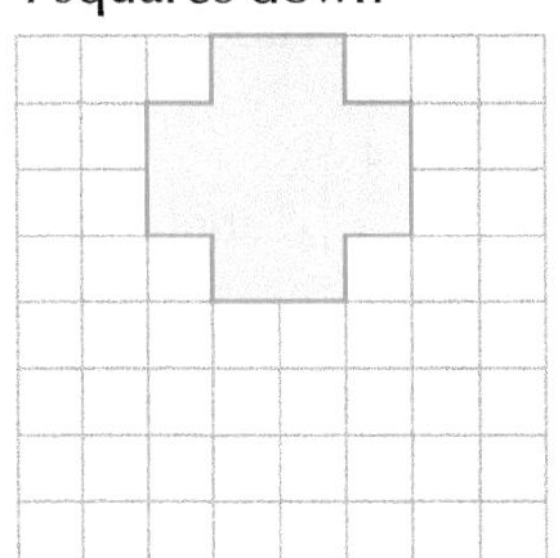

d 4 squares left

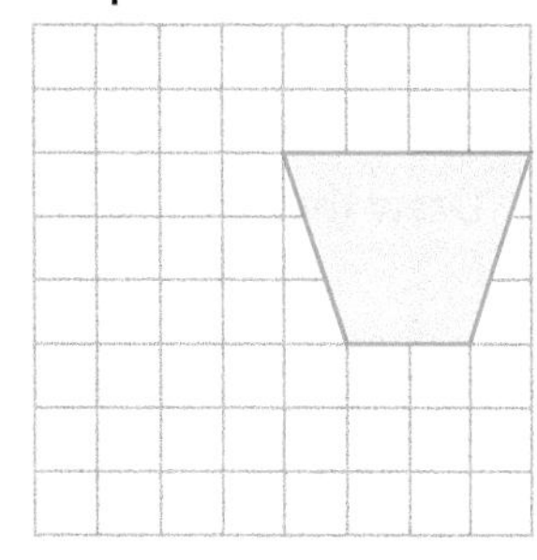

2 Copy each shape onto squared paper and draw its image after the given translation.

a 3 squares right and 2 squares down

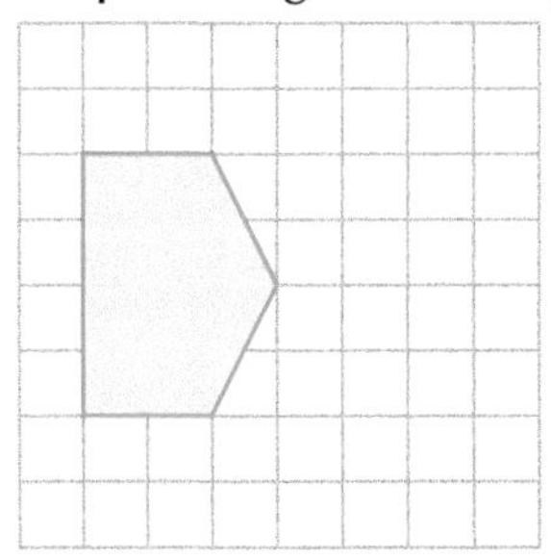

b 3 squares right and 4 squares up

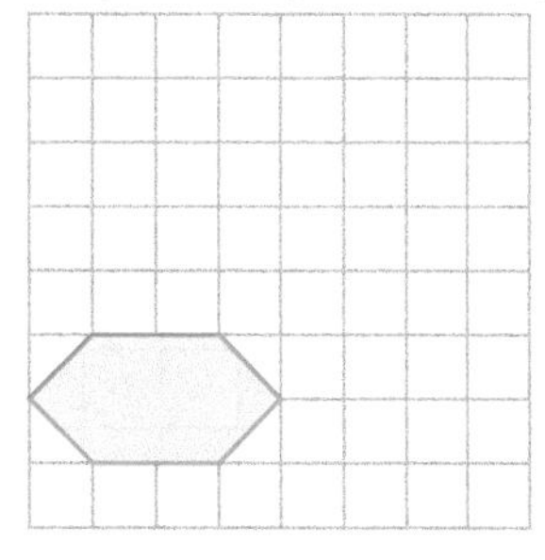

c 3 squares left and 3 squares down

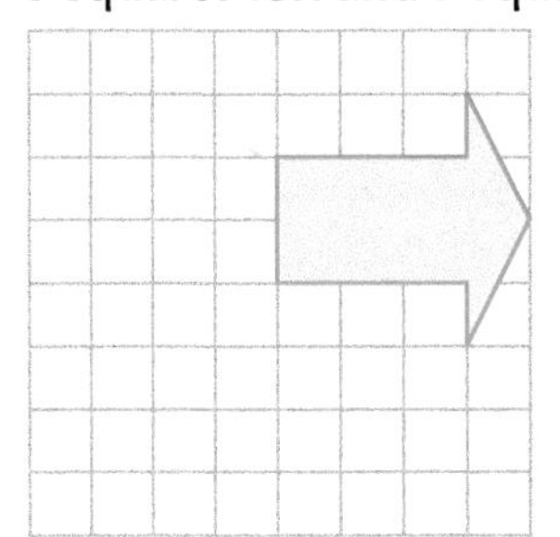

d 4 squares left and 1 square up

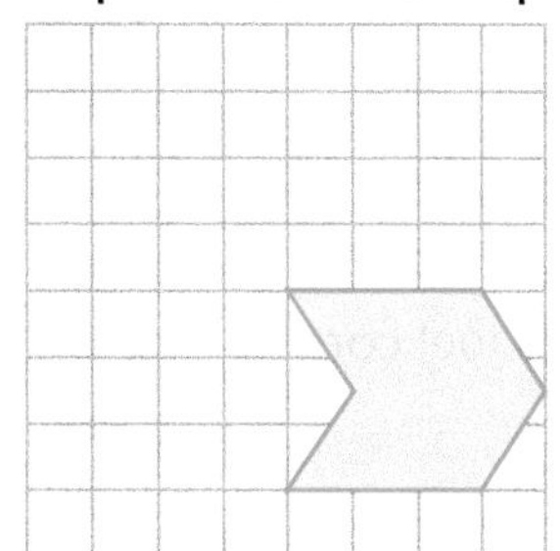

3 Look at the triangles on this grid.

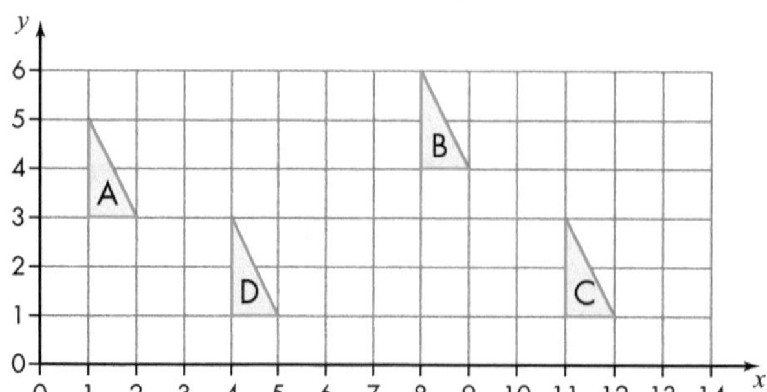

Use vectors to describe these translations.

i A to B **ii** A to C **iii** A to D **iv** B to A **v** B to C **vi** B to D

4 **a** Draw coordinate axes with values of x and y from 0 to 10. Draw a triangle with coordinates A(4, 4), B(5, 7) and C(6, 5).

b Draw the image of ABC after a translation with vector $\begin{pmatrix} 3 \\ 2 \end{pmatrix}$. Label this P.

c Draw the image of ABC after a translation with vector $\begin{pmatrix} 4 \\ -3 \end{pmatrix}$. Label this Q.

d Draw the image of ABC after a translation with vector $\begin{pmatrix} -4 \\ 3 \end{pmatrix}$. Label this R.

e Draw the image of ABC after a translation with vector $\begin{pmatrix} -3 \\ -2 \end{pmatrix}$. Label this S.

5 Write down a series of translations that will take you from the Start/finish, around the shaded square without touching it, and back to the Start/finish. Make as few translations as possible.

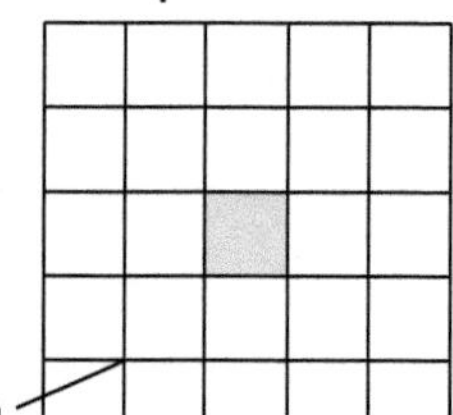

6 Joel says that if the translation from a point X to a point Y is described by the vector $\begin{pmatrix} -3 \\ 2 \end{pmatrix}$, then the translation from the point Y to the point X is described by the vector $\begin{pmatrix} 2 \\ -3 \end{pmatrix}$.

Is Joel correct? Explain how you decide.

12.3 Reflections

Homework 12C

1 Copy these shapes and mirror lines onto squared paper. Draw the reflection of each shape in the given mirror line.

a

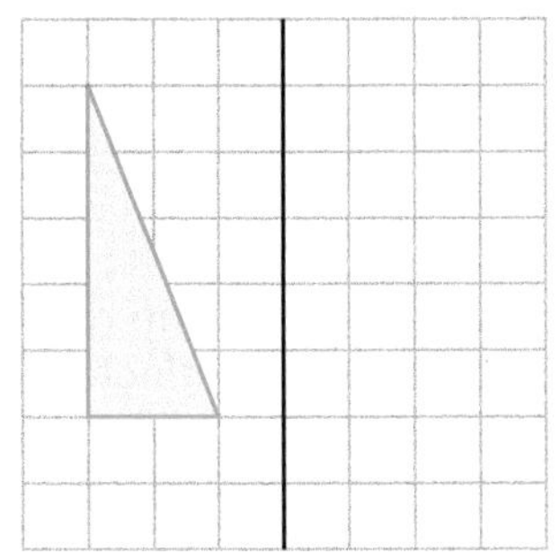

b

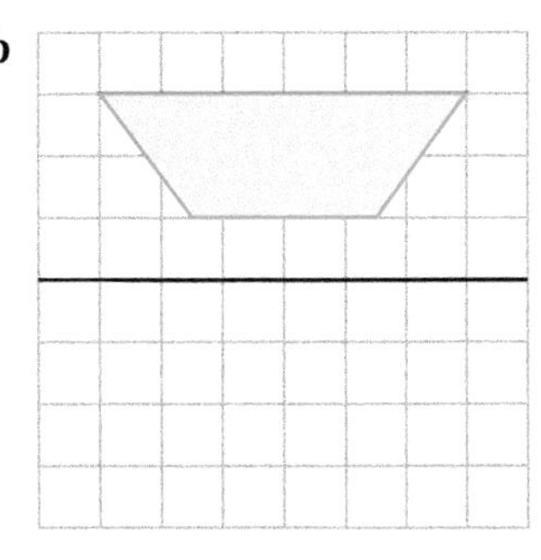

c

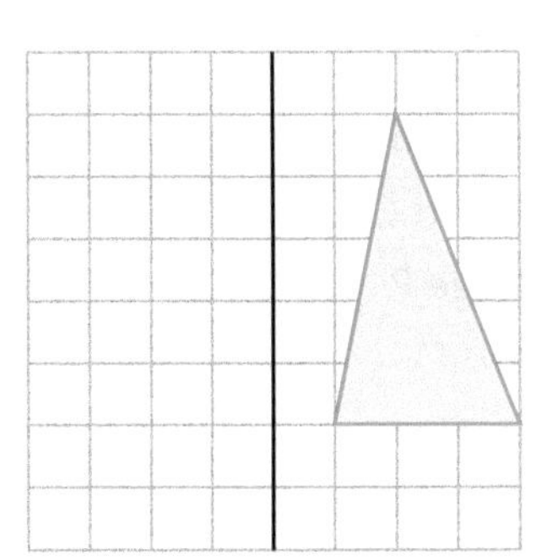

d

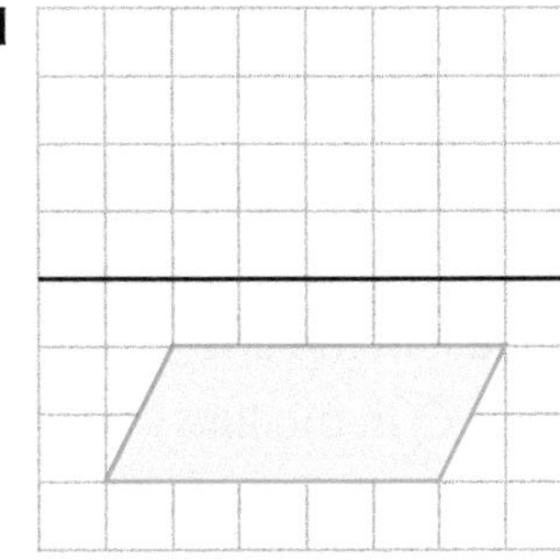

2 Copy these shapes and mirror lines onto squared paper. Draw the reflection of each shape in the given mirror line.

a

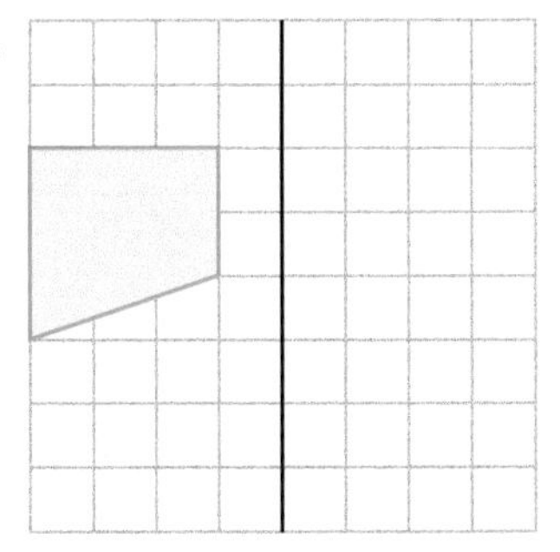

b

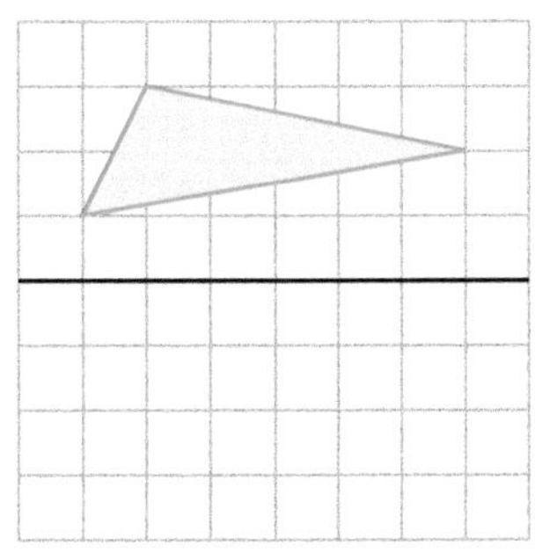

c

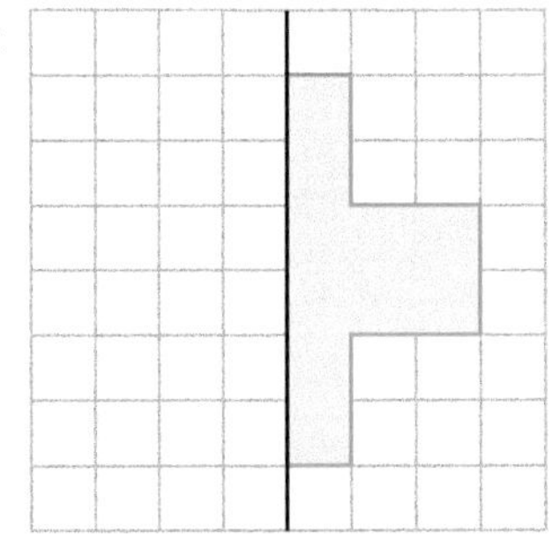

d

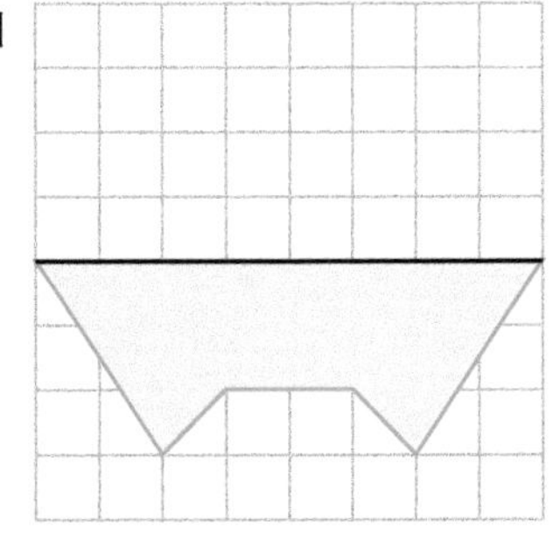

3 Copy this diagram onto squared paper.

a Reflect triangle ABC in the x-axis.

Label the image R.

b Reflect triangle ABC in the y-axis.

Label the image S.

c What special name is given to figures that are exactly the same shape and size?

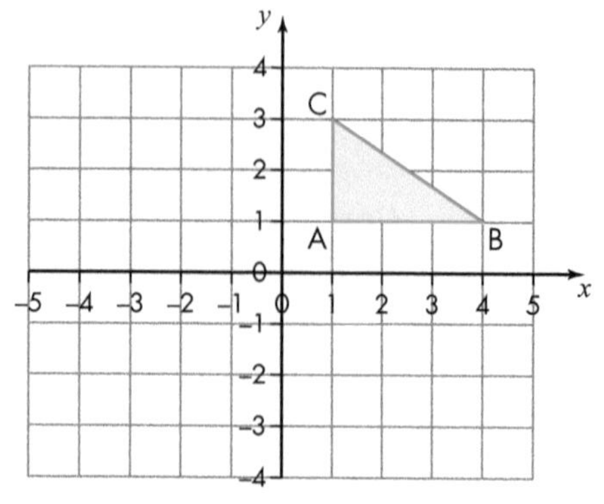

4 There are five capital letters that can form another capital letter, when reflected. Name these letters and the letters that they make when they are reflected.

5 There is a triangle that, when you draw a reflection on each side, creates the net of a tetrahedron.

What is the name of this triangle?

6 **a** Draw a pair of axes. Label the x-axis from –5 to 5 and the y-axis from –5 to 5.

b Draw the triangle with coordinates A(2, 2), B(3, 4) and C(2, 4).

c Reflect triangle ABC in the line $y = x$. Label the image P.

d Reflect triangle P in the line $y = -x$. Label the image Q.

e Reflect triangle Q in the line $y = x$. Label the image R.

f Describe the reflection that will transform triangle ABC to triangle R.

12.4 Rotations

Homework 12D

You may use tracing paper for this exercise.

1 Copy each diagram onto squared paper. Draw the image of each triangle after the given rotation about the centre, A.

a $\frac{1}{2}$ turn **b** $\frac{1}{4}$ turn clockwise **c** $\frac{1}{4}$ turn anticlockwise **d** $\frac{3}{4}$ turn clockwise

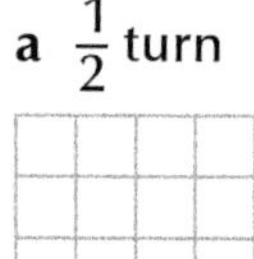
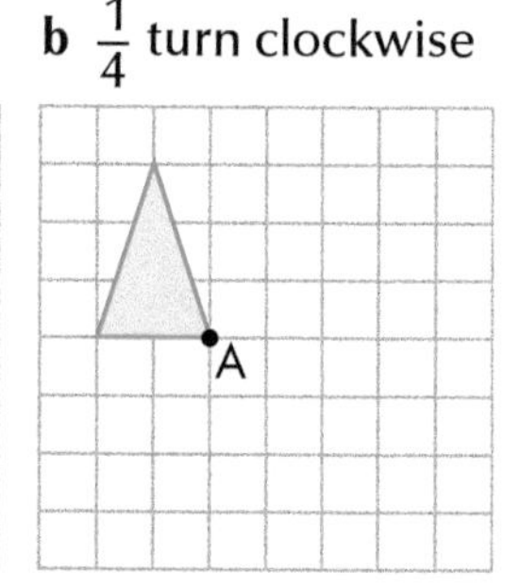

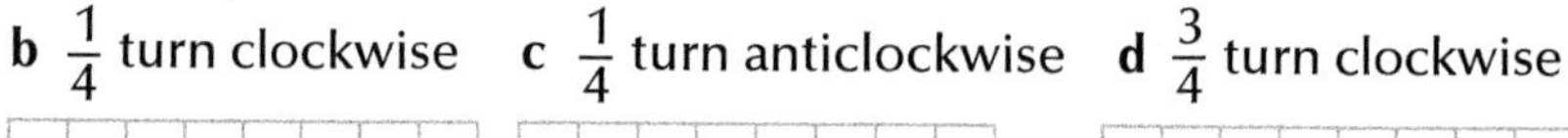
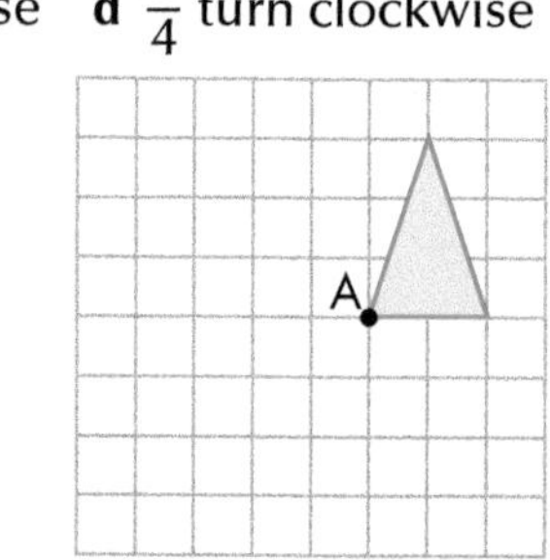

2 Copy each diagram onto squared paper. Draw the image of each flag after the given rotation about the centre, A.

a 180° turn

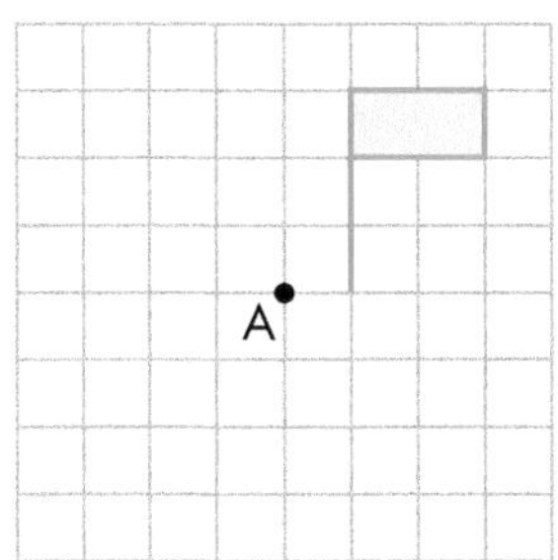

b 90° turn clockwise

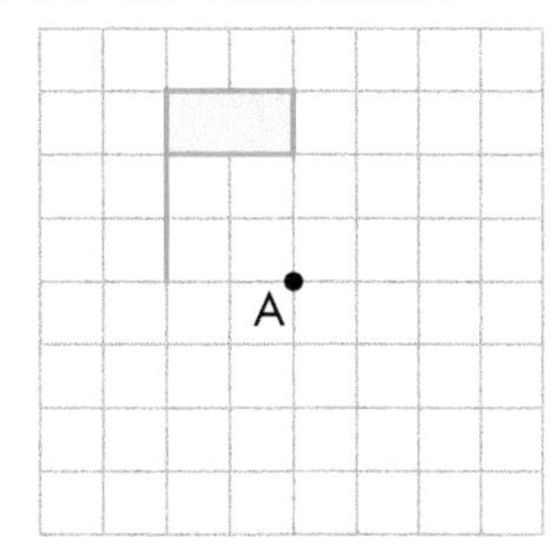

c 90° turn anticlockwise

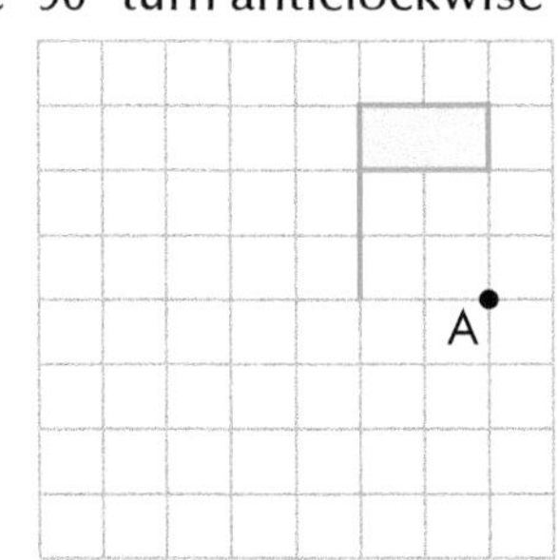

d 270° turn clockwise

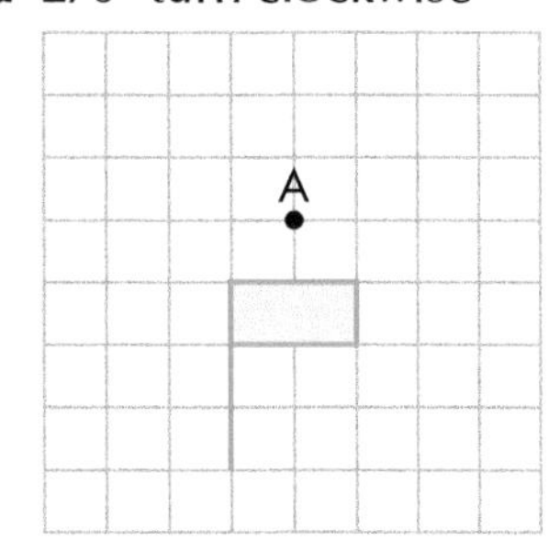

3 Copy this diagram onto squared paper.

a Rotate the shape 90° clockwise about (0, 0). Label the image P.

b Rotate the shape 180° about (0, 0). Label the image Q.

c Rotate the shape 90° anticlockwise about (0, 0). Label the image R.

d What rotation takes R back to the original shape?

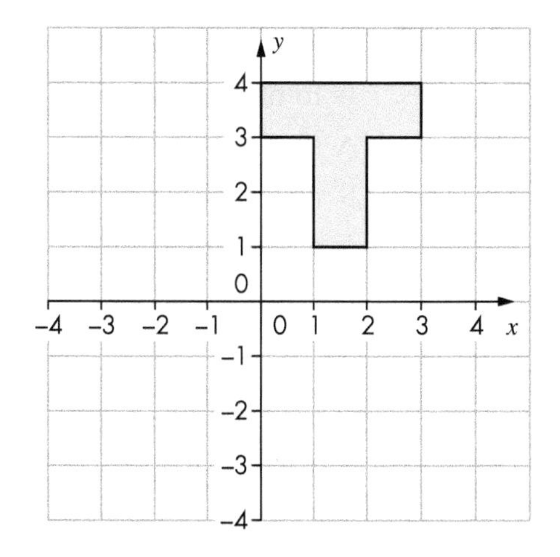

4 A graphic designer made this routine for creating a design.

- Start with a rectangle ABCD.
- Reflect the rectangle in the line AC.
- Rotate the whole shape 90° clockwise about the centre point of line AC.

Draw any rectangle on squared paper and create a design using the above routine.

5 Choose any one of the triangles below as a starting triangle (ABC). Describe how to keep rotating the triangle to make the shape in the diagram.

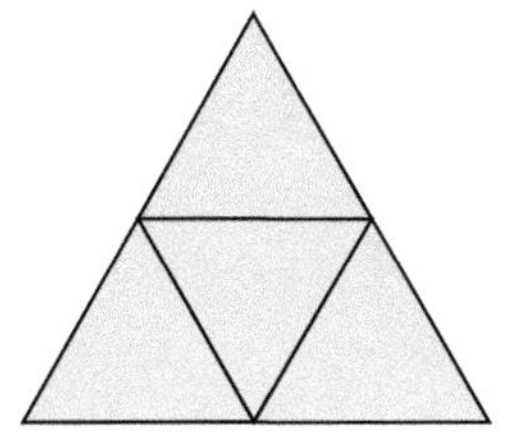

6 Copy the diagram and rotate the given triangle as described.

a 90° clockwise about (0, 0)

b 180° about (0, –2)

c 90° anticlockwise about (–1, –1)

d 180° about (0, 0)

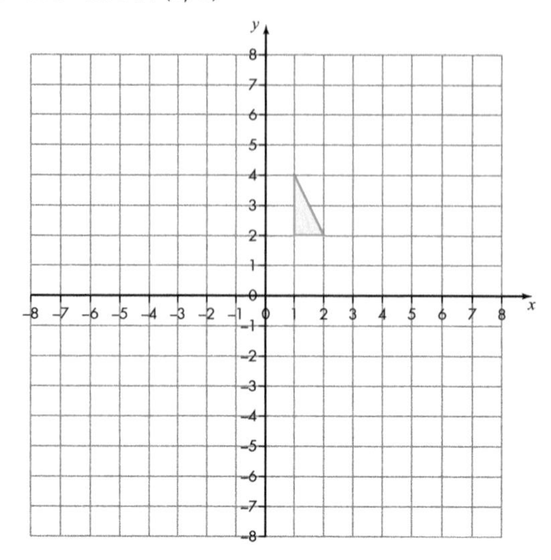

7 Tom said: 'If I rotate a shape, then the image is always congruent.'

Is Tom's statement:

A sometimes true

B never true

C always true?

12.5 Enlargements

Homework 12E

1 Copy each shape with its centre of enlargement onto squared paper. Use the ray method to enlarge it by the given scale factor.

a

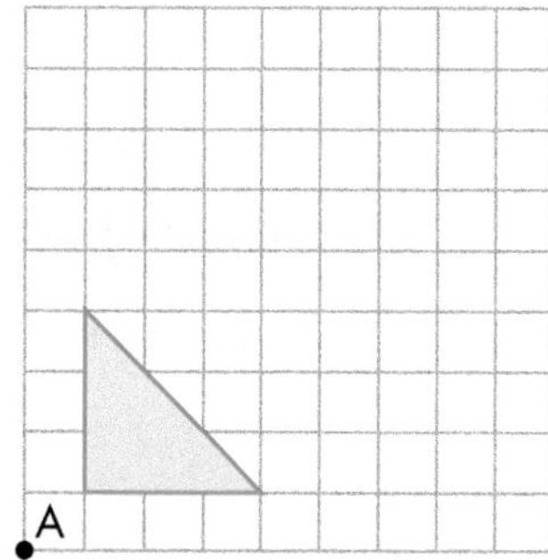

Scale factor 2

b

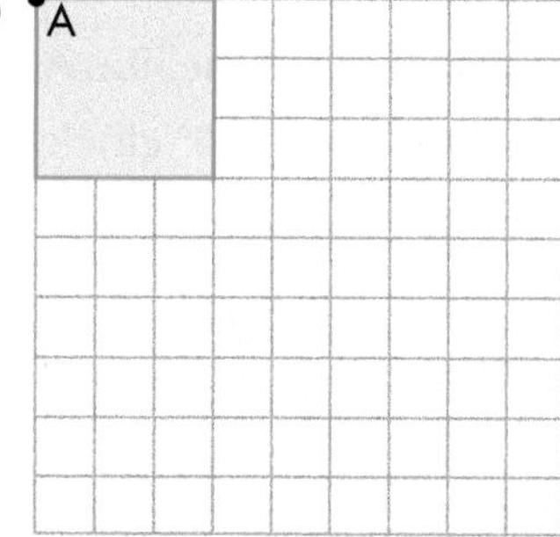

Scale factor 3

2 Copy each shape and grid onto squared paper. Enlarge each one by scale factor 2 using the origin as the centre of enlargement.

a

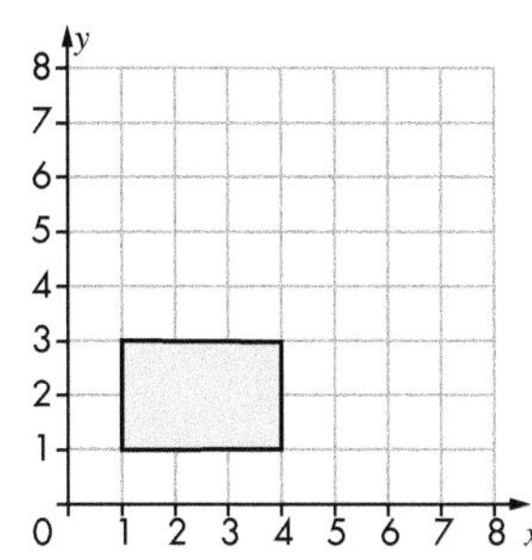

b

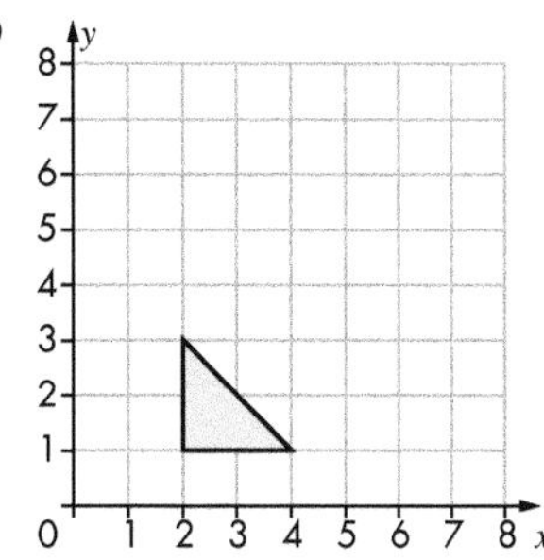

c

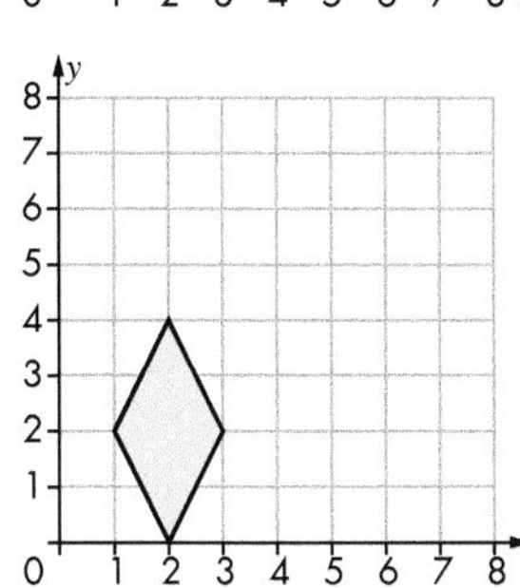

d

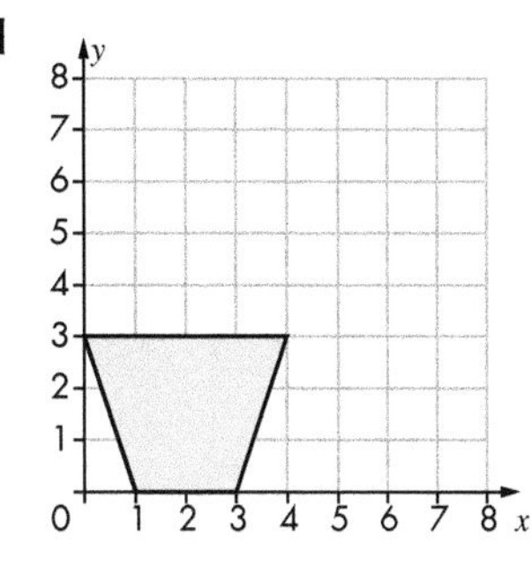

3 Copy each shape and its centre of enlargement. Then enlarge each shape by the given scale factor using the counting squares method.

a

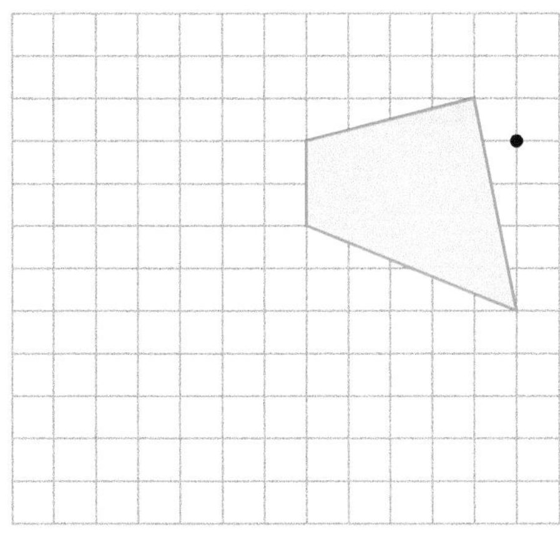

Scale factor 2

b

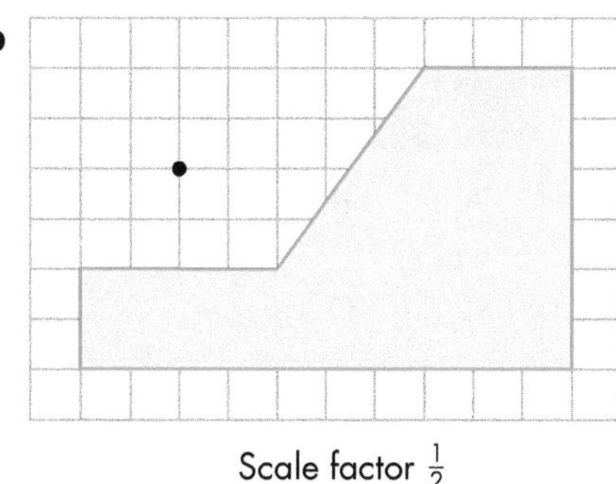

Scale factor $\frac{1}{2}$

4 A designer creates a logo using this routine.

- Start with an octagon in the shape of a letter T. Reflect the T in the small line on the bottom of the T.
- Rotate this whole shape about the midpoint, M, of the small line from the previous step.
- Enlarge the whole shape by scale factor 2, centre of enlargement point M.

Start with a T-shape of your choice and create your own logo using this routine.

5 Tina enlarged a shape and found the image was congruent to the original. Explain how this might have happened.

6 If I enlarge a shape by scale factor 4, by what scale factor will the area of the shape increase?

12.6 Using more than one transformation

Homework 12F

1 Copy and complete the table by describing fully the transformations that will give the required movements.

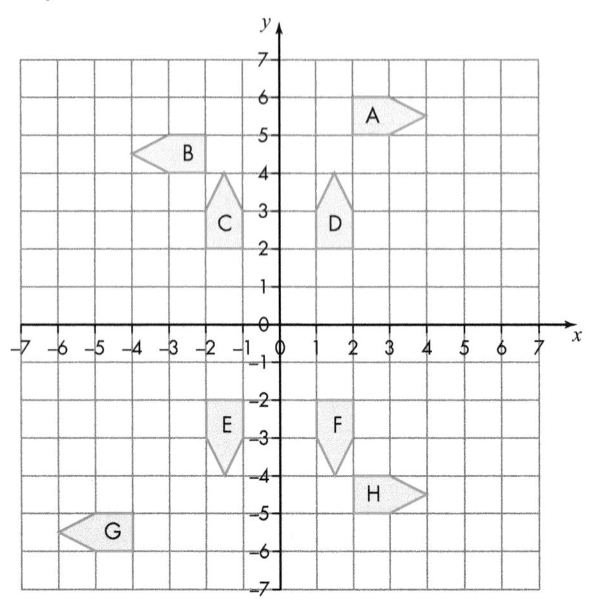

From	To	Transformation
A	B	
C	D	
D	F	
E	F	
G	H	

12.7 Vectors

Homework 12G

1 Draw lines to represent these vectors. (Remember to include the arrow.)

$\mathbf{a} = \begin{pmatrix} 3 \\ 4 \end{pmatrix}$ $\mathbf{b} = \begin{pmatrix} -3 \\ 4 \end{pmatrix}$ $\mathbf{c} = \begin{pmatrix} 3 \\ -4 \end{pmatrix}$ $\mathbf{d} = \begin{pmatrix} -3 \\ -4 \end{pmatrix}$

2 **a** Draw the points X(0, 2), Y(4, 5) and Z(–2, –6) on a coordinate grid.

b Write these as column vectors.

i $\overrightarrow{XY}$ **ii** $\overrightarrow{YZ}$ **iii** $\overrightarrow{XZ}$

3 C is the point (2, 3). $\overrightarrow{CD} = \begin{pmatrix} -3 \\ -4 \end{pmatrix}$. Write down the coordinates of point D.

4 K is the point (–3, 7). $\overrightarrow{KJ} = \begin{pmatrix} 5 \\ -9 \end{pmatrix}$ and $\overrightarrow{KL} = \begin{pmatrix} 7 \\ -4 \end{pmatrix}$.

a Write down the coordinates of J and L.

b What is the column vector for $\overrightarrow{JL}$?

5 Copy each part onto squared paper and show the position of each shape after the given translation.

a $\begin{pmatrix} -5 \\ -2 \end{pmatrix}$ **b** $\begin{pmatrix} -3 \\ 1 \end{pmatrix}$ **c** $\begin{pmatrix} -6 \\ 0 \end{pmatrix}$ **d** $\begin{pmatrix} 3 \\ 1 \end{pmatrix}$

a

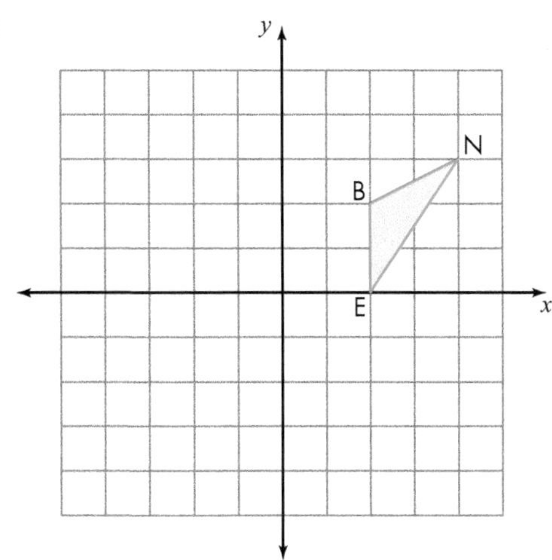

b

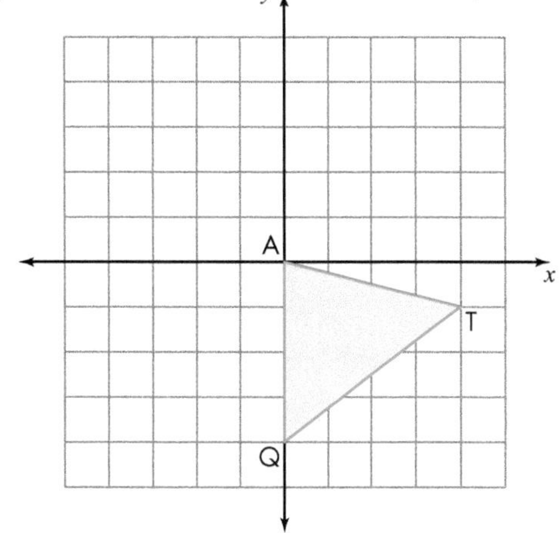

c

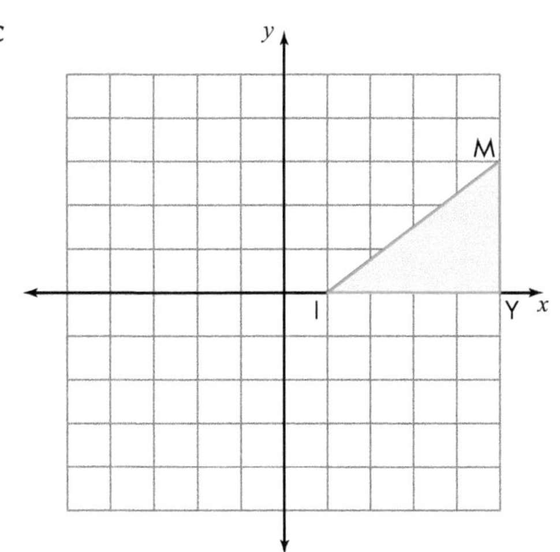

d

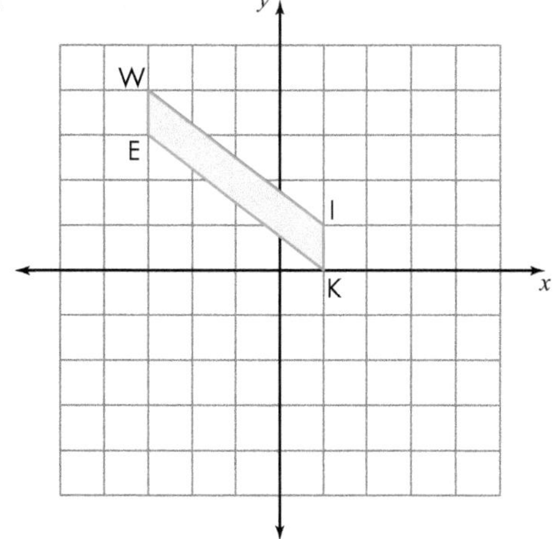

6 For parts **a** to **d**, state the translation that gives the following.

i A to A′

ii B to B′

a

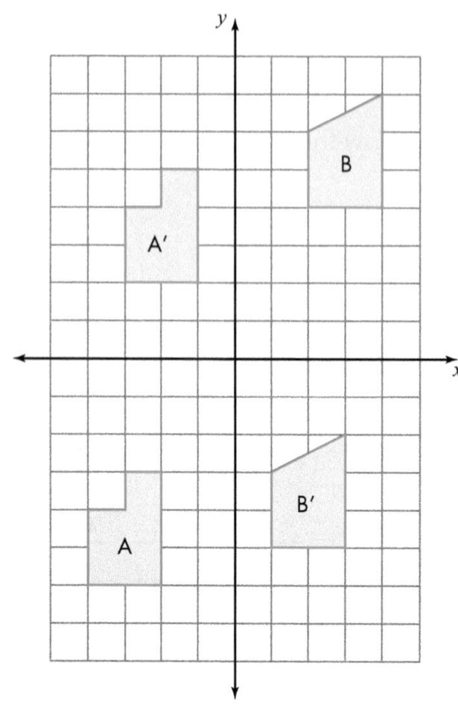

b

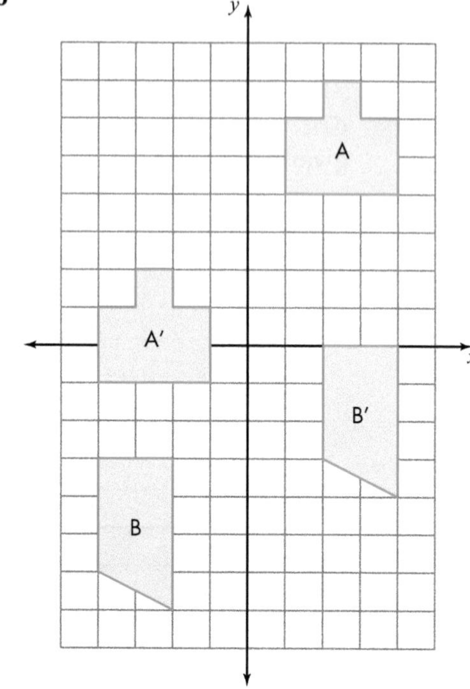

c

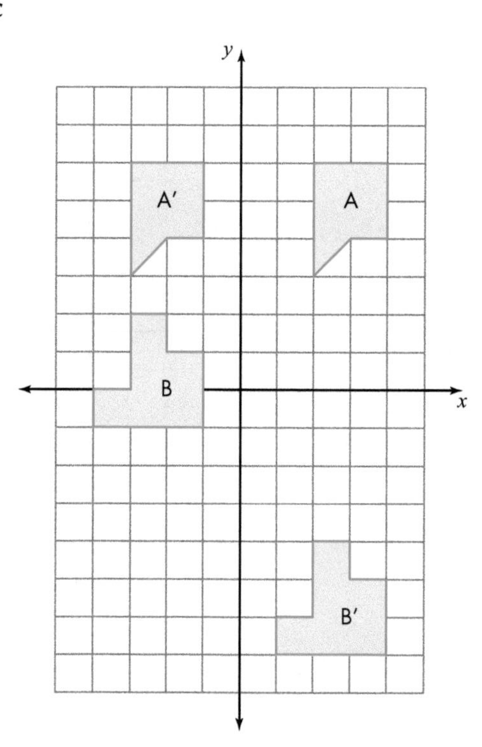

d

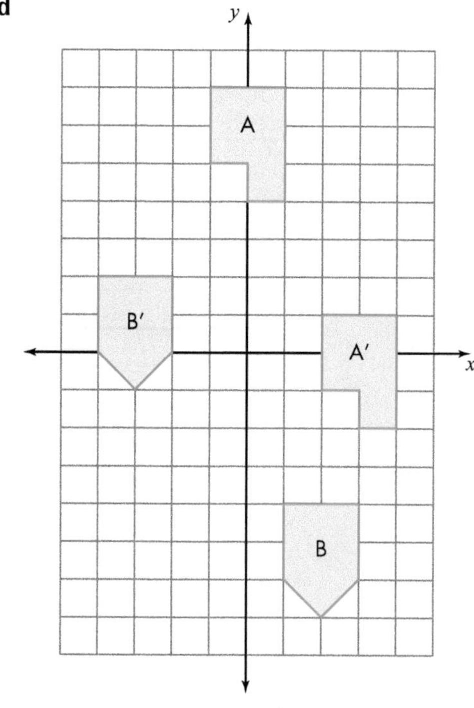

1 ABCDEF is a regular hexagon. $\overrightarrow{AB} = \mathbf{a}$, $\overrightarrow{BC} = \mathbf{b}$ and $\overrightarrow{CD} = \mathbf{b} - \mathbf{a}$.

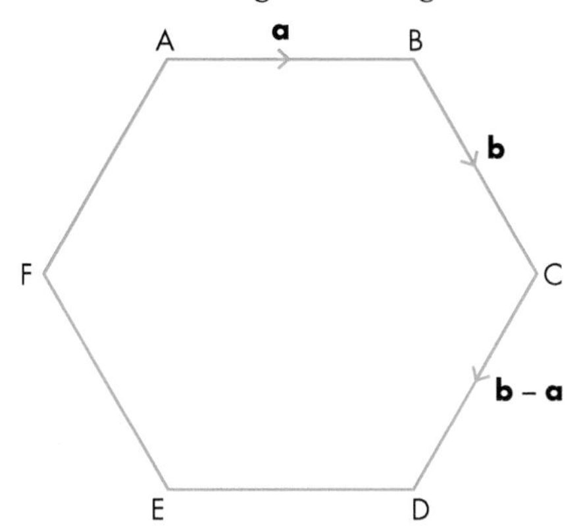

Write each of these vectors in terms of **a** and **b**.

i $\overrightarrow{DE}$ **ii** $\overrightarrow{EF}$ **iii** $\overrightarrow{FA}$ **iv** $\overrightarrow{AD}$

v $\overrightarrow{BE}$ **vi** $\overrightarrow{CF}$ **vii** $\overrightarrow{AE}$

2 The diagram shows a triangle.

$\overrightarrow{OA} = \mathbf{a}$ and $\overrightarrow{OB} = \mathbf{b}$.

C is a point on the line $\overrightarrow{OB}$ such that $\overrightarrow{OC} : \overrightarrow{CB} = 2 : 1$.

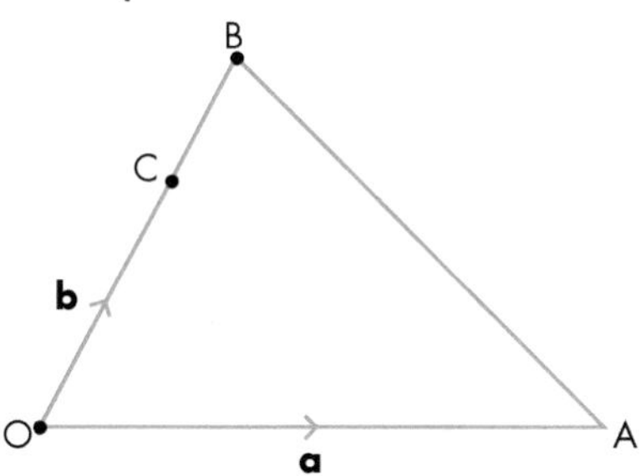

Write each of these vectors in terms of **a** and **b**.

i $\overrightarrow{OC}$ **ii** $\overrightarrow{AC}$ **iii** $\overrightarrow{BC}$

13 Probability: Probability and events

13.1 Calculating probabilities

Homework 13A

Example

A bag contains five red balls and three blue balls. A ball is taken out at random.

What is the probability that it is:

a red **b** blue **c** green?

a There are five red balls out of a total of eight, so P(red) = $\frac{5}{8}$.

b There are three blue balls out of a total of eight, so P(blue) = $\frac{3}{8}$.

c There are no green balls, so P(green) = 0.

1 Osamu takes a card from an ordinary pack. What is the probability of each outcome? Remember to cancel the probability fraction if possible.

a taking an ace **b** taking a picture card
c taking a diamond **d** taking a queen or a king
e taking the ace of spades **f** taking a red jack
g taking a club or a heart

2 Ten cards numbered from 1 to 10 (inclusive) are placed in a hat. Irene takes a card out of the hat without looking. What is the probability that she takes out:

a the number 10 **b** an odd number **c** a number greater than 4
d a prime number **e** a number between 5 and 9?

3 A bag contains two blue balls, three red balls and four green balls. Frank takes a ball from the bag without looking. What is the probability that he takes out:

a a blue ball **b** a red ball **c** a ball that is not green
d a yellow ball?

4 In a prize raffle, 50 tickets are sold. Ten are red, ten are blue and the rest are white. What is the probability that the first ticket drawn out is:

a red **b** blue **c** white
d red or white **e** not blue?

5 A bag contains 15 coloured balls. Three are red, five are blue and the rest are black. Paul takes a ball at random from the bag.

a Write down:

i P(he takes a red) **ii** P(he takes a blue) **iii** P(he takes a black).

b Add together the three probabilities from part **a**. What do you notice?

c Explain your answer to part **b**.

6 Adam, Ewa, Katie, Daniel and Maria are all in the same class. Their teacher asks two of these students, at random, to tidy a cupboard.

a Write down all the ways of choosing two students.

b How many pairs give two girls?

c What is the probability she chooses two girls?

d How many pairs give a girl and a boy?

e What is the probability she chooses a boy and a girl?

f What is the probability she chooses two boys?

7 The table shows some information about the year groups in one school.

Year	Y7		Y8		Y9		Y10		Y11	
	Boys	Girls	Boys	Girls	Boys	Girls	Boys	Girls	Boys	Girls
Pets	7	8	8	9	10	9	8	9	8	11
No pets	4	5	4	5	6	8	5	6	5	4

A representative is chosen at random from each year.

Which year has the highest probability of having a male representative with a pet?

13.2 Probability that an outcome will not happen

Homework 13B

Example

What is the probability of not taking an ace from a pack of cards?

First, calculate the probability of taking an ace. P(taking an ace) is $\frac{4}{52} = \frac{1}{13}$

So, P (not taking an ace) is $1 - \frac{1}{13} = \frac{12}{13}$

1 Here are the probabilities of some events happening. Write down the probabilities of the events not happening:

a P(h) = 0.3 **b** P(h) = 0.4 **c** P(h) = 0.52

d P(h) = 0.21 **e** P(h) = 25% **f** P(h) = 98%

g P(h)= 55.5% **h** P(h) = $\frac{2}{5}$ **i** P(h) = $\frac{6}{10}$

j P(h) = $\frac{12}{15}$

2 a The probability of winning a prize in a tombola is $\frac{1}{25}$. What is the probability of not winning a prize in the tombola?

b The probability that it will rain tomorrow is 65%. What is the probability that it will not rain tomorrow?

c The probability that Josie wins a game of tennis is 0.8. What is the probability that she loses a game?

d The probability of getting two sixes when throwing two dice is $\frac{1}{36}$. What is the probability of not getting two sixes?

3 Harvinder takes a card from an ordinary pack of cards. Write down the probability that the card she takes is:

a i a king ii a card that is not a king

b i a spade ii a card that is not a spade

c i a 9 or a 10 ii neither a 9 nor a 10.

4 These letter cards are put into a bag.

a Stan takes a letter card at random. What is the probability that:

i he takes a letter A ii he does not take a letter A?

b Stan takes an R from the original set of cards and keeps it. Eliza now takes a letter from those remaining.

i What is P(A) now? ii What is P(not A) now?

5 This is the starting section from a board game.

START	Chelsea Park	Take a chance	London Road	Carter Road	Pay tax £500	Banner Road	Struen Road	Rest area

You throw a dice and move, from the start, the number of places shown by the dice. What is the probability of *not* landing on:

a a blue square b the Pay tax square c a coloured square?

6 Elijah and Harris are playing a board game. Elijah will have to miss a turn if the next dice he rolls shows an odd number. Harris will miss a turn if the next dice he rolls shows a 1 or a 2.

Who has the better chance of *not* missing a turn the next time they roll the dice?

7 Arran is told that the chance he loses a game is 0.1. He says, 'So my chance of winning is 0.9.'

Give a reason why Arran might be wrong.

13.3 Mutually exclusive and exhaustive outcomes

Homework 13C

1. Write down which of these pairs of outcomes are mutually exclusive.
   a Winning a rugby match and drawing the same match
   b Wearing one spotty sock and one stripy sock
   c Eating toast for breakfast and pizza for lunch
   d Being on time for school and being late for school on the same day

2. A bag contains six fudges, four toffees and 10 caramels. I take one sweet, at random, from the bag. What is the probability I take:
   a a toffee **b** a fudge **c** a caramel
   d a toffee or a caramel **e** a fudge or a caramel?

3. Paul knows that when he plays a game of chess, he has a 65% chance of winning a game and a 15% chance of losing a game. What is the probability that he draws a game?

4. In a game of darts, what is the probability of throwing a single dart and it landing on:
   a 20
   b an even number
   c a prime number
   d an even number and multiple of 3
   e an odd number and multiple of 5?

5. A bag contains red, green and purple balls. For each set of balls, P(red ball) and P(green ball) are given. Work out P(purple ball) for each set.

   a

Red	Green	Purple
0.1	0.05	

   b

Red	Green	Purple
0.62	0.21	

   c

Red	Green	Purple
0.15	0.02	

6. George is playing a game of tennis.
   His sister says, 'He either wins or loses, so he must have an even chance of winning.' Explain why George's sister might not be correct.

7. Alan and Alice play noughts and crosses. The probability that Alan wins is 0.31 and the probability that Alice wins 0.45. What is the probability that they draw?

13.4 Experimental probability

Homework 13D

1 Katrina throws two fair six-sided dice and records the number of doubles that she gets after various numbers of throws. The table shows her results.

Number of throws	10	20	30	50	100	200	600
Number of doubles	2	3	6	9	17	35	102

a Calculate the experimental probability of scoring a double at each stage that Katrina recorded her results.

b What is the theoretical probability of throwing a double with two dice?

2 Victoria made a six-sided spinner, like the one shown in the diagram. She used it to play a board game with her friend Katie.

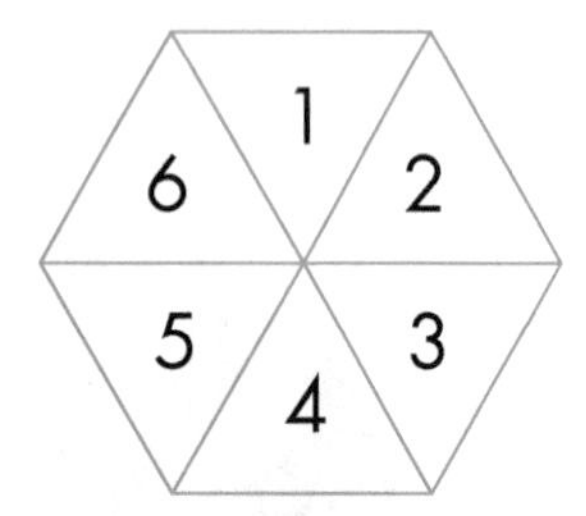

The girls thought that the spinner was not very fair as it seemed to land on some numbers more than others. They spun the spinner 120 times and recorded the results, as shown in the table.

Number spinner lands on	1	2	3	4	5	6
Number of times	22	17	21	18	26	16

a Work out the relative frequency of the spinner landing on each number.

b How many times would you expect the spinner to land on each number if the spinner is fair?

c Do you think that the spinner is fair? Give a reason for your answer.

3 In a game at the fairground, a player rolls a coin onto a grid with coloured squares. If the coin lands completely within one of the blue, green or red squares, the player wins a prize.

The table below shows the probabilities of the coin landing completely within a winning colour.

Colour	Blue	Green	Red
Probability	0.3	0.2	0.1

a On one afternoon, 300 games were played. How many coins would you expect to land on:

i a blue square

ii a green square

iii a red square?

b What is the probability that a player loses a game?

4 A survey was carried out for one week on all route 79 buses to find out how many of the passengers were pensioners.

	Mon	Tue	Wed	Thu	Fri
Passengers	950	730	1255	796	980
Pensioners	138	121	168	112	143

For each day, calculate the probability that the 400th passenger to board the bus was a pensioner.

5 Joseph made a six-sided spinner. He tested it out to see if it was fair.

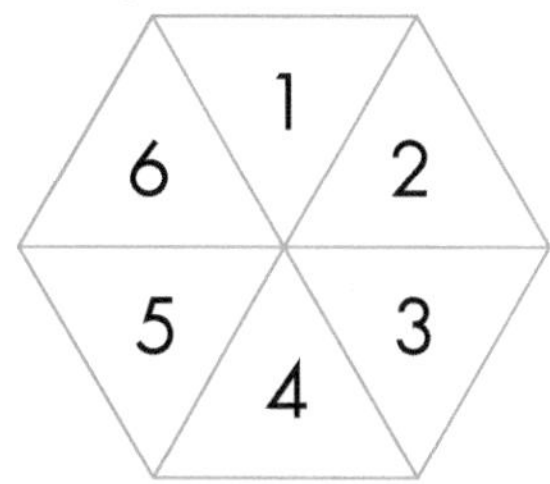

He spun the spinner 240 times and recorded the results in a table.

Number spinner lands on	1	2	3	4	5	6
Frequency	43	38	32	41	42	44

Do you think the spinner is fair?

Give reasons for your answer.

6 Aleena tossed a coin 50 times.

She said: 'If this is a fair coin, I should get exactly 25 tails.'

Explain why she is wrong.

13.5 Expectation

Homework 13E

1 I roll an ordinary fair dice 600 times. How many times can I expect to get a score of 1?

2 I throw a coin 500 times. How many times can I expect to get a tail?

3 I take a card at random from an ordinary pack and replace it. I do this 104 times. How many times would I expect to get:

a a red card

b a queen

c a red seven

d the jack of diamonds?

4 The ball in a roulette wheel can land in one of 37 spaces that are marked in numbers from 0 to 36 inclusive. I always bet on the same block of numbers, 1–6.

If I play all evening and there are exactly 111 spins of the wheel in that time, how many times could I expect the ball to land on the numbers 1–6?

5 I have five tickets for a raffle. The probability that I win the only prize is 0.003. How many tickets were sold altogether in the raffle?

6 A bag contains 20 balls. Ten are red, three are yellow and seven are blue. A ball is taken out at random and then replaced. This is done 200 times. How many times would I expect to get:

a a red ball

b a yellow or blue ball

c a ball that is not blue

d a green ball?

7 A headteacher is told that the probability of any student being left-handed is 0.14.

She has 1200 students in her school. Explain how she can work out how many of her students she should expect to be left-handed.

8 A bag contains black and white balls. Lucas knows that the probability of getting a black ball is 0.4. He takes 200 samples. How many of them would he expect to give a white ball?

9 Kara rolls two six-sided dice 200 times.

a How many times would she expect to roll a double?

b How many times would she expect to roll a total score greater than 7?

10 The ball in a roulette wheel can land in one of 37 spaces that are marked with numbers from 0 to 36 inclusive. I always bet on a prime number.

If I play the game 100 times, how many times could I expect the ball to land on a prime number?

11 An opinion poll uses a sample of 200 voters in one area. 112 of them said they would vote for Party A.

a There are a total of 50 000 voters in the area. If they all voted, how many would you expect to vote for Party A?

b The poll is accurate to within 10%. Can Party A be confident of winning?

12 **a** Franz is about to take his driving test. The chance that he passes is $\frac{1}{3}$. His sister says: 'You are sure to pass within three attempts because $3 \times \frac{1}{3} = 1$.' Explain why his sister is wrong.

b If Franz does fail, would you expect the chance that he passes next time to increase or decrease? Explain your answer.

13.6 Choices and outcomes

Homework 13F

1 A café offers a lunch deal.

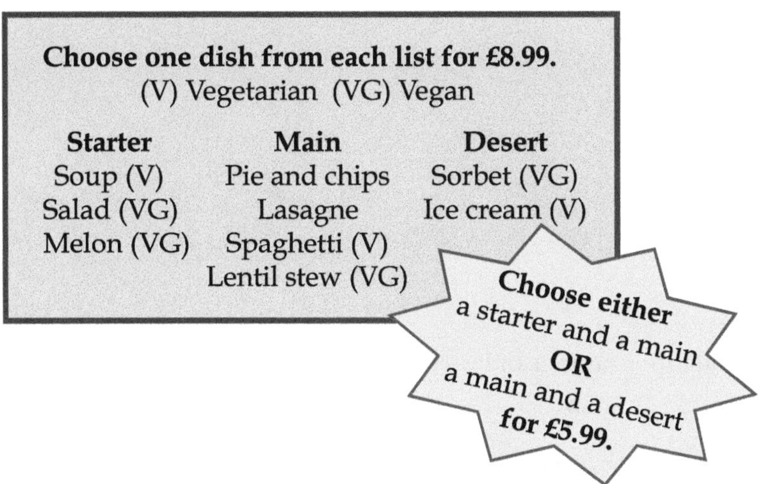

a How many different £8.99 lunch deals are possible?

b How many different £5.99 lunch deals are possible?

c Naz is vegan. How many different £5.99 lunch deals could he choose?

2 The first three places in a local 10K race were filled by the Wilson brothers: John, Keith and Len.

Work out the probability that the finishing order was 1st Keith, 2nd Len and 3rd John.

3 Alfie takes two of these letter cards without looking.

a Explain why there are 20 possible ways of taking two letters.

b What is the probability that both of the letters he took were vowels?

4 Martyn writes down all the numbers from 100 to 200.

100 101 102 ... 197 198 199 200

How many times does he write the digit 9?

5 A combination lock has four digits with numbers from 0 to 9 inclusive.

a How many different combinations are possible?

b Pablo knows that his combination starts with 4 and that there are no repeated numbers in the combination. He uses this to guess its combination.

Work out the probability he guesses the correct combination on his first try.

14 Geometry and measures: Volumes and surface areas of prisms

14.1 3D shapes

Homework 14A

1 For each shape, state the number of **i** vertices, **ii** edges and **iii** faces.

a

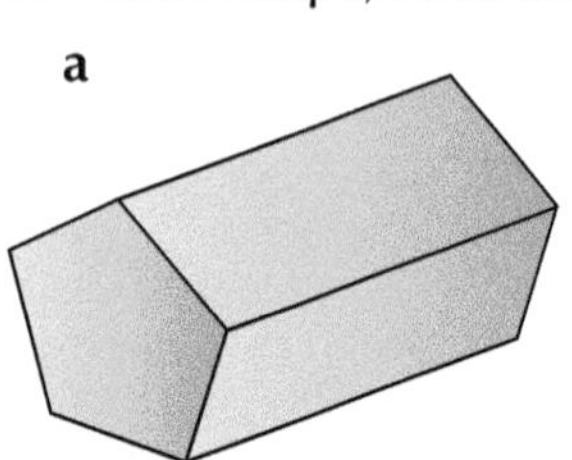

b

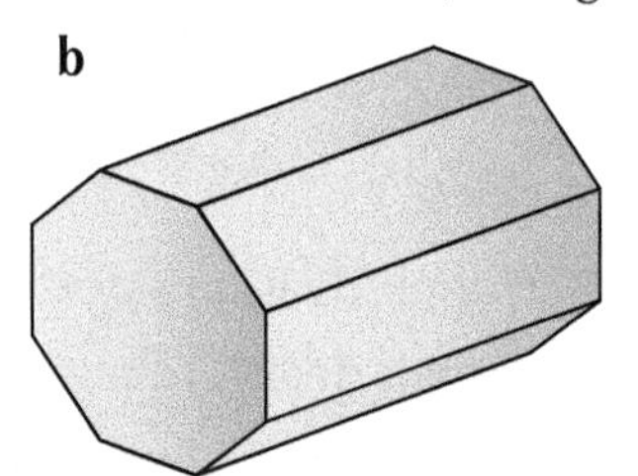

c

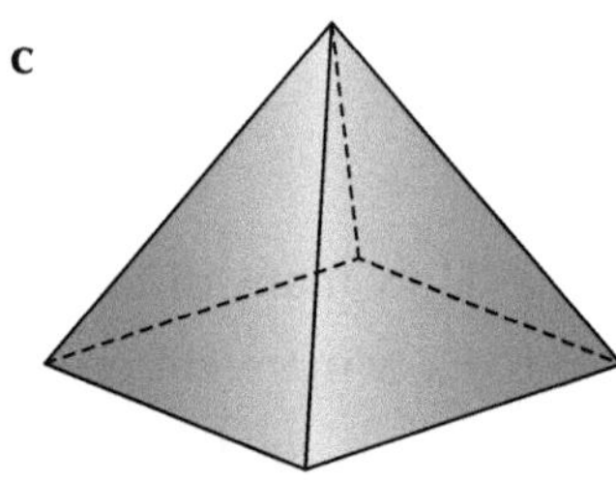

2 These shapes are made from unit cubes with an edge length of 1 cm. For each shape, work out:

i the volume

ii the surface area.

a

b

c

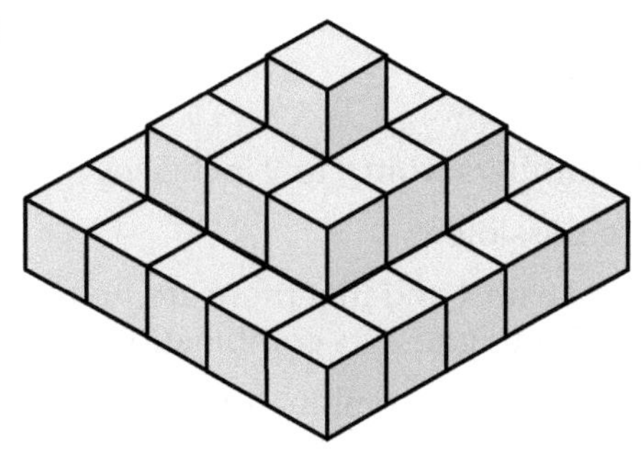

d

e

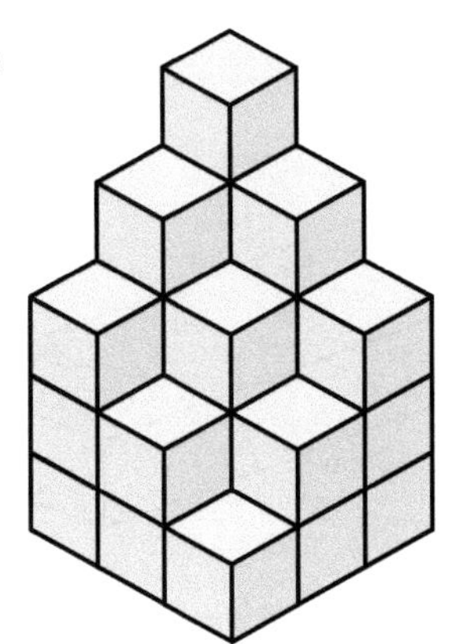

f

g

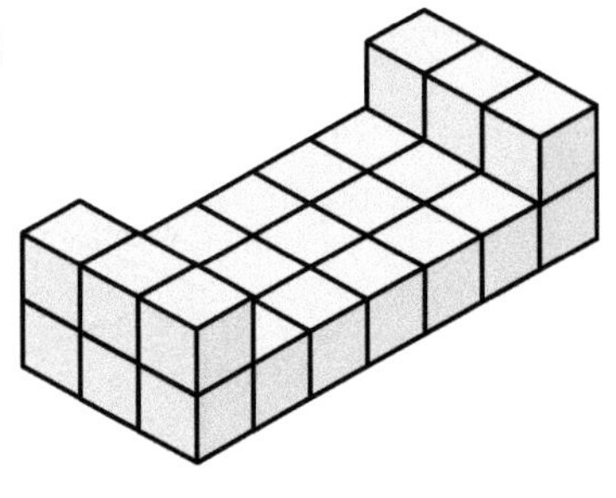

3 **a** Work out the volume and surface area of each shape.

A

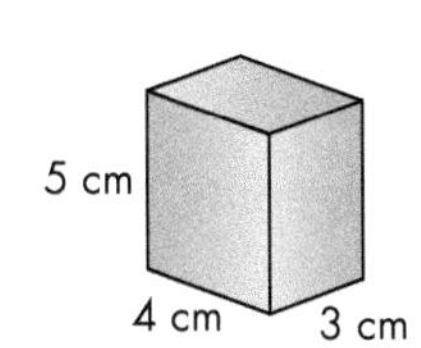

B

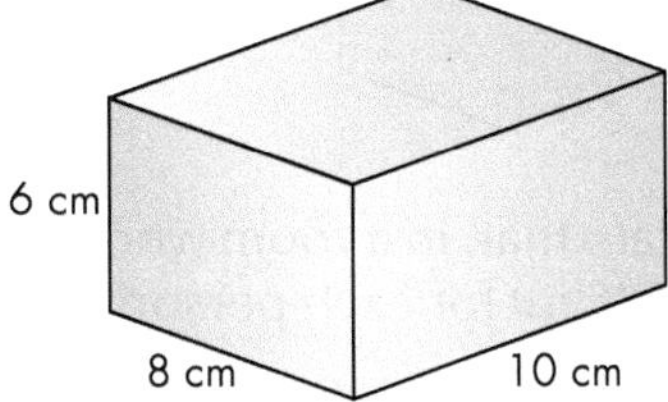

b Copy and complete these sentences.

i The length, width and height of shape B is __ times as big as the length, width and height of shape A.

ii The surface area of shape B is __ times as big as the surface area of shape A.

iii The volume of shape B is __ times as big as the volume of shape A.

14.2 Volume and surface area of a cuboid

Homework 14B 

Example

Calculate the volume and surface area of this cuboid.

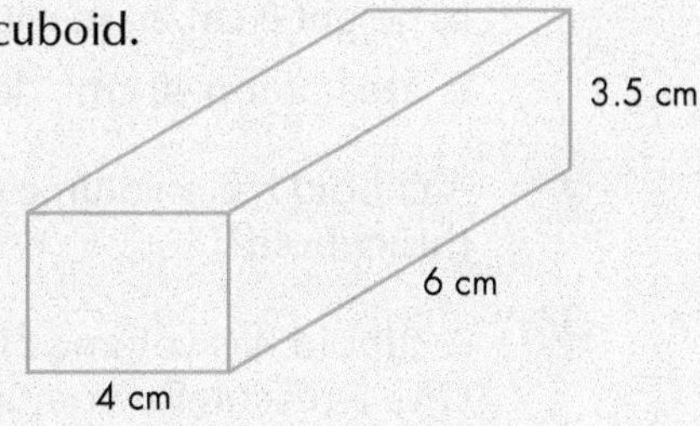

Volume = $6 \times 4 \times 3.5 = 84\ \text{cm}^3$

Surface area = $(2 \times 6 \times 4) + (2 \times 3.5 \times 4) + (2 \times 3.5 \times 6)$

$= 48 + 28 + 42$

$= 118\ \text{cm}^2$

1 Calculate the capacity of a swimming pool with length 12 m, width 5 m and depth 1.5 m.

2 For each cuboid, calculate:

i the volume **ii** the surface area.

a

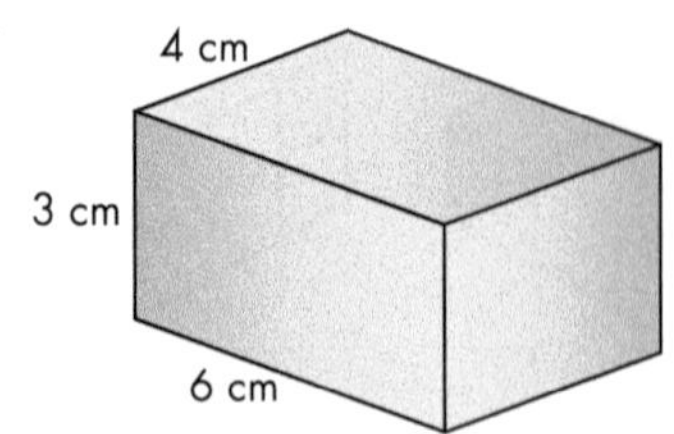

b

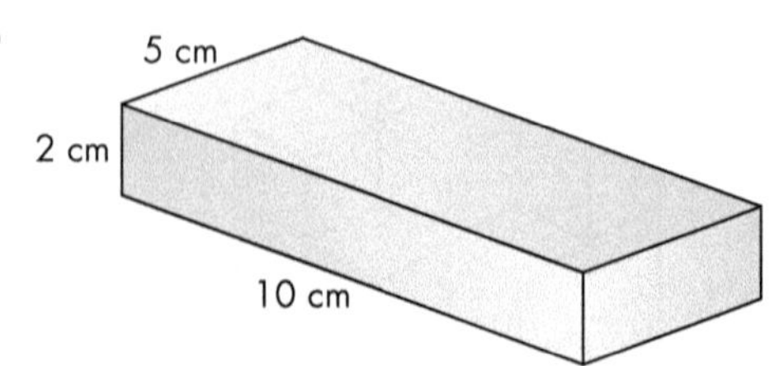

c

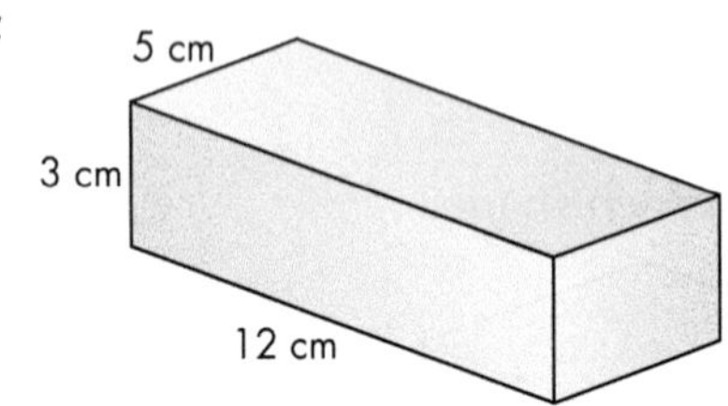

d

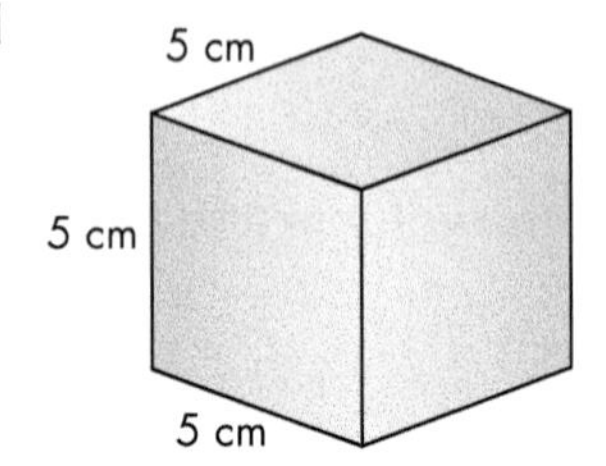

3 Safety regulations state that, in a room where people are sleeping, there should be a minimum volume of 18 m^3 for each person.

A dormitory is 15 m long, 12 m wide and 3.5 m high.

What is the maximum number of people who can safely sleep in this dormitory?

4 Copy and complete the table for cuboids **a** to **d**.

	Length	Width	Height	Volume
a	4 cm	3 cm	2 cm	
b		3 cm	3 cm	45 cm^3
c	8 cm		4 cm	160 cm^3
d	6 cm	6 cm		216 cm^3

5 Calculate the volume of cuboids with these dimensions.

a base area 20 cm^2, height 3 cm

b height 8 cm, base with one side 4 cm and the other side 1 cm longer

c area of top 40 cm^2, depth 3 cm

6 A cuboid has a volume of 512 cm^3. What is the smallest possible surface area of this cuboid?

7 A cuboid has volume 216 cm^3 and a total surface area of 216 cm^2. Could it be a cube? Give a reason for your answer.

14.3 Volume and surface area of a prism

Homework 14C

1 Calculate the volume and the total surface area of this prism.

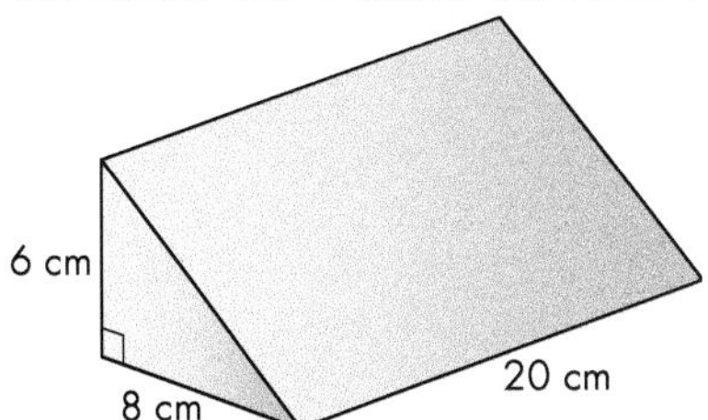

2 For each prism, calculate:

i the area of the cross-section **ii** the volume.

a

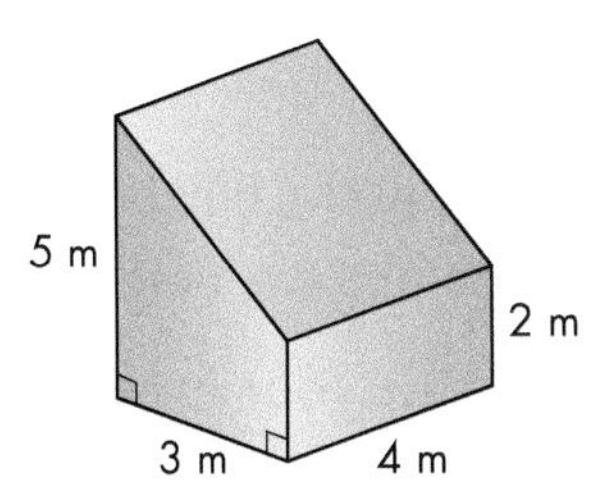

b

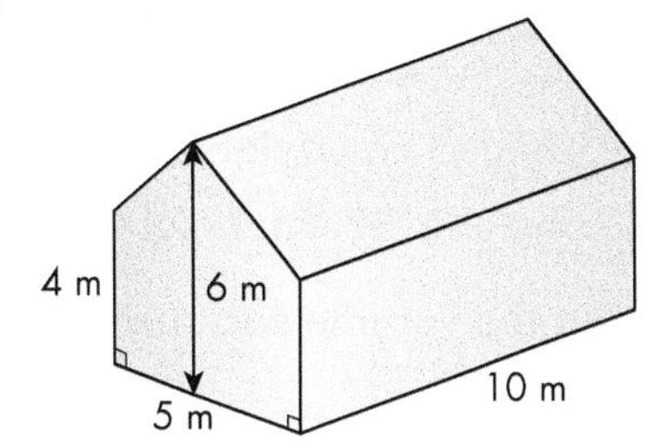

3 Calculate the mass of each prism.

a

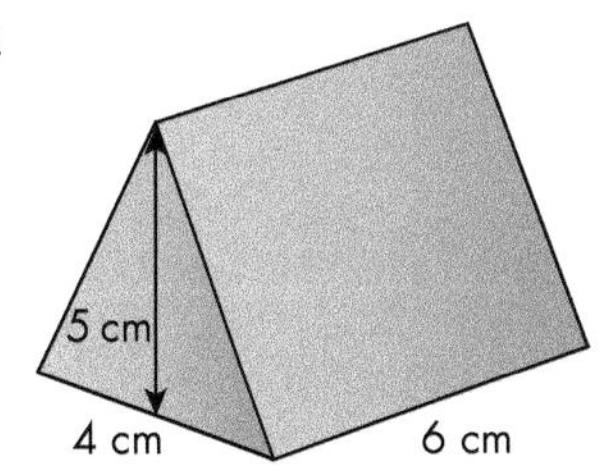

1 cm^3 has a mass of 3.13 g

b

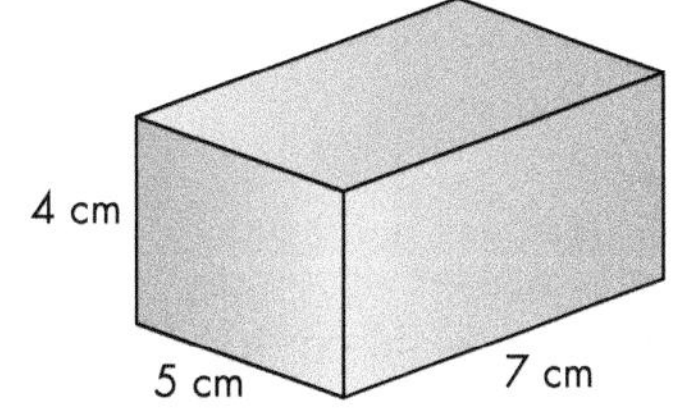

1 cm^3 has a mass of 1.35 g

4 A 100-m trench is to be dug as part of a construction job. The trench will be in the shape of a trapezium that is 2.4 m wide at the top, 1.9 m wide at the bottom and 1.6 m deep.

a How much earth needs to be removed?

b One lorry can carry a maximum load of 15 tonnes of earth. 1 cm^3 of earth weighs 2.5 g. How many lorry loads will be needed to transport the earth?

5 A girl has cubes that each have edge length 2 cm. She builds a single large cube out of 27 of these cubes.

How many more 2-cm cubes would she need to build a single cube with edge length 2 cm longer than the first one?

6 Imagine you have a large glass bottle and you need to mark on the outside the levels 1 litre, 2 litres, 3 litres, 4 litres, etc.

Explain how you could do this if you only had one 2-litre jug and one 5-litre jug.

14.4 Volume and surface area of cylinders

Homework 14D

Example

Calculate the volume and surface area of a cylinder with a radius of 4 cm and a height of 10 cm.

Volume $= \pi r^2 h$

$= \pi \times 4^2 \times 10$

$= 502.7\ \text{cm}^3$ (to one decimal place)

Surface area $= 2\pi rh + 2\pi r^2$

$= 2 \times \pi \times 4 \times 10 + 2 \times \pi \times 4^2$

$= 251.3 + 100.5$

$= 351.8\ \text{cm}^2$

1 Calculate the volume and surface area of cylinders with these dimensions. Give your answers to one decimal place.

a base radius 5 cm, height 7 cm

b base radius 10 cm, height 8 cm

c base diameter 12 cm, height 20 cm

d base diameter 9 cm, height 9 cm

2 Work out the volume and surface area of each cylinder. Give your answers to one decimal place.

a

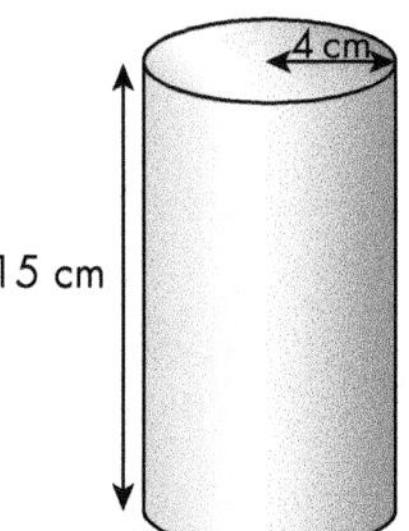

b

6 cm

5 cm

c

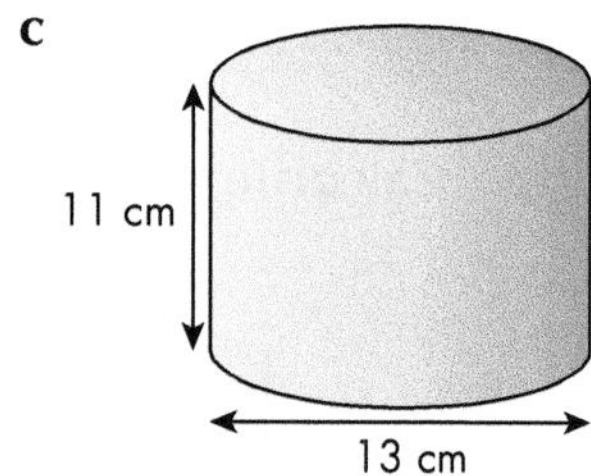

3 A solid iron bar is 40 cm long and has a radius of 2 cm. 1 cm^3 of iron weighs 8 g. Work out the mass of the iron bar in kilograms.

4 Work out the volume of cylinders with these dimensions. Give your answers in terms of π.

a radius 4 cm, height 11 cm

b diameter 16 cm, height 18 cm

5 Andrea's mum is hosting a party and is working out how much juice to order.

Her glasses are cylindrical in shape. They have radius 1.5 cm and height 5 cm.

The juice cartons are in the shape of a cuboid that measures 30 cm by 15 cm by 18 cm.

a Calculate the amount of juice in a full carton.

b Calculate the capacity of one glass.

c How many glasses can be 'half filled' from a carton of juice?

d There are going to be 30 guests at the party. If Andrea's mum serves each guest three half-filled glasses of juice, how many cartons of juice should she order?

6 A tunnel with a semi-circular cross-section is cut through a hillside. The diameter of the semicircle is 15 m and the length of the tunnel is 250 m.

One lorry can take away 8 m^3 of earth.

How many lorry loads are needed to move all the earth that is dug out to create the tunnel?

15 Algebra: Linear equations

15.1 Solving linear equations

Homework 15A

1 Solve the following equations.

a $x + 2 = 8$ **b** $y - 4 = 3$ **c** $s + 7 = 10$ **d** $t - 7 = 4$

e $3p = 12$ **f** $5q = 15$ **g** $\frac{k}{2} = 4$ **h** $4n = 20$

i $\frac{a}{3} = 2$ **j** $b + 1 = 2$ **k** $c - 7 = 7$ **l** $\frac{d}{5} = 1$

2 **a** Rafiq ran 26.9 miles less than Kathryn last week. Rafiq ran 11.1 miles. How many miles did Kathryn run?

b Terry and nine of his friends went out for a meal. They split the bill equally and each paid £10.48. What was the total bill?

3 Set up an equation to represent the following. Use x for the variable.

My mother is twice as old as me. She is 38 years old. How old am I?

4 Set up an equation to represent the following. Use y for the variable.

Ten litres of petrol cost £9.50. How much is one litre?

Homework 15B

1 Use inverse flow diagrams to solve these equations. Check that each answer works in the original equation.

a $2x + 5 = 13$ **b** $3x - 2 = 4$ **c** $2x - 7 = 3$ **d** $3y - 9 = 9$

e $5a + 1 = 11$ **f** $4x + 5 = 21$ **g** $6y + 6 = 24$ **h** $5x + 4 = 9$

i $8x - 10 = 30$ **j** $\frac{x}{2} + 1 = 4$ **k** $\frac{a}{2} - 2 = 3$ **l** $\frac{c}{3} + 2 = 8$

m $\frac{x}{3} - 3 = 1$ **n** $\frac{m}{3} - 1 = 2$ **o** $\frac{z}{5} + 6 = 10$

2 The reverse flow diagram shows the solution to an equation.

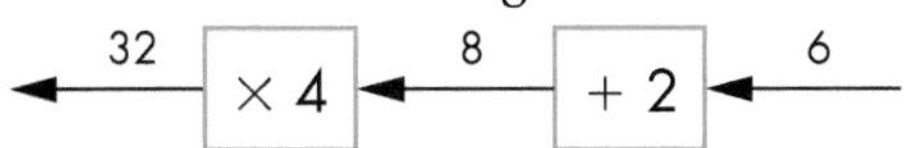

What is the equation?

3 The diagram shows a two-step number machine.

a If x is the input, what is the output in its simplest form?

b Draw a reverse flow diagram to show that this is correct.

Homework 15C

Example

Solve the equation $3x - 5 = 16$ by balancing.

Balancing means doing the same to both sides.

Add 5 to both sides. $3x - 5 = 16$

$$3x - 5 + 5 = 16 + 5$$

$$3x = 21$$

Divide both sides by 3. $\frac{3x}{3} = \frac{21}{3}$

$$x = 7$$

1 Solve these equations by balancing. Remember to check that each answer works in the original equation.

a $x + 5 = 6$ **b** $y - 3 = 4$ **c** $x + 5 = 3$

d $2y + 4 = 12$ **e** $3t + 5 = 20$ **f** $2x - 4 = 12$

g $6b + 3 = 21$ **h** $4x + 1 = 5$ **i** $2m - 3 = 4$

2 Solve these equations using any method. Remember to check that each answer works in the original equation.

a $2x + 1 = 7$ **b** $2t + 5 = 13$ **c** $3x + 5 = 17$

d $4y + 7 = 27$ **e** $2x - 8 = 12$ **f** $5t - 3 = 27$

g $8 - x = 2$ **h** $13 - 2k = 3$ **i** $6 - 3z = 0$

Homework 15D

1 Solve these equations.

a $\frac{x}{2} + 3 = 6$ **b** $\frac{p}{3} + 2 = 3$ **c** $\frac{x}{2} - 3 = 5$

d $\frac{x}{2} - 5 = 2$ **e** $\frac{a}{3} + 3 = 6$ **f** $\frac{z}{5} - 1 = 1$

2 The solution to the equation $\frac{x}{3} + 8 = 16$ is $x = 24$.
Make up two different equations of the form $\frac{x}{a} + b = c$, where a, b and c are positive whole numbers, for which the answer is also 24.

3 Here are two students' solutions to the equation $\frac{x}{2} + 3 = 5$.

Student 1	**Student 2**
$\frac{x}{2} + 3 = 5$	$\frac{x}{2} + 3 = 5$
$\frac{x}{2} + 3 - 3 = 5 - 3$	$\frac{x}{2} + 3 + 3 = 5 + 3$
$\frac{x}{2} = 2$	$\frac{x}{2} = 8$
$\frac{x}{2} \times 2 = 2 \times 2$	$\frac{x}{2} \div 2 = 8 \div 2$
$x = 4$	$x = 4$

a Which student used the correct method?

b Explain the mistakes the other student made.

4 Solve these equations.

a $\frac{x + 2}{3} = 4$ **b** $\frac{y - 4}{5} = 2$ **c** $\frac{z + 4}{8} = 5$

5 A teacher read out the text below to her class:

'I am thinking of a number. I multiply it by 2 and then subtract 3 from the result. The answer is 12. What number did I think of to start with?'

a What was the number the teacher thought of?

b Ben misunderstood the instructions and got the operations the wrong way round.
What number did Ben think the teacher started with?

6 Six boxes of apples, each holding A apples, are delivered to a supermarket. 18 of the apples are thrown away because they are rotten.

The rest are packed into 45 trays with six apples in each before being put on the shelves. How many apples, A, are in each box?

15.2 Solving equations with brackets

Homework 15E 

Solve $3(2x - 7) = 15$.

First multiply out the brackets. $6x - 21 = 15$

Add 21 to both sides. $6x = 36$

Divide both sides by 6. $x = 6$

1 Solve these equations. Some of the answers may be fractions or negative numbers. Remember to check that each answer works for its original equation. You can use your calculator if necessary.

a $2(x + 1) = 8$ **b** $3(x - 3) = 12$ **c** $3(t + 2) = 9$ **d** $2(x + 5) = 20$

e $2(2y - 5) = 14$ **f** $2(3x + 4) = 26$ **g** $4(3t - 1) = 20$ **h** $2(t + 5) = 6$

i $2(x + 4) = 2$ **j** $2(3y - 2) = 5$ **k** $4(3k - 1) = 11$ **l** $5(2x + 3) = 26$

2 Mike has been asked to solve the equation $a(bx + c) = 40$.

He knows that the values of a, b and c are 2, 4 and 5, but he doesn't know which is which.

What are the correct values of a, b and c given that $x = 1.5$?

3 As the class are coming in for the start of a mathematics lesson, the teacher is writing some equations on the board.

So far she has written:

$5(2x + 3) = 13$

$2(5x + 3) = 13$

Zak says, 'That's easy! Both equations have the same solution, $x = 2$.'

Is Zak correct? If not, what mistake has he made? What are the correct answers?

15.3 Solving equations with the variable on both sides

Homework 15F 

Example

Solve the equation $5x + 4 = 3x + 10$.

Subtract $3x$ from both sides.	$2x + 4 = 10$
Subtract 4 from both sides.	$2x = 6$
Divide both sides by 2.	$x = 3$

1 Solve these equations.

a $2x + 1 = x + 3$ **b** $3y + 2 = 2y + 6$ **c** $5a - 3 = 4a + 4$

d $5t + 3 = 3t + 9$ **e** $7p - 5 = 5p + 3$ **f** $6k + 5 = 3k + 20$

g $6m + 1 = m + 11$ **h** $5s - 1 = 2s - 7$ **i** $4w + 8 = 2w + 8$

j $5x + 5 = 3x + 10$

2 Amber is thinking of a number. She multiplies it by 3 and subtracts 6.

Callum is thinking of a number. He multiplies it by 5 and adds 2.

Amber and Callum discover that they both thought of the same original number and both got the same final answer.

What number did they think of?

3 Solve these equations.

a $5(t - 2) = 4t - 1$ **b** $4(x + 2) = 2(x + 1)$

c $5p - 2 = 5 - 2p$ **d** $2(2x + 3) = 3(x - 4)$

4 The triangle shown is isosceles. What is the perimeter of the triangle?

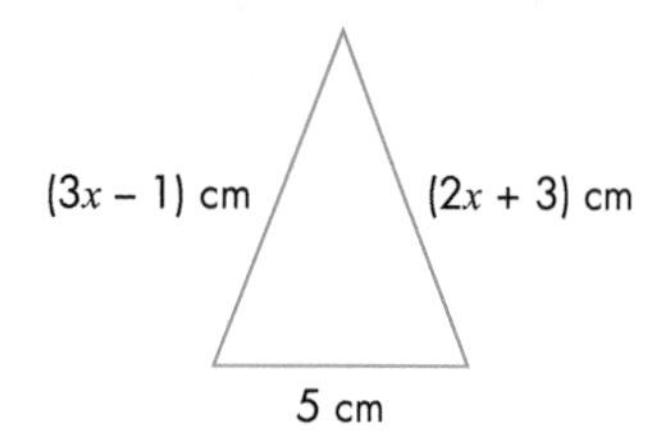

5 A teacher asks her class to think of a starting number and subtract 6 from it.

Alan said: 'My starting number was 9.'

Vikram said: 'My final answer was –2.'

a What was Alan's final answer?

b What was Vikram's starting number?

6 A class of 24 students had a collection to buy some chocolates for their teacher's birthday. Each student gave p pence and the teaching assistant gave a pound.

The chocolates cost £10.60.

a Which of the following equations represents this situation?

$24p + 1 = 10.6$ $\quad$ $24p + 100 = 10.6$ $\quad$ $24p + 100 = 1060$

b How much did each student contribute?

7 A girl is Y years old. Her father is 23 years older than she is. The sum of their ages is 37. How old is the girl?

8 A boy is X years old. His sister is twice as old as he is. The sum of their ages is 24. How old is the boy?

9 The diagram shows a rectangle. Work out the value of x if the perimeter is 24 cm.

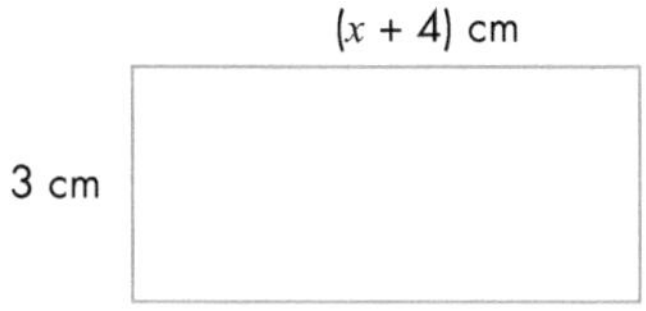

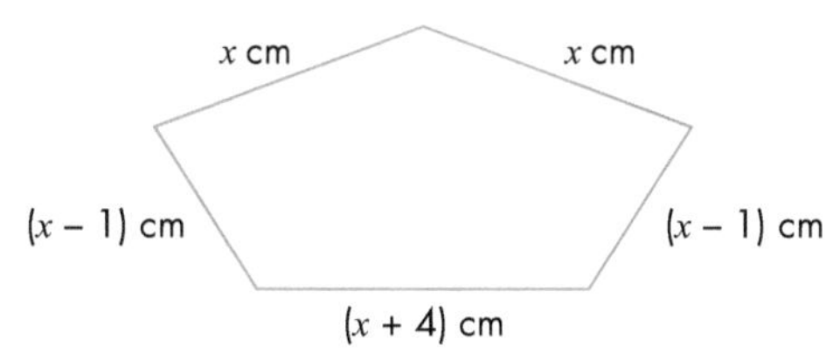

10 This pentagon has a perimeter of 32 cm. Work out the length of each side of the pentagon.

11 On a bookshelf there are $2b$ crime novels, $3b – 2$ science fiction novels and $b + 7$ romance novels. There are 65 books altogether. How many of each type of book are there?

12 Could this triangle be equilateral? Explain your answer.

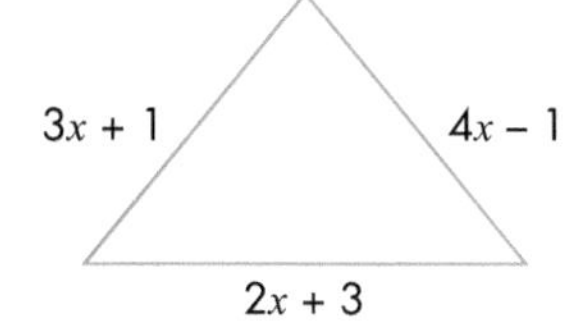

16 Ratio and proportion and rates of change: Percentages and compound measures

16.1 Equivalent fractions, percentages and decimals

Homework 16A

Example

As a fraction 32% = $\frac{32}{100}$ and this simplifies (or cancels) to $\frac{8}{25}$.

Example

As a decimal 65% = 65 ÷ 100

= 0.65

1 Write each percentage as a fraction in its simplest form.

a 10% **b** 40% **c** 25% **d** 15% **e** 75% **f** 35%

g 12% **h** 28% **i** 56% **j** 18% **k** 42% **l** 6%

2 Write each percentage as a decimal.

a 87% **b** 25% **c** 33% **d** 5% **e** 1% **f** 72%

g 58% **h** 17.5% **i** 8.5% **j** 68.2% **k** 150% **l** 132%

3 Copy and complete the table to show the equivalent values.

Percentage	Fraction	Decimal
10%		
20%		
30%		
		0.4
		0.5
		0.6
	$\frac{7}{10}$	
	$\frac{8}{10}$	
	$\frac{9}{10}$	

4 If 45% of students walk to school, what percentage do not walk to school?

5 If 84% of the families in a village own at least one car, what percentage of the families do not own a car?

6 In a local election 48% of people voted for Mrs Slater, 29% voted for Mr Rhodes and the remainder voted for Mr Mulley. What percentage voted for Mr Mulley?

7 From his gross salary, Mr Hardy pays 20% Income Tax, 6% pension contributions and 5% National Insurance. What percentage is his net (take-home) pay?

8 Approximately what percentage of each can is filled with oil?

a 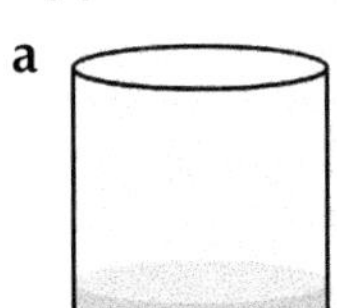

b 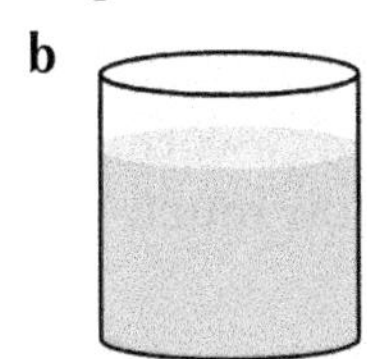

c 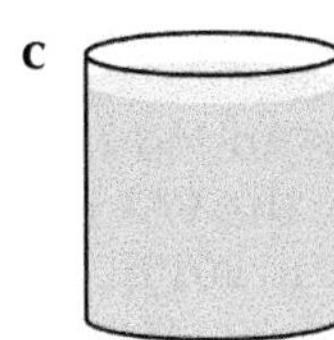

9 Write each fraction as a percentage.

a $\frac{3}{4}$ **b** $\frac{2}{5}$ **c** $\frac{7}{20}$ **d** $\frac{3}{25}$ **e** $\frac{43}{50}$ **f** $\frac{3}{8}$

10 Write each decimal as a percentage.

a 0.23 **b** 0.87 **c** 0.09 **d** 0.235 **e** 1.8 **f** 2.34

11 Tom scored 68 marks out of 80 in a geography test.

a Write his score as a fraction in its simplest form.

b Write his score as a decimal.

c Write his score as a percentage.

d His next test is marked out of 50.

How many marks out of 50 does he need to improve his percentage score?

16.2 Calculating a percentage of a quantity

Homework 16B 

Example

Calculate 12% of 54 kg.

Method 1 $\frac{12}{100} \times 54 = 6.48$ kg

Method 2 $0.12 \times 54 = 6.48$ kg

1 Write down the multiplier for each percentage.

a 23% **b** 70% **c** 4% **d** 120%

2 Write down the percentage that is equivalent to each multiplier.

a 0.38 **b** 0.8 **c** 0.07 **d** 1.5

3 Calculate these quantities.

a 25% of £200 **b** 10% of £120 **c** 53% of 400 kg **d** 75% of 84 cm

e 22% of £84 **f** 71% of 250 g **g** 24% of £3 **h** 95% of 320 m

i 6% of £42 **j** 17.5% of £56 **k** 8.5% of 160 litres **l** 37.2% of £800

4 During one week at a test centre, 320 people took their driving test and 65% passed. How many people passed?

5 A school has 250 students in each year and the attendance record on one day for each year group is shown below.

Year 7 240, Year 8 230, Year 9 210, Year 10 220, Year 11 200

The school has a target of 90% attendance overall. Did the school meet its target?

6 A certain type of stainless steel is 84% iron, 14% chromium and 2% carbon (by mass). How many tonnes of each element are in 450 tonnes of stainless steel?

7 VAT (Value Added Tax) is a tax that the Government adds to the price of goods sold. Until 2011, the VAT rate was 17.5%. How much VAT would have been added on to these bills in 2010?

a a restaurant bill for £40 **b** a telephone bill for £82 **c** a car repair bill for £240

8 An insurance firm sells house insurance with an annual premium of 0.5% of the value of the house. What will be the annual premium for a house valued at £120 000?

16.3 Increasing and decreasing quantities by a percentage

Homework 16C 

Example

Increase £6 by 5%.

Method 1 Work out 5% of £6: $(5 \div 100) \times 6 = £0.30$

Add £0.30 to the original amount: $£6 + £0.30 = £6.30$

Method 2 $1.05 \times 6 = £6.30$

Example

Decrease £6 by 5%.

Method 1 Work out 5% of £6: $(5 \div 100) \times 6 = £0.30$

Subtract £0.30 from the original amount: $£6 - £0.30 = £5.70$

Method 2 $0.95 \times 6 = £5.70$

1 Write down the multiplier you would use to increase a quantity by each percentage.

a 15% **b** 17.5% **c** 22% **d** 8%

2 Write down the multiplier you would use to decrease a quantity by each percentage.

a 9% **b** 14% **c** 84% **d** 37%

3 Increase each amount by the given percentage.

a £80 by 5% **b** £150 by 10% **c** 800 m by 15% **d** 320 kg by 25%

e £42 by 30% **f** £24 by 65% **g** 120 cm by 18% **h** £32 by 46%

i 550g by 85% **j** £72 by 72%

4 Decrease each amount by the given percentage.

a £20 by 10% **b** £150 by 20% **c** 90 kg by 30% **d** 500 m by 12%

e £260 by 5% **f** 80 cm by 25% **g** 400 g by 42% **h** £425 by 23%

i 48 kg by 75% **j** £63 by 37%

5 Mr Sables was given a pay rise of 4%. His salary before the rise was £132 500. What is his new salary?

6 Mrs Denghali buys a new car from a garage for £18 400. The garage owner tells her that her car will lose 24% of its value after one year. Calculate the value of the car after one year.

7 In 2010 the population of a village was 2400. In 2014 the population had decreased by 12%. What was the population of the village in 2014?

8 VAT (Value Added Tax) is a tax that the Government adds to the price of goods sold. Until 2011, the VAT rate was 17.5%. In 2010, Mrs Moat purchased these items from a gift catalogue. The prices shown below are before VAT of 17.5% was added.

Gift	Pre-VAT price
Travel alarm clock	£18.00
Ladies' purse wallet	£15.20
Pet's luxury towel	£12.80
Silver-plated bookmark	£6.40

She estimated that the total cost would be £60. Was this a good estimate? Show how you decide.

9 A dining table costs £300 before VAT is added.

If the rate of VAT increases from 15% to 20%, by how much will the cost of the dining table increase?

10 A travel agent is offering a 15% discount on holidays. How much will the advertised holiday now cost?

NEW YEAR'S SALE:
All prices reduced by 15%

11 A shop increases all its prices by 10%.

One month later it advertises 10% off all marked prices.

Are the goods cheaper, the same price or more expensive than before the increase?

Show how you work out your answer.

12 Shop A increases its prices by 5% and then a month later by another 5%. Shop B increases its prices by 10%. Which shop increases its prices by the greatest percentage?

13 Show that a 20% decrease followed by a 20% increase is equivalent to a 4% decrease.

16.4 Expressing one quantity as a percentage of another

Homework 16D

Example

Express £6 as a percentage of £40.

Set up the fraction: $\frac{6}{40}$

Multiply it by 100: 6 ÷ 40 = 15%

1 Write the first quantity as a percentage of the second. Give your answers to one decimal place where necessary.

a £8, £40 **b** 20 kg, 80 kg **c** 5 m, 50 m

d £15, £20 **e** 400 g, 500 g **f** 23 cm, 50 cm

g £12, £36 **h** 18 minutes, 1 hour **i** £27, £40

j 5 days, 3 weeks

2 What percentage of these shapes are shaded?

a

b

3 In a class of 30 students, 18 are girls.

a What percentage of the class are girls?

b What percentage of the class are boys?

4 The area of a farm is 820 hectares. The farmer uses 240 hectares for pasture.

What percentage of the farm land is used for pasture? Give your answer to one decimal place.

5 **a** Work out the percentage profit on each item. Give your answers to one decimal place.

	Item	Wholesale price paid by the shop (£)	Selling price (£)
i	Micro hi-fi system	150	250
ii	Mp3 player	60	90
iii	CD player	30	44.99
iv	Cordless headphones	18	29.99

b The shopkeeper wants to make over 40% profit on each item. Does he succeed?

6 These are the results from two tests taken by Paul and Val. Both tests are out of the same mark.

	Test A	Test B
Paul	30	40
Val	28	39

Whose result has the greater percentage increase from test A to test B?

Show your working.

7 A small train is carrying 48 passengers.

At a station, more passengers get on so that all the seats are filled and no one is standing.

At the next station, 70% of the passengers leave the train and 30 new passengers get on.

There are now 48 passengers on the train again.

How many seats are on the train?

8 In a secondary school, 30% of students have a younger brother or sister at a primary school.

20% of students have two younger brothers or sisters at the primary school. None of them have more than two siblings.

Altogether there are 700 brothers and sisters at the primary school.

How many students are at the secondary school?

16.5 Compound measures

Homework 16E 

Hints and tips The relationship between total pay, hourly rate and hours worked can be expressed in three ways.

total pay = hours worked × hourly rate

$$\text{hours worked} = \frac{\text{total pay}}{\text{hourly rate}} \qquad \text{hourly rate} = \frac{\text{total pay}}{\text{hours worked}}$$

1 Work out the total pay for each person:

a 40 hours at £2.64 per hour

b $37\frac{1}{2}$ hours at £24.51 per hour

c $47\frac{1}{2}$ hours at £17.52 per hour

d 12 hours at £6.50 per hour

2 Work out the hourly rate for each payment:

a £189 for 18 hours' work

b £207 for 12 hours' work

c £439.28 for 19 hours' work

d £476.16 for 24 hours' work

3 Work out the number of hours worked for each job:

a £104.47 at £6.74 per hour

b £154.47 at £8.13 per hour

c £92.87 at £2.51 per hour

d £802.28 at £12.94 per hour

4 Jonathan works for 39 hours at an hourly rate of £12.13. He pays 20% of his pay in income tax. He also pays National Insurance at a rate of x%. This leaves him with £340.61. What is the value of x? Show your working.

Homework 16F

1 **a** The mass of 5 m^3 of copper is 44 800 kg.

Work out the density of copper.

b The density of zinc is 7130 kg/m^3.

Work out the mass of 5 m^3 of zinc.

2 An ice hockey puck is a cylinder with a radius of 3.8 cm and a depth of 2.5 cm. It is made out of rubber with a density of 1.5 g/cm^3.

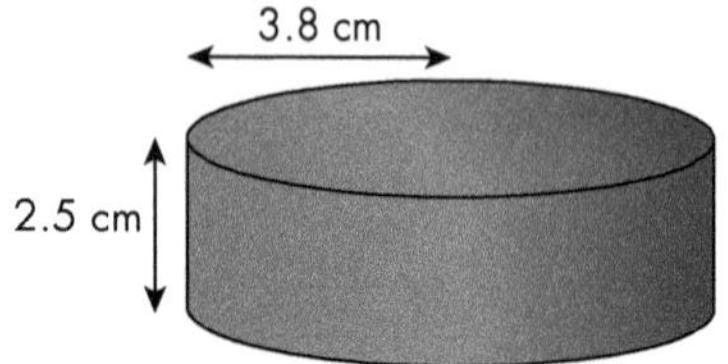

Work out the mass of the ice hockey puck.

Give your answer correct to three significant figures.

3 The diagram shows a solid cuboid. It is made from wood with a density of 0.6 g/cm^3. Work out the mass of the cuboid.

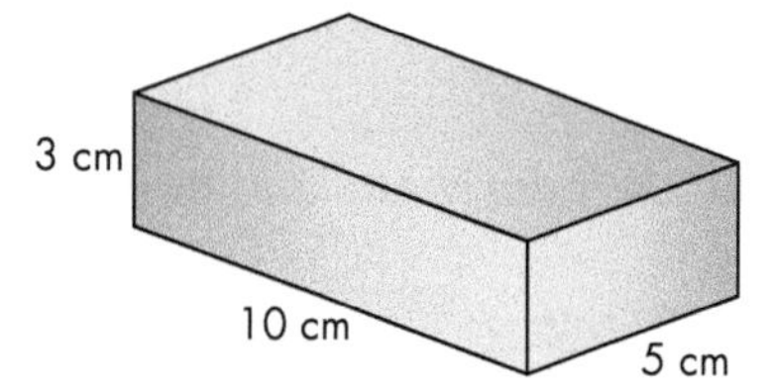

4 Metal A has a mass of 1000 g and a volume of 20 cm^3. Metal B has a mass of 1050 g and a density of 50 g/cm^3. Which piece of metal has a larger volume?

5 A steel block is in the shape of a cuboid that measures 30 cm by 35 cm by 25 cm. Which face should it stand on to exert most pressure?

6 Josh works 8 hours per week and earns £h. Mia works 14 hours per week. They both get paid exactly the same weekly pay. What is Mia's hourly rate? Give your answer in terms of h.

7 A cuboid has a mass of 75 kg and a volume of 0.1 m^3. When placed on each of its faces the pressures exerted are 50 Pa, 20 Pa and 100 Pa.

Take g = 10 m/s^2.

Work out the area of each face of the cuboid.

17 Ratio and proportion and rates of change: Percentages and variation

17.1 Compound interest and repeated percentage change

Homework 17A

1. A bank pays simple interest of 8% per year on money invested in a savings account. Miss Pettica invests £2000 in this account. How much will she have in her account after:

   a 1 year **b** 2 years **c** 3 years?

2. I invest £3000 in a savings account that pays compound interest of 4%. How much will I have in the bank after 6 years if I do not make any withdrawals?

3. My dad put all the money I was sent when I was born into a savings account. He invested £470 in an account that pays compound interest of 11.2% for as long as the money is in the account. How much will I have when I turn 18?

4. I take out a loan for £14 499 to buy a car. The loan is for 5 years and has an interest rate of 6.9%. What will my total loan repayments be?

5. Veronika buys a mobile phone for £250. It depreciates at a rate of 11.3% each year. Amelia buys a mobile phone for £320. It loses value at a rate of 8.9% each year. Scarlett buys a phone for £650. It depreciates by 22% each year.

   Whose phone is worth the most after three years?

6. A chick increases its mass by 3% each day for the first six weeks of life. The mass of the chick was 85 g at birth. What will its mass be after:

   a 1 day

   b 5 days

   c 10 days

   d 20 days?

7. The population of a small village in Lincolnshire is projected to increase at a steady rate of 9% each year. In 2010 the population was 19 102.

   a In which year is the population:

   i 22 695

   ii 29 390

   iii 45 221

   iv 107 055?

   b When the population of the village reaches 55 000, the council will need to build a new school. In which year will they begin to build the school?

17.2 Reverse percentage (working out the original value)

Homework 17B

1. A laptop is reduced by 15% to £520 in a sale. What was its original price?
2. A carpenter increases his charges by 6% to £25.44 per hour. How much did he charge before the increase?
3. A pair of shoes are reduced by 30% to £105. How much did they cost before the reduction?
4. I bought a sofa in the January sales with 25% discount. I paid £330. What is the full price of the sofa?
5. 13% more tickets were sold for Z Festival this year than last year. This year 58 082 tickets were sold. How many tickets were sold last year?
6. Steve reduced his personal best marathon time by 4% to 3 hours and 36 minutes. What was his previous personal best?
7. Every year the amount of rubbish sent to landfill increases by 7%. This year 25 265 696 tonnes of rubbish will be sent to landfill. How much rubbish was sent to landfill last year?
8. Ciara has reduced her calorie intake by 5% and now consumes 1995 calories per day. What was her original calorie intake?
9. Every year 20% of university students drop out of Course B at CleberClobs University. This year 44 students dropped out of Course B. How many students started the course?
10. Aaron is demoted at work and his salary decreases by 15% to £38 250. How much was he earning before his demotion?

17.3 Direct proportion

Homework 17C

Hints and tips

There is direct proportion between two variables when one is always the same multiple of the other.

The cost of nachos is directly proportional to the number of bags bought.

Number of bags	1	2	3	4
Cost (£)	2	4	6	8

When two variables are in direct proportion, they produce a straight-line graph that starts at the origin.

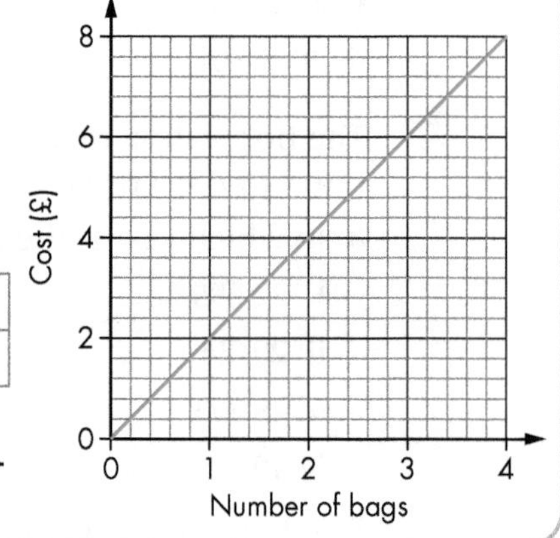

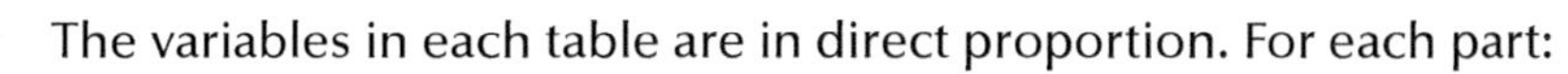

1 The variables in each table are in direct proportion. For each part:

i Work out the multiplier.

ii Copy and complete the table of values.

iii Draw a direct proportion graph.

iv Read the required value from the graph.

a

x	2	4	20	
y	14			56

What is the value of y when $x = 10$?

b

p	4	20	1	
q	20			75

What is the value of q when $p = 2$?

c

x	5		20
y	30	6	

What is the value of y when $x = 10$?

d

x	2	6	
y	15	45	90

What is the value of y when $x = 3$?

e

a	2	8	
b	7		105

What is the value of b when $a = 11$?

f

x	6	12	30	
y	24			4

What is the value of y when $x = 10$?

g

x	4		20
y	10	30	

What is the value of y when $x = 1$?

h

a	4	12	
b	18		90

What is the value of b when $a = 5$?

i

x	12	8	
y	18		6.9

What is the value of y when $x = 5.5$?

j

Pounds (P)	3	12	25	
Dollars (D)	4.50			180.00

How many dollars are equivalent to £40?

2 A is directly proportional to r. When $r = 4$, $A = 6$. Work out the value of:

a A when $r = 12$ **b** r when $A = 30$.

3 C is directly proportional to p. When $p = 4$, $C = 18$. Work out the value of:

a C when $p = 20$ **b** p when $C = 63$.

4 Do these graphs show direct proportion? Give a reason for each answer.

a Revenue against quantity of items produced

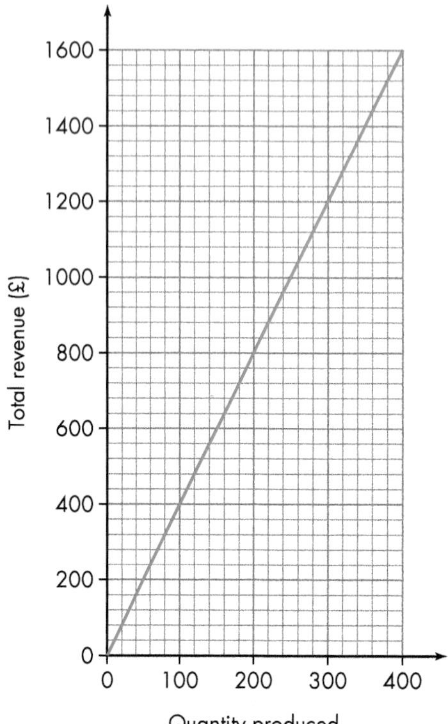

b Postage cost against distance

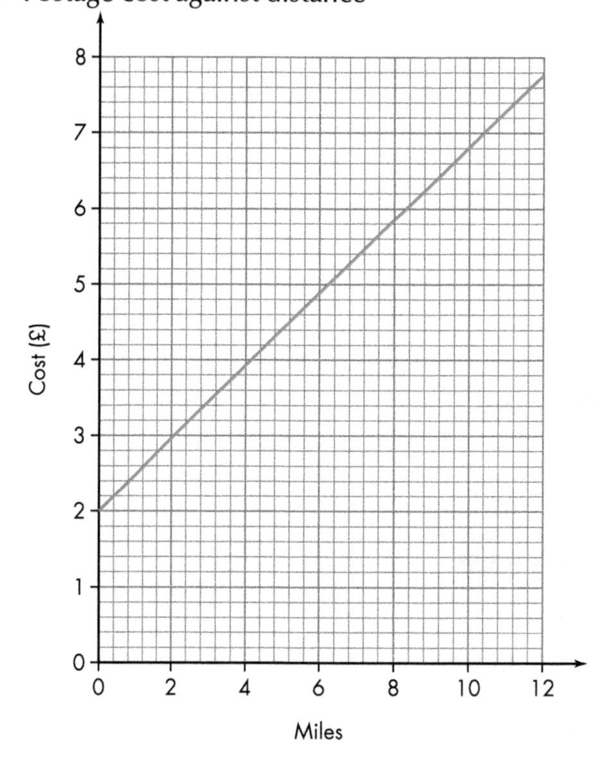

c Distance travelled against time

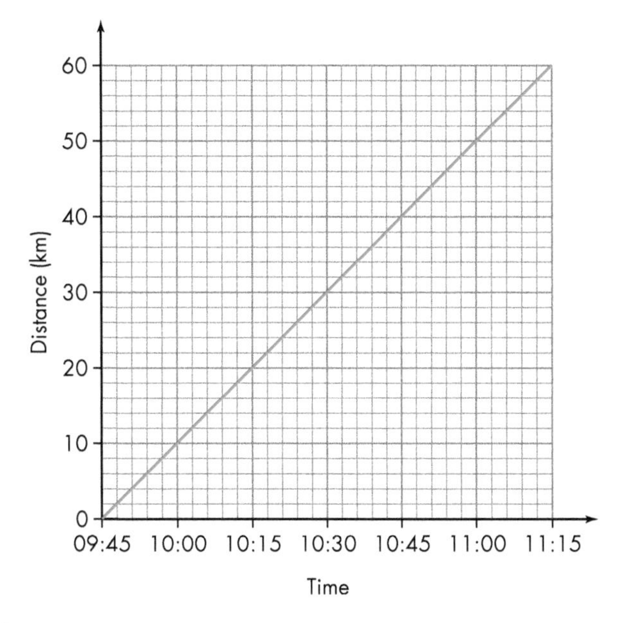

d Temperature of a hot drink against time

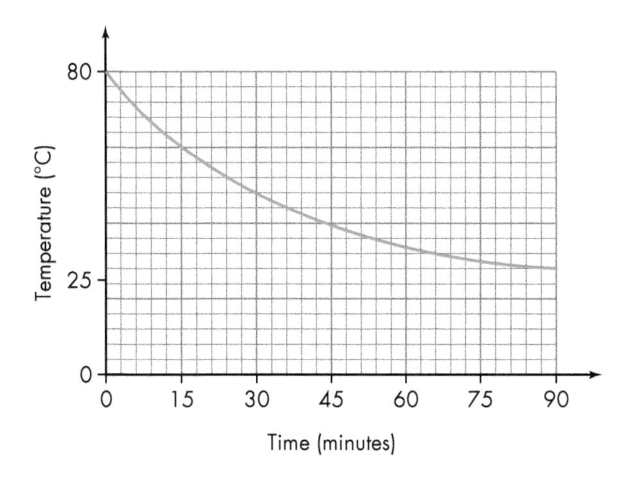

17.4 Inverse proportion

Homework 17D

1. Twelve men can dig a pond in 8 days. How many men are needed to dig a pond in 6 days?

2. A hostel has enough food for 125 students for 16 days. For how long will the food last 200 students?

3. A fort has enough provisions to last 80 soldiers for 60 days. After 15 days, 20 more soldiers arrive at the fort. How long will the food last altogether?

4. Eight taps (with the same rate of flow) fill a tank in 27 minutes. How long will it take for six taps to fill the tank?

5. y is inversely proportional to x. When $x = 6$, $y = 12$. Work out the value of y when $x = 8$.

6. h is inversely proportional to k. When $k = 2$, $h = 24$. Work out the value of h when $k = 8$.

7. y is inversely proportional to x. When $x = 1$, $y = 12$. Work out the value of y when $x = 4$.

8. y is inversely proportional to the square of x. When $x = 4$, $y = 3$. Work out the value of y when $x = 12$.

9. z is inversely proportional to the square of w. When $w = 4$, $z = 32$. Work out the value of z when $w = 2$.

10. The speed of a car (S) and the time (T) it takes to go around a circuit are inversely proportional.

    The proportionality equation is $S = \frac{50}{T}$.

    Copy and complete the table and plot the graph. Remember to join the points with a smooth curve.

S	50	25	20	12.5		5		
T	1	2		4	5	10	15	20

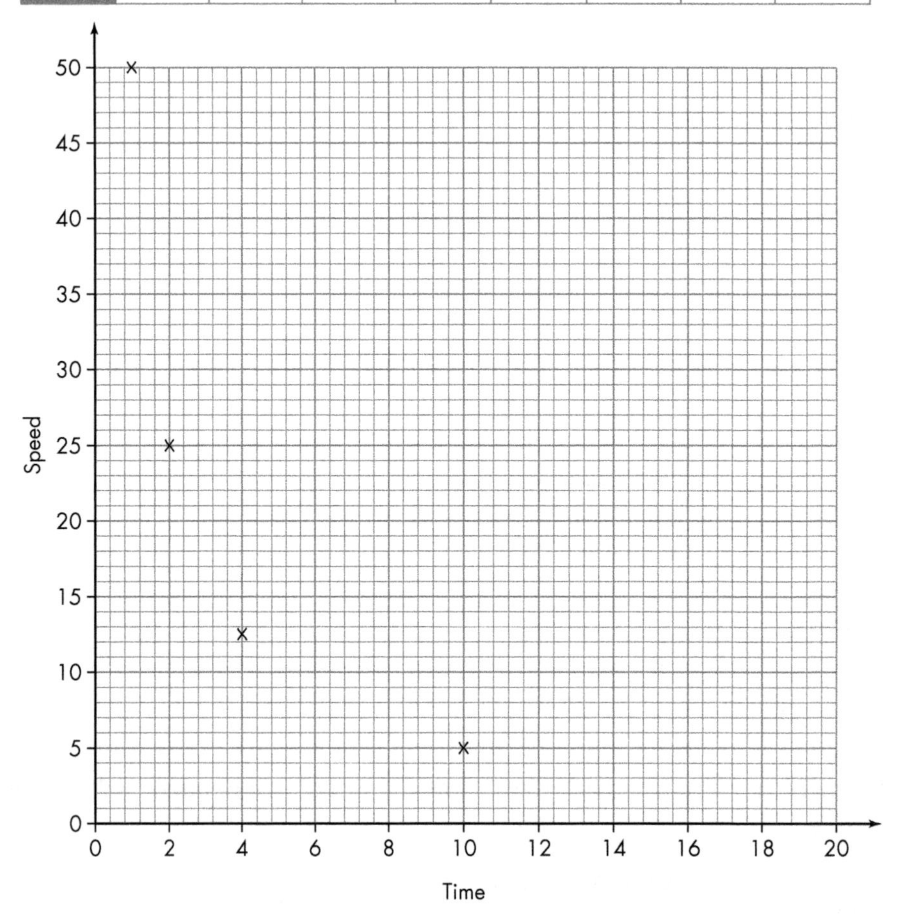

18 Statistics: Representation and interpretation

18.1 Sampling

Homework 18A

1. Isabelle wants to find out who uses the local shop. She asks the first 10 customers who use the shop when it opens at 8:30 their age. Is her sample representative? Give reasons for your answer.

2. Jenson wants to know which flavour drinks the school should stock in the vending machine. He asks a random sample of Year 11 students at his school.

   a Explain why this sample may not be a good one.

   b There are 1000 students in his school. Jenson wants to use a sample of 50 students. Describe how Jenson could select a suitable sample.

3. Molly carries out a survey to find out the favourite band of students in her school year.

   There are 160 girls and 40 boys in her school year.

   She takes a random sample containing 20 girls and 30 boys.

   Is her sample representative? Give reasons for your answer.

18.2 Pie charts

Homework 18B

1. The table shows the times taken by 60 people to travel to work.

Time in minutes	10 or less	Between 11 and 30	31 or more
Frequency	8	19	33

   Draw a pie chart to illustrate the data.

2. The table shows the numbers of GCSE passes that 180 students obtained.

GCSE passes	9 or more	7 or 8	5 or 6	4 or less
Frequency	20	100	50	10

   Draw a pie chart to illustrate the data.

3 For a statistics project, Tom asks 36 of his school friends about their main use of computers and records the results in the table shown below.

Main use	E-mail	Internet	Word processing	Games
Frequency	5	13	3	15

a Draw a pie chart to illustrate his data.

b What conclusions can you draw from his data?

c Give reasons why Tom's data is not really suitable for his project.

4 In a survey, a TV researcher asks 120 people at a leisure centre to name their favourite type of television programme. The results are given in the table.

Type of programme	Comedy	Drama	Films	Soaps	Sport
Frequency	18	11	21	26	44

a Draw a pie chart to illustrate the data.

b Do you think the sample chosen by the researcher is representative of the population? Give a reason for your answer.

5 Marion is writing a magazine article about healthy living. She asked a sample of people the question: "How often do you consider your health when planning your diet?" The pie chart shows the results of her survey.

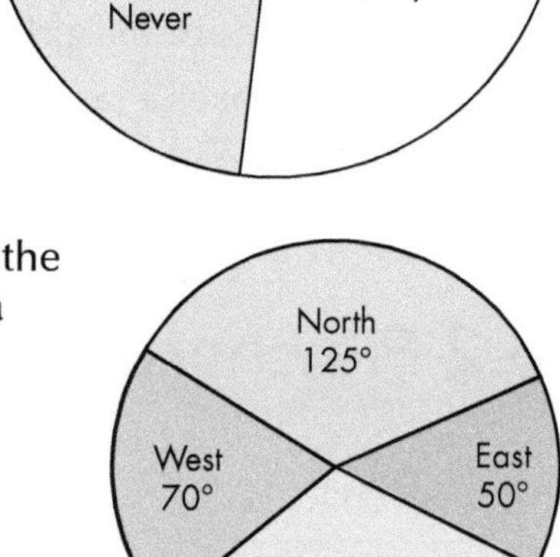

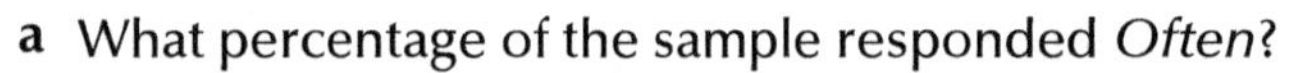

a What percentage of the sample responded *Often*?

b What response was given by about a third of the sample?

c Can you tell how many people there were in the sample? Give a reason for your answer.

d What other questions could Marion ask?

6 A nationwide survey was carried out to canvas opinion on the friendliest region of England. What is the probability that a person picked at random from this survey answered *East*?

7 You are asked to draw a pie chart to represent the different breakfasts that students have one morning.

What data would you need to obtain in order to do this?

18.3 Scatter diagrams

Homework 18C

1 The table shows the heights and masses of 12 students in a class.

a Plot the data on a scatter diagram.

b Draw a line of best fit.

c Chloe was absent from the class, but is 152 cm tall. Use the line of best fit to estimate her mass.

d A new girl, with a mass of 45 kg, joined the class. What height would you expect her to be?

Student	Mass (kg)	Height (cm)
Arianna	41	143
Bea	48	145
Caroline	47.5	147
Dhiaan	52	148
Emma	49.5	149
Fiona	55	149
Gill	55	153
Hanna	55.5	155
Imogen	61	157
Jasmine	65.5	160
Keira	60	163
Laura	68	165

2 The table shows the marks for 10 students in their mathematics and music examinations.

Student	Alex	Ben	Chris	Dom	Ellie	Ffion	Giordan	Hannah	Isabel	Jemma
Maths	52	42	65	60	77	83	78	87	29	53
Music	50	52	60	59	61	74	64	68	26	45

a Plot the data on a scatter diagram. Use the horizontal axis for the mathematics scores and mark it from 20 to 100. Use the vertical axis for the music scores and mark it from 20 to 100.

b Draw a line of best fit.

c One of the students was ill when they took the mathematics examination. Which student was it most likely to have been?

d Another student, Kris, was absent for the music examination. He scored 45 in mathematics. What mark would you expect him to have scored in music?

e Lexie was absent for the mathematics examination but scored 78 in music. What mark would you expect her to have scored in mathematics?

3 Twelve students took part in a maths challenge. The table shows their scores for two tests: a mental test and a problem-solving test.

Mental	25	32	32	43	47	50	55	58	61	65	68	72
Problem-solving	32	30	36	40	50	59	52	53	62	60	48	73

Another student, Harry, scored 53 in the mental test but his problem-solving test went missing. How many marks would you expect him to get in the problem-solving test?

4 Describe what you would expect the scatter graph to look like if someone said that it showed no correlation.

18.4 Grouped data and averages

Homework 18D

1 For each set of data:

i write down the modal group

ii calculate an estimate for the mean.

a The daily wages of 94 factory employees

Wages (£)	0–20	21–40	41–60	61–80	81–100
Frequency	9	13	21	34	17

b The manufacturing costs for 88 pieces of self-assembly furniture

Cost (£)	0.00–10.00	10.01–20.00	20.01–30.00	30.01–40.00	40.01–60.00
Frequency	9	17	27	21	14

2 A hospital has to report the average waiting time for patients in the Accident and Emergency department. During one shift, a survey was made to see how long these patients had to wait before seeing a doctor. The table summarises the results. The times are rounded to the nearest minute.

Time (minutes)	0–10	11–20	21–30	31–40	41–50	51–60	61–70
Frequency	1	12	24	15	13	9	5

a How many patients were seen by a doctor during this shift?

b Estimate the mean waiting time.

c Which average would the hospital use to represent the waiting time?

3 The table shows the average speed of cyclists in Lincoln along a stretch of the Newark Road.

Speed (mph)	0–9	10–19	20–29	30–39	40–49
Frequency	5	6	9	4	1

Chris noticed that one response had been recorded in the wrong part of the table. When corrected, the mean was 20.1.

Which response was recorded in the wrong part of the table?

4 The profit, to the nearest pound, made each week by a sandwich shop is shown in the table below.

Profit	£0–£200	£201–£500	£501–£750	£751–£1000
Frequency	21	26	11	2

Explain how you would estimate the mean profit made each week.

19 Geometry and measures: Constructions and loci

19.1 Constructing triangles

Homework 19A

1 Construct each of these triangles accurately. Measure the sides and angles not marked in the diagrams.

a

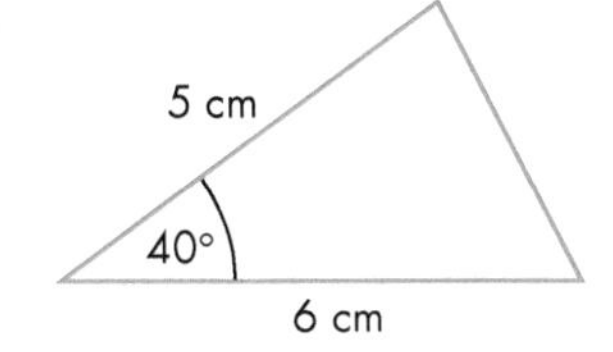

b

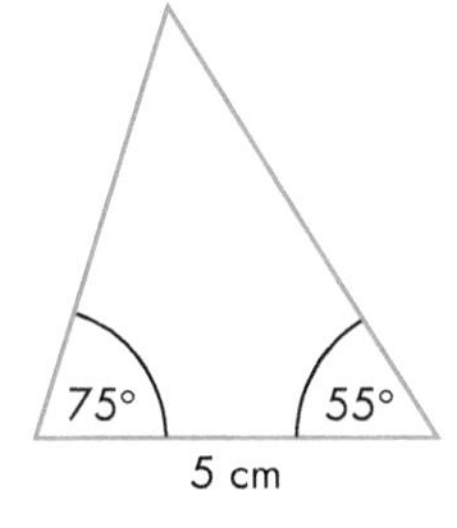

c

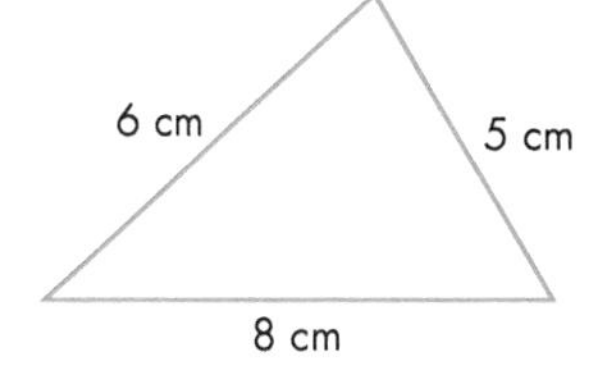

d

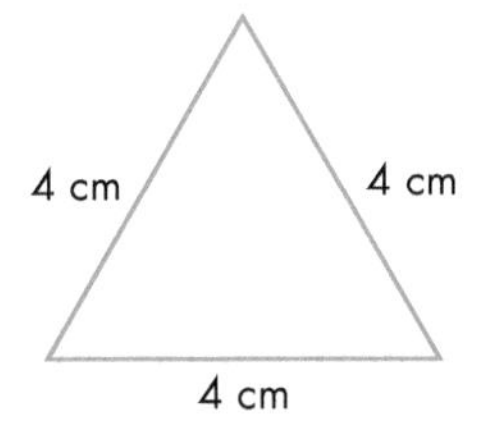

e

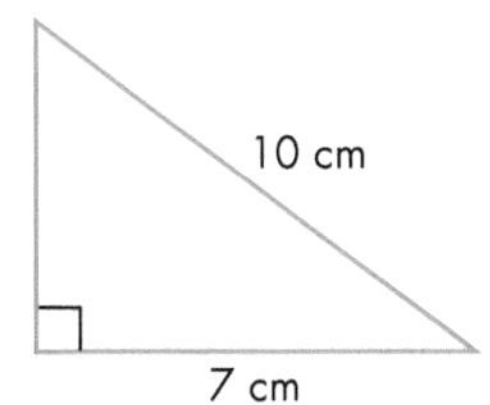

2 Can you draw this triangle accurately? Give reasons for your answer.

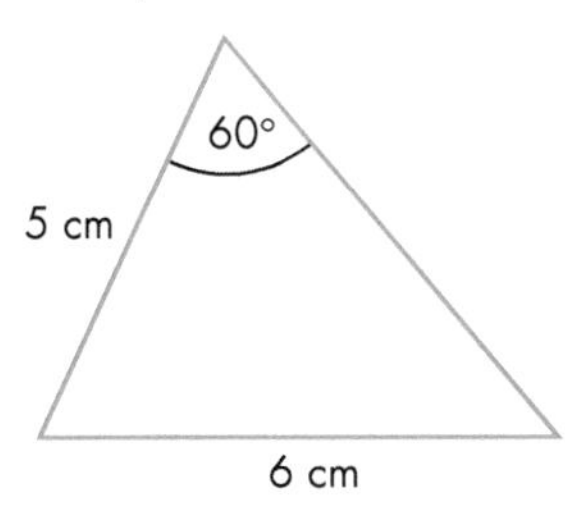

3 **a** Accurately draw the shape below.

b What is the name of the shape you have drawn?

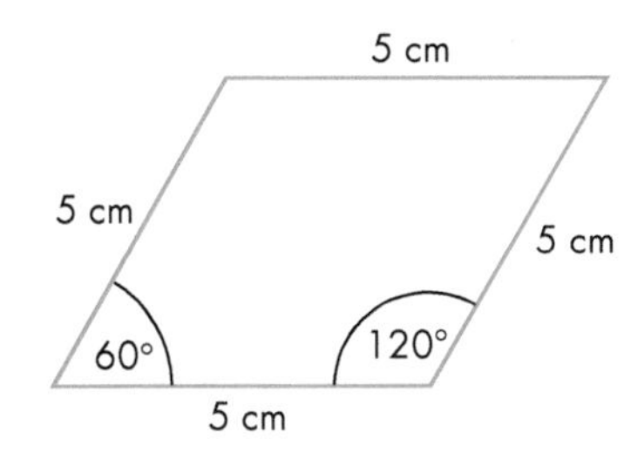

4. Chloé says, 'As long as I know two sides of a triangle and one angle I can draw it.' Is Chloé correct? Give a reason for your answer.

5. You are asked to construct a triangle with one side 9 cm, one side 10 cm and an angle of 60°. Sketch all the possible triangles that match this description.

19.2 Bisectors

Homework 19B

1. Draw a line 8 cm long. Construct the perpendicular bisector of the line. Check your accuracy by measuring each half.

2. **a** Draw any triangle.

   b Construct the perpendicular bisector of each side. The bisectors should all intersect at the same point.

   c Draw a circle with this point as the centre so that it fits exactly inside the triangle.

3. **a** Construct a circle with a radius of about 4 cm.

   b Draw a quadrilateral inside the circle with the corners of the quadrilateral on the circumference of the circle.

   c Construct the perpendicular bisector of two sides of the quadrilateral. The bisectors should meet at the centre of the circle.

4. **a** Draw any angle.

   b Construct the angle bisector.

   c Check your accuracy by measuring each half.

5. The diagram shows a park with two ice-cream sellers, A and B. People always go to the ice-cream seller nearest to them. Copy the diagram and shade the region of the park from which people go to ice-cream seller B.

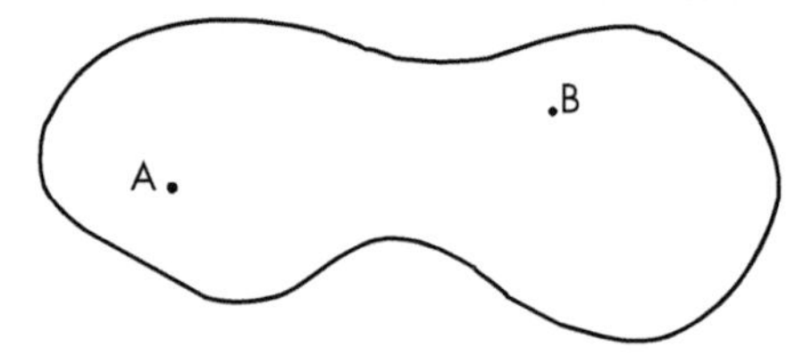

6. Using only a pencil, a straight edge and a pair of compasses, construct:

   a an angle of 15 degrees

   b an angle of 75 degrees.

7. If I construct all the angle bisectors of any triangle, they will meet at a point.

   Explain why, if I draw a circle with this as the centre, the circle will just touch each side of the triangle.

19.3 Defining a locus

Homework 19C

1 A is a fixed point. Sketch the locus of the point P where AP > 3 cm and AP < 6 cm.

2 A and B are two fixed points 4 cm apart. Draw the locus of the point P for these situations:

a AP < BP

b P is always within 3 cm of A and within 2 cm of B.

3 A fly is tethered by a length of spider's web that is 1 m long. Describe the locus of the fly's movement.

4 ABC is an equilateral triangle of side 4 cm. In each of the following loci, the point P moves only inside the triangle. Sketch the locus in each case.

a AP = BP

b AP < BP

c CP < 2 cm

d CP > 3 cm and BP > 3 cm

5 A wheel rolls around the inside of a square. Sketch the locus of the centre of the wheel.

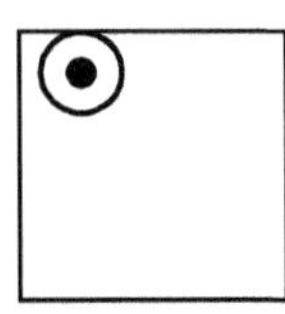

6 The same wheel rolls around the outside of the square. Sketch the locus of the centre of the wheel.

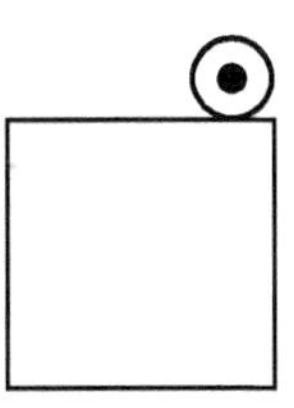

7 In each case, sketch the locus of point X and describe it in words. A point X moves on a flat surface so that it is:

a always 6 cm from a fixed point P

b always less than 6 cm from a fixed point P

For parts **c** and **d**, use this diagram.

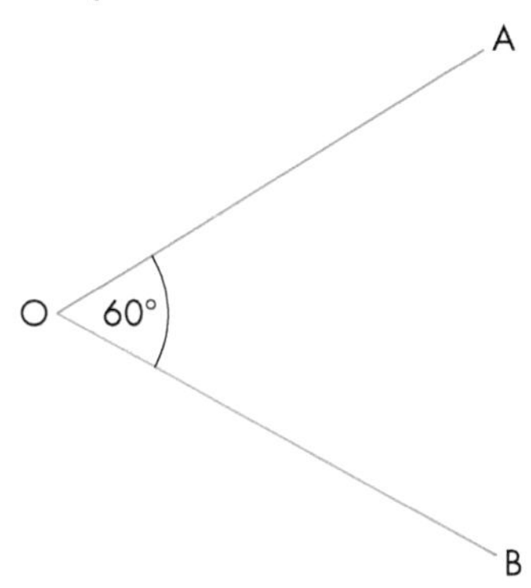

c the same distance from line OA as it is from line OB

d nearer to line OA than it is to line OB.

8. On a piece of plain paper, mark three points A, B and C, about 5 to 7 cm away from each other.

   Draw the locus of point P when:

   a P is always closer to point A than point B

   b P is always the same distance from points B and C.

9. Sketch the locus of a point on the rim of a bicycle wheel as it makes three revolutions along a flat road.

19.4 Loci problems

Homework 19D

For Questions **1** to **3**, you should start by sketching the picture given in each question on centimetre-squared paper. The scale for each question is given.

1. A goat is tied by a rope, 10 m long, to a stake that is 2 m from each side of a field. What is the locus of the area that the goat can graze? Use a scale of 1 cm to 2 m.

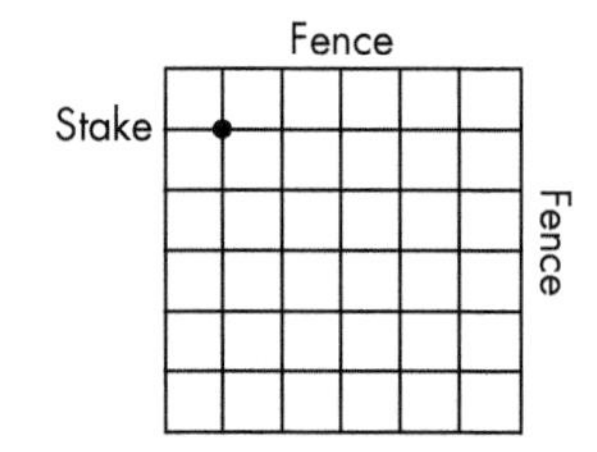

2. A cow is tied to a rail at the top of a fence 4 m long. The rope is 4 m long. Sketch the area that the cow can graze. Use a scale of 1 cm to 2 m.

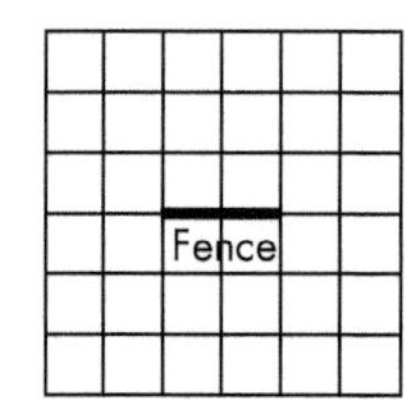

3. A horse is tied to a corner of a shed, 3 m by 1 m. The rope is 4 m long. Sketch the area that the horse can graze. Use a scale of 1 cm to 1 m.

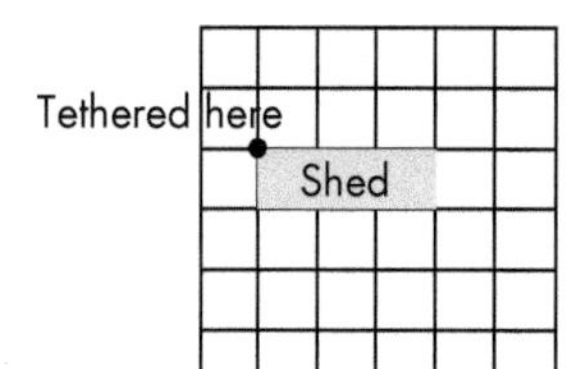

For Questions **4** to **7**, you should use a copy of this map. For each question, trace the map and mark on those points that are relevant to that question.

4 A radio station broadcasts from Birmingham with a range that is just far enough to reach York. Another radio station broadcasts from Glasgow with a range that is just far enough to reach Newcastle.

a Sketch the area to which each station can broadcast.

b Will the Birmingham station broadcast as far as Norwich?

c Will the two stations interfere with each other?

5 An air traffic control centre is to be built in Newcastle. If it has a range of 200 km, will it cover all the area of Britain north of Sheffield and south of Glasgow?

6 There are plans to build a new radio transmitter so that it is the same distance from Exeter, Norwich and Newcastle.

a Draw the perpendicular bisectors of the lines joining these three places and so find its proposed location.

b The radio transmitter will cause problems if it is built within 50 km of Birmingham. Will the proposed location cause problems?

7 Three radio stations pick up a distress call from a boat in the North Sea.

The station at Norwich can tell from the strength of the signal that the boat is within 150 km of the station. The station at Sheffield can tell that the boat is between 100 and 150 km from Sheffield.

If these two reports are correct, state the least and greatest possible distance of the boat from the helicopter station at Newcastle.

8 Two ships, A and B, are 7 km apart. They both hear a distress signal from a fishing boat. The fishing boat is less than 4 km from ship A and less than 4.5 km from ship B.

A helicopter pilot sees that the fishing boat is nearer to ship A than to ship B. On a scale drawing, shade the region that contains the fishing boat.

9 AB and BC are two walls of a house.

Bob wants to pave the part of the garden that is:

- less than 3 metres from wall AB

  and

- less than 3 metres from point P (along wall BC).

  Make a scale drawing and shade the part of the garden that Bob will pave.

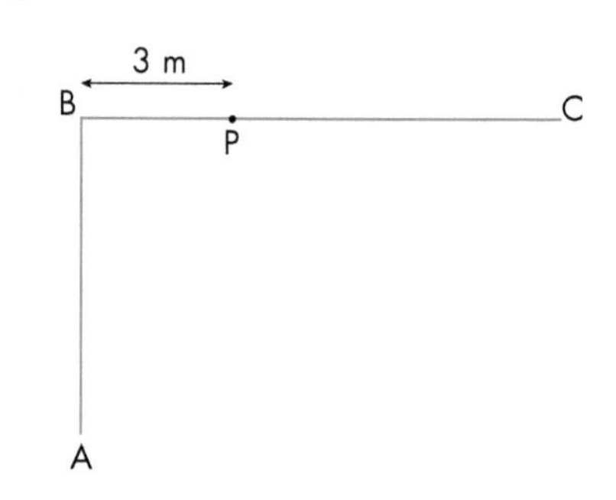

10 Copy and draw the set of points inside square ABCD which are both:

- nearer to side AB than to side AD

  and

- nearer to D than to C.

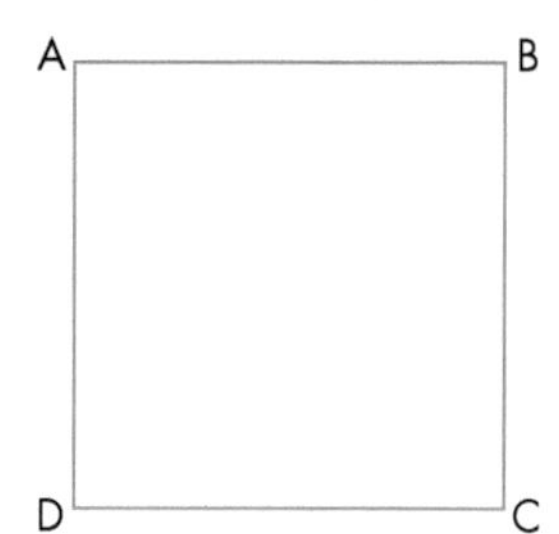

20 Geometry and measures: Curved shapes and pyramids

20.1 Sectors

Homework 20A

1 For each sector, calculate the length of the arc.

a

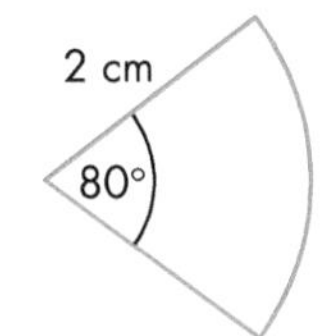

b

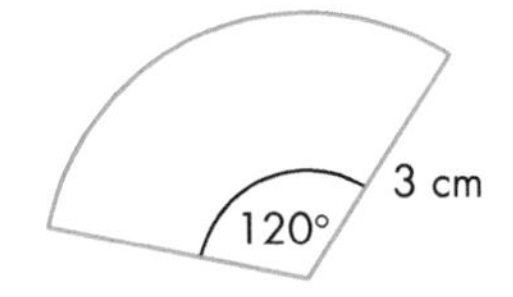

c

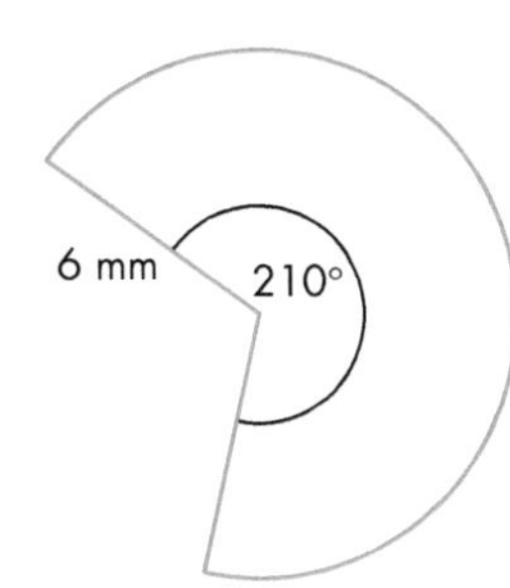

d

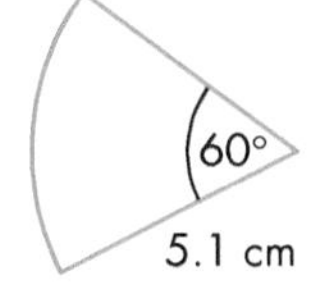

e

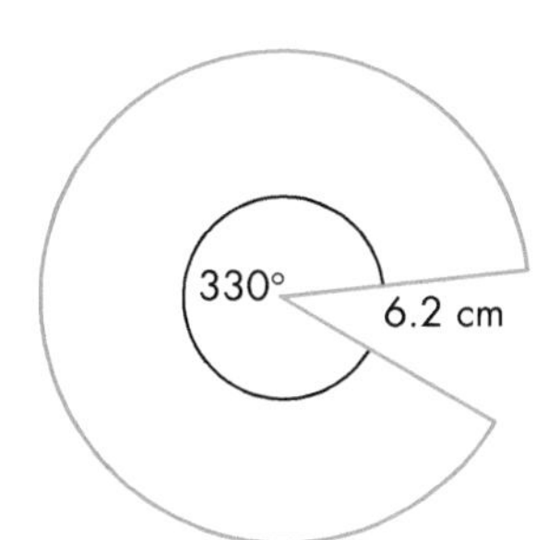

f

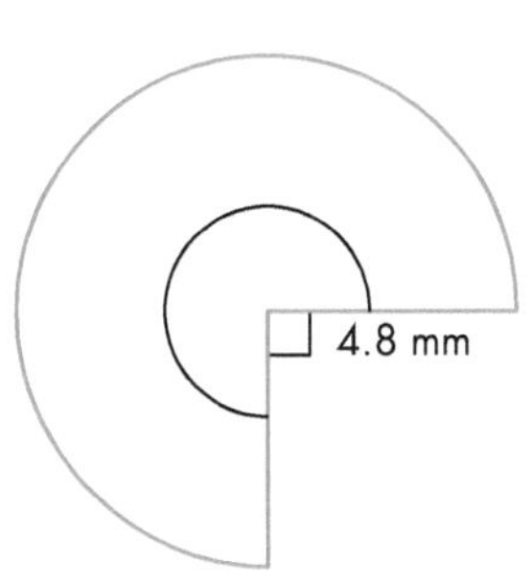

2 Calculate the area of each sector.

a

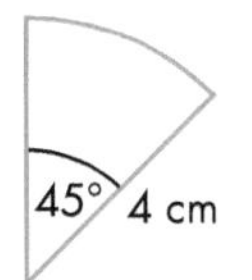

b

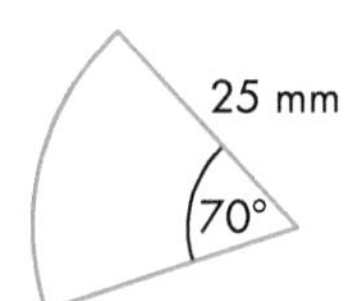

c

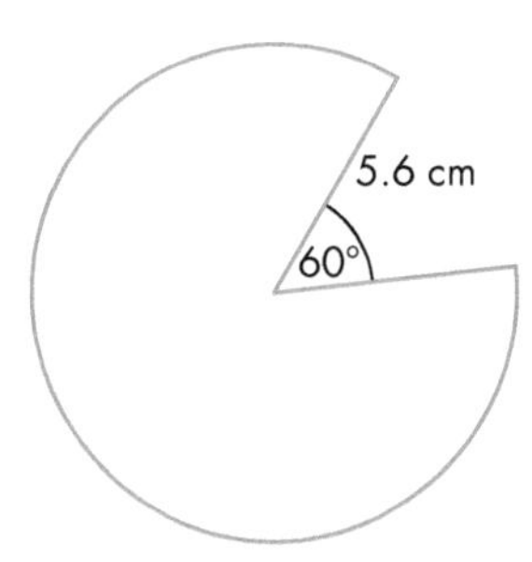

d

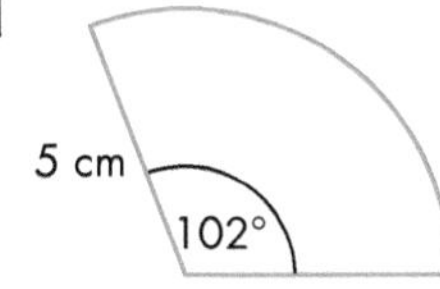

e

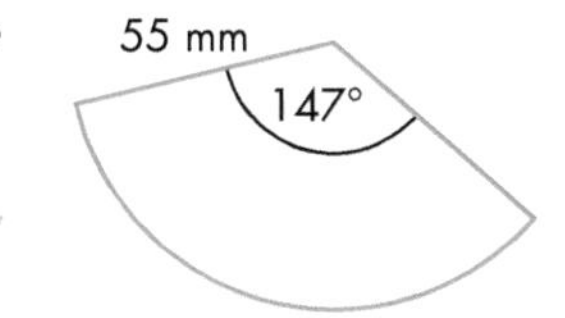

f

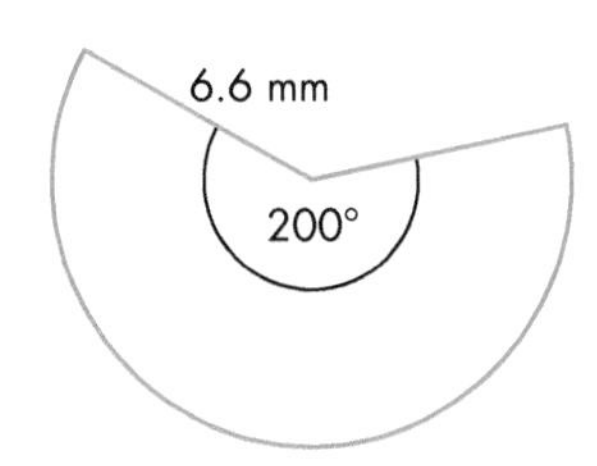

3. Calculate **i** the length of the arc and **ii** the area of a sector with these measures.

   a sector angle 72°, radius 2 cm

   b sector angle 105°, diameter = 19 cm

   c sector angle 270°, radius = 6 cm

   d sector angle 18°, diameter = 11 cm

4. A pie chart shows how students travel to school. The sector for those using a car has an angle of 120°.

   a What fraction of all the students use a car? Give your answer in its lowest terms.

   b The diameter of the pie chart is 10 cm. What is the area of the sector?

5. When full, a circular box of radius 4 cm contains six cheese sectors.

   Four of the sectors have been eaten. What area of the base of the box is visible when the lid is taken off?

6. A sector (of a circle) has an angle of 30° at the centre. The length of the arc is 10 cm. Calculate the radius of the circle.

7. A school has a shot put area that is the shape of a sector of a circle, radius 9 m. It has an area of 98 m^2. Work out the angle of the sector.

20.2 Pyramids

Homework 20B

1. Calculate the volumes of these rectangular-based pyramids.

   a

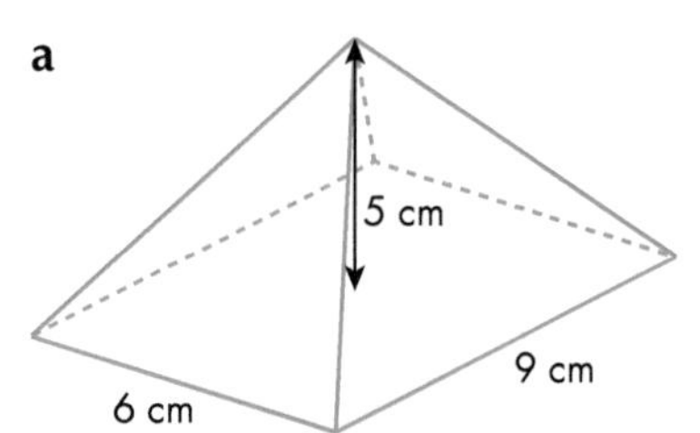

   b

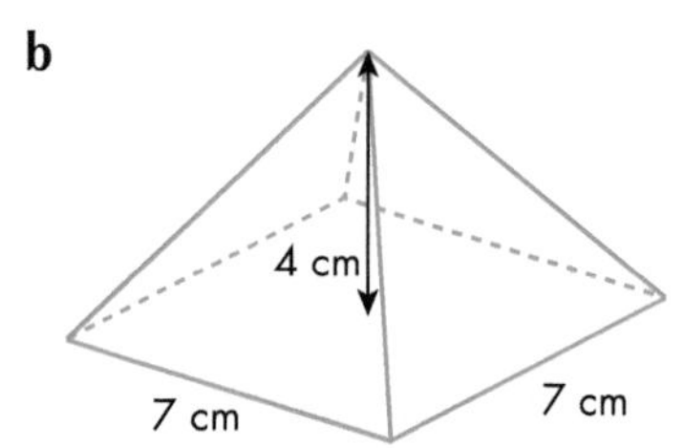

2. A pyramid has a square base of side 4 cm and a vertical height of 12 cm. Calculate the volume of the pyramid.

3. A pyramid has a square base of side 12 cm and a vertical height of 8 cm. Calculate the volume of the pyramid.

4. A pyramid has a 4 cm by 5 cm rectangular base and a vertical height of 10 cm. Calculate the volume of the pyramid.

5. A pyramid has a 5.6 cm by 5.1 cm rectangular base and a vertical height of 8.2 cm. Calculate the volume of the pyramid.

6. A pyramid has a square base of side 4.3 cm and a vertical height of 1.2 cm. Calculate the volume of the pyramid.

7. A pyramid has a 2.9 cm by 5.3 cm rectangular base and a vertical height of 5.8 cm. Calculate the volume of the pyramid.

Homework 20C

1. The diagram shows the net of a square-based pyramid.

   Calculate its surface area.

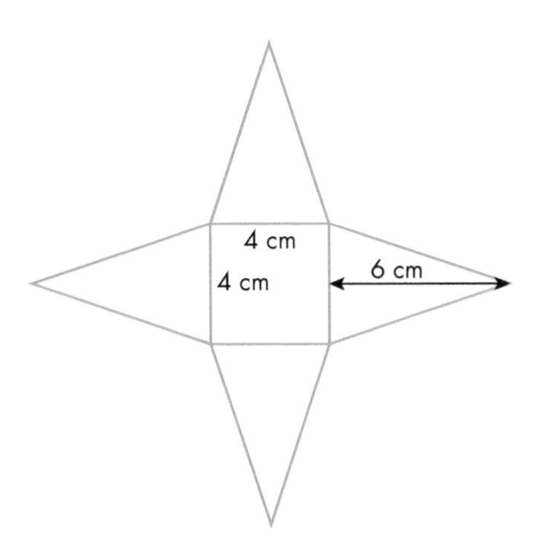

2. A pyramid has a square base of side 10 cm and a total surface area of 360 cm^2. Calculate its vertical height.

3. Calculate the total surface area of each pyramid.

a

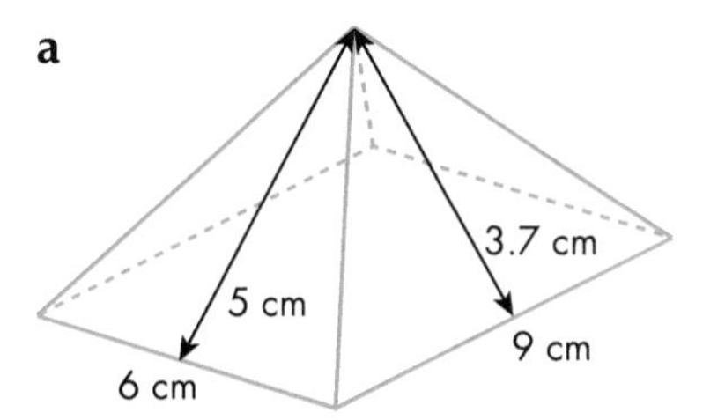

b

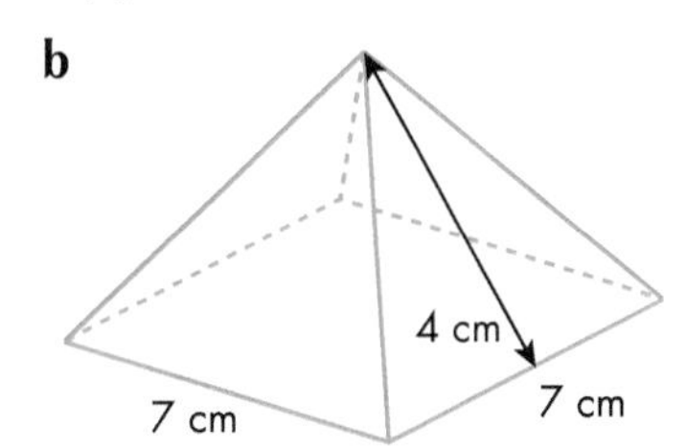

4. A pyramid has a square base of side 3 cm and a total surface area of 360 cm^2. Calculate its vertical height.

20.3 Cones

Homework 20D

1 Work out the area of each sector. Give your answers to three significant figures.

A

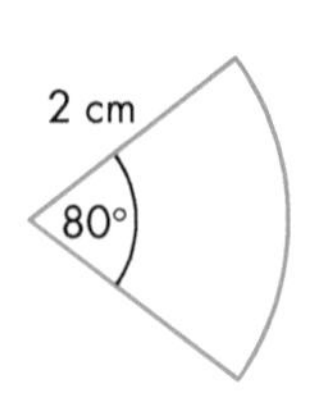

B

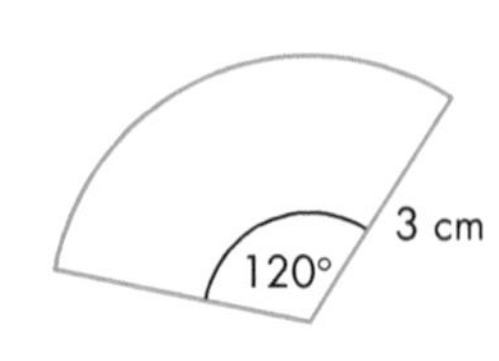

C

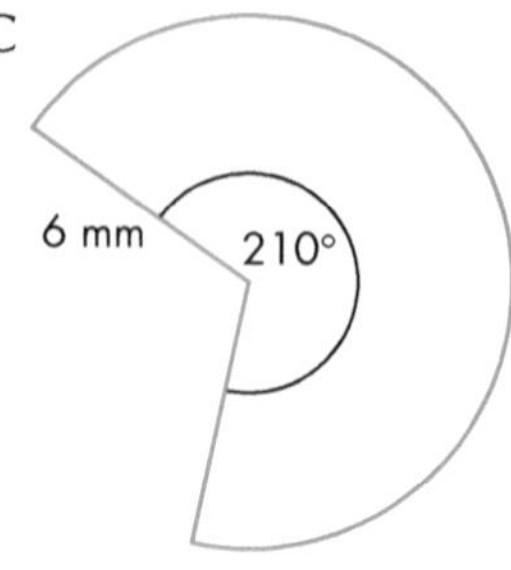

D

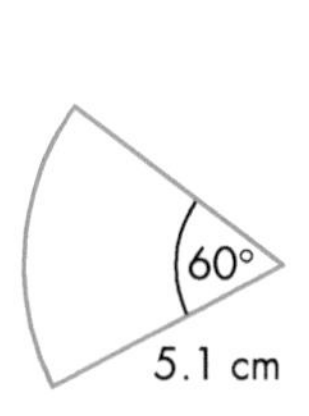

E

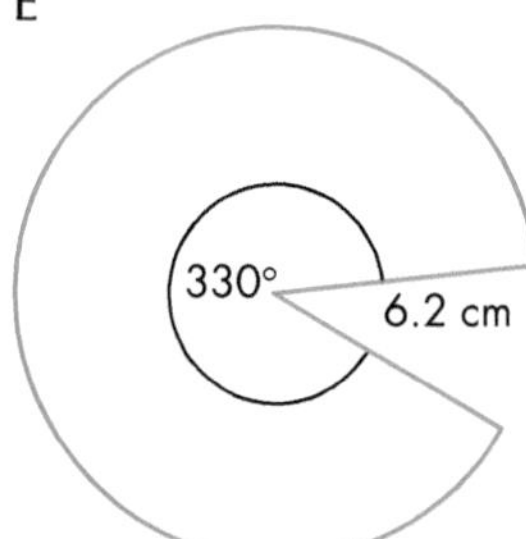

F

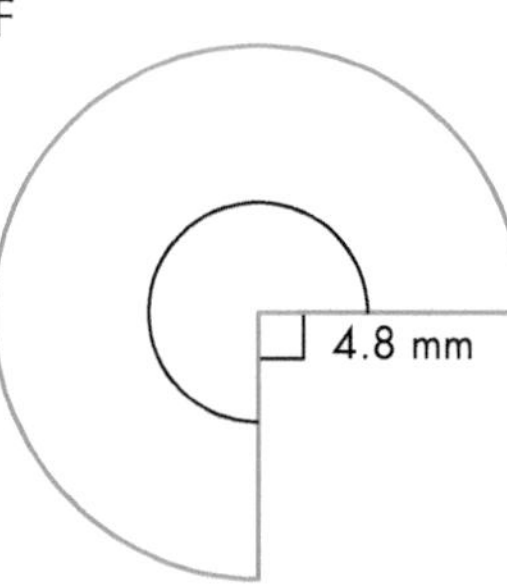

2 **a** Make a drawing of each sector from question **1** and measure the length of each arc.

b Calculate the length of each arc using $\frac{\theta}{360} \times 2 \times \pi \times \text{radius}$.

c Compare your answers to parts **a** and **b**.

3 Imagine making a cone from each sector in question **1** by taping the edges together.

a Write down the circumference of the base of each cone.

b Using the formula circumference = $2\pi r$, write down the radius of each cone.

4 Copy and complete the table using your answers to questions 1 to 3.

Sector	Area of sector	Length of arc	Radius of cone, r	Slant height, l	$\pi \times r \times l$
A				2	
B				3	
C				6	
D				5.1	
E				6.2	
F				4.8	

Homework 20E

1 Calculate the curved surface area of each shape.

a

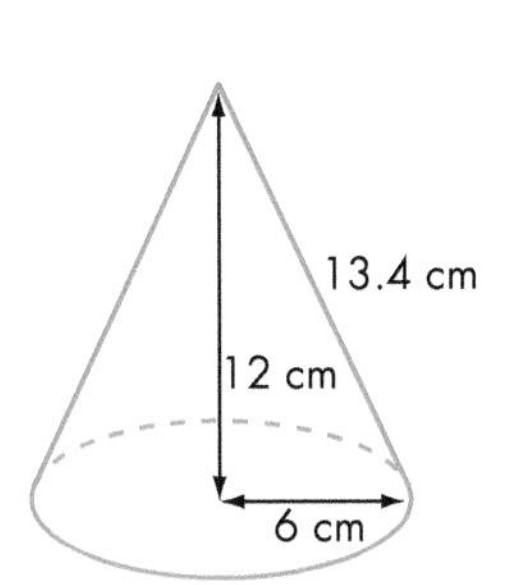

b

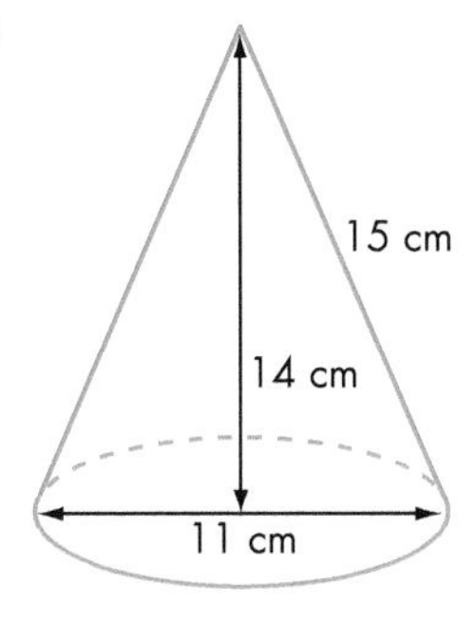

c

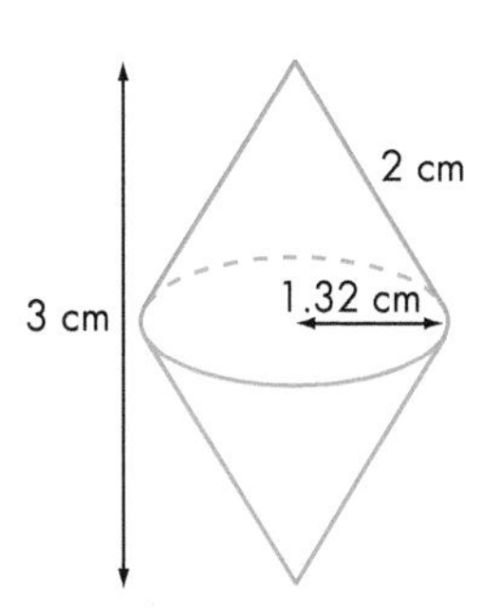

2 Calculate the total surface area of cones with these measures.

a base radius 10 cm, slant height 10 cm

b base radius 5 cm, slant height 16 cm

3 Work out the volume of an ice-cream cone with these measures.

a radius 2 cm, vertical height 12 cm

b radius 3 cm, vertical height 15 cm

4 Calculate **i** the volume and **ii** the surface area of each shape.

a

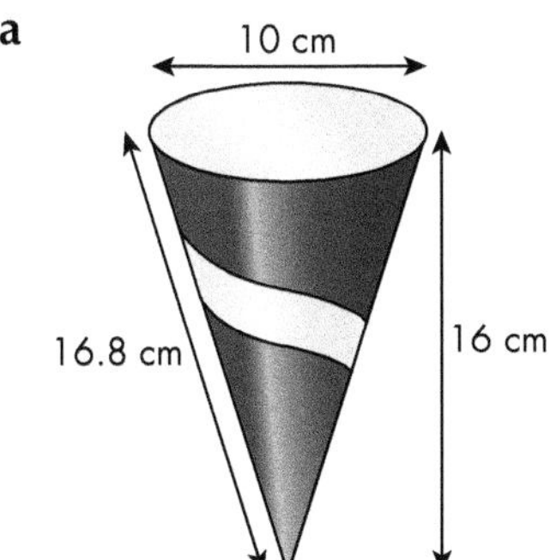

b

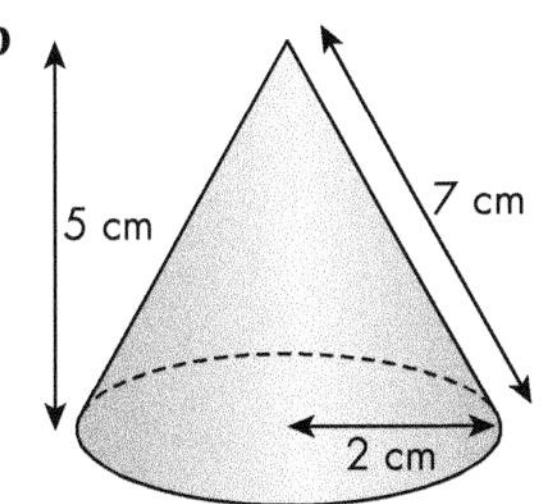

c

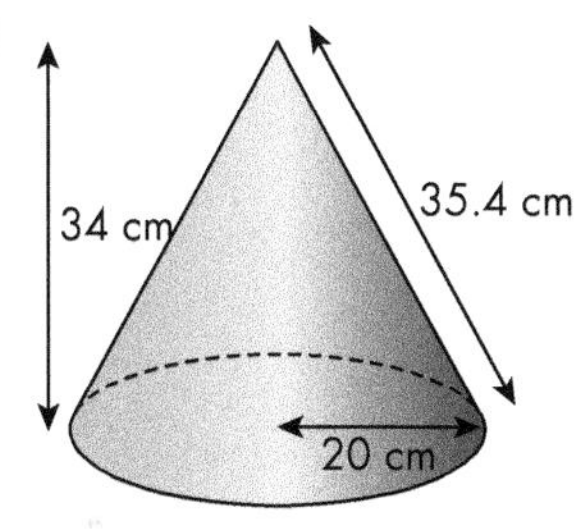

d

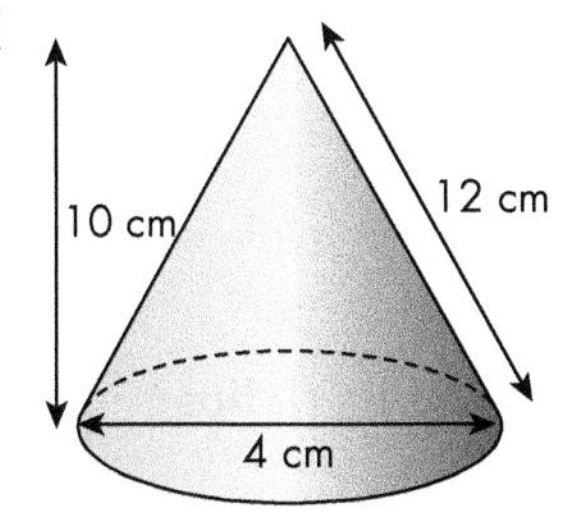

e

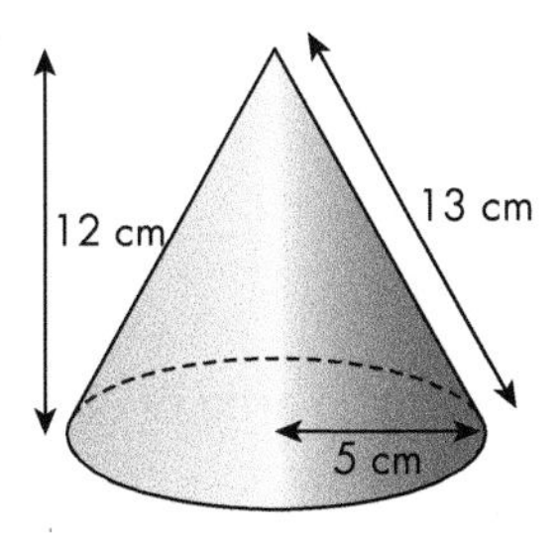

20.4 Spheres

Homework 20F

1 Calculate **i** the volume and **ii** the surface area of each sphere.

a

7 cm

b

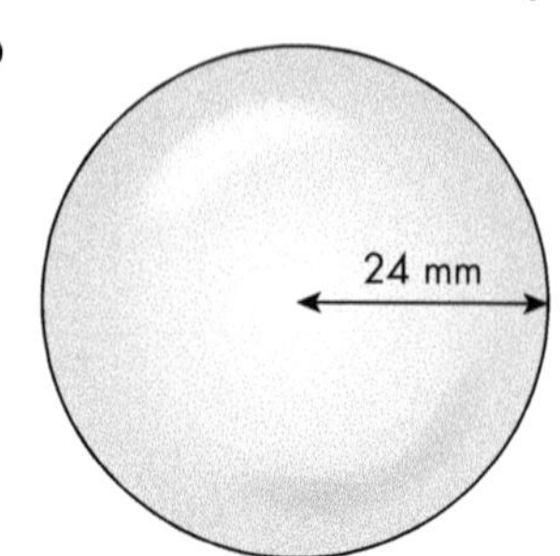

c

6.3 cm

d

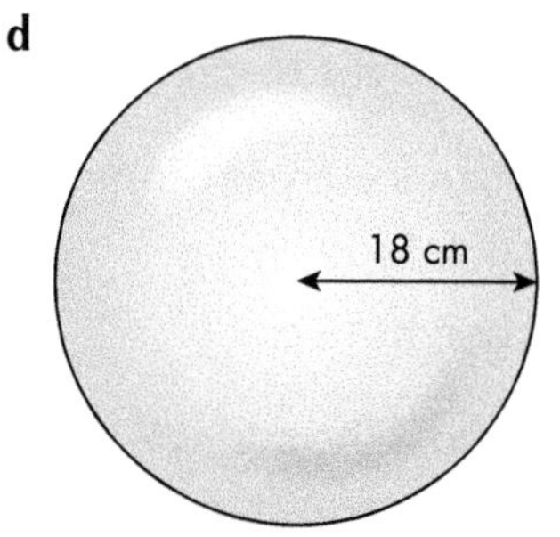

e

25.6 cm

f

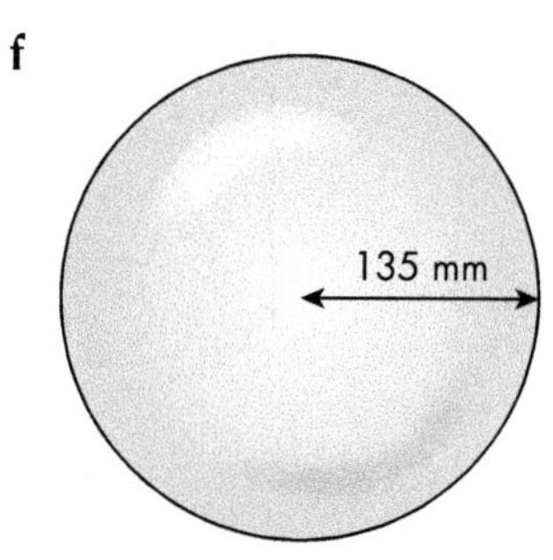

2 Calculate the surface area of:

a a tennis ball with radius 5 cm

b a volleyball with radius 8 cm.

3 A globe of the Earth has a surface area of 3000 cm^2. What is its radius?

4 The surface area of a football is 1560 cm^2. Calculate its circumference.

5 Work out the volume of a ball bearing with these measures.

a radius 0.5 cm

b radius 0.135 cm

21 Algebra: Number and sequences

21.1 Patterns in number

Homework 21A

In questions **1** to **4**, copy the number sentences, look for the pattern and then write down the next two lines. Then check your predictions with a calculator.

1
$1 \times 8 + 1 = 9$
$12 \times 8 + 2 = 98$
$123 \times 8 + 3 = 987$
$1234 \times 8 + 4 = 9876$

2
$9 \times 9 + 7 = 88$
$98 \times 9 + 6 = 888$
$987 \times 9 + 5 = 8888$
$9876 \times 9 + 4 = 88\,888$

3
$7 \times 11 \times 13 \times 2 = 2002$
$7 \times 11 \times 13 \times 3 = 3003$
$7 \times 11 \times 13 \times 4 = 4004$
$7 \times 11 \times 13 \times 5 = 5005$

4
$3 \times 7 \times 13 \times 37 \times 2 = 20\,202$
$3 \times 7 \times 13 \times 37 \times 3 = 30\,303$
$3 \times 7 \times 13 \times 37 \times 4 = 40\,404$
$3 \times 7 \times 13 \times 37 \times 5 = 50\,505$

Use your observations on the number patterns in questions **3** and **4** to answer questions **5** to **9** without using a calculator.

5 $7 \times 11 \times 13 \times 9 =$

6 $3 \times 7 \times 13 \times 37 \times 8 =$

7 $7 \times 11 \times 13 \times 15 =$

8 $3 \times 7 \times 13 \times 37 \times 15 =$

9 $3 \times 7 \times 13 \times 37 \times 99 =$

10 This is the calendar for January 2016.

January 2016						
Mon	**Tue**	**Wed**	**Thu**	**Fri**	**Sat**	**Sun**
28	29	30	31	1	2	3
4	5	6	7	8	9	10
11	12	13	14	15	16	17
18	19	20	21	22	23	24
25	26	27	28	29	30	31

a Take any 3 by 3 square, for example:

7	8	9
14	15	16
21	22	23

Add up the three numbers in the central column, then add up the central row and the two diagonals.

b What do you notice about the four totals?

c How is your answer to part **b** connected to the middle number?

d Choose a different 3 by 3 square. Without adding any numbers, predict the totals of the three numbers in the central column, the central row and the two diagonals.

11 Copy the number sentences, look for the pattern and then write down the next two lines. Check your predictions with a calculator.

a $3 \times 5 = 4^2 - 1 = 15$

$4 \times 6 = 5^2 - 1 = 24$

$5 \times 7 = 6^2 - 1 = 35$

$6 \times 8 = 7^2 - 1 = 48$

b $3 \times 7 = 5^2 - 4 = 21$

$4 \times 8 = 6^2 - 4 = 32$

$5 \times 9 = 7^2 - 4 = 45$

$6 \times 10 = 8^2 - 4 = 60$

21.2 Number sequences

Homework 21B

1 Write down the next three terms in each sequence and explain the pattern.

a 4, 6, 8, 10, …

b 3, 6, 9, 12, …

c 2, 4, 8, 16, …

d 5, 12, 19, 26, …

e 3, 30, 300, 3000, …

f 1, 4, 9, 16, …

2 Write down the next two terms in each sequence and explain the pattern.

a 1, 1, 2, 3, 5, 8, 13, 21, …

b 2, 3, 5, 8, 12, 17, …

3 For each number sequence, find the rule and write down the next three terms.

a 7, 14, 28, 56, …
b 3, 10, 17, 24, 31, …
c 1, 3, 7, 15, 31, …
d 40, 39, 37, 34, …
e 3, 6, 11, 18, 27, …
f 4, 5, 7, 10, 14, 19, …
g 4, 6, 7, 9, 10, 12, …
h 5, 8, 11, 14, 17, …
i 5, 7, 10, 14, 19, 25, …
j 10, 9, 7, 4, …
k 200, 40, 8, 1.6, …
l 3, 1.5, 0.75, 0.375, …

4 Write down the first five terms of the sequences formed by these rules.

a $3n + 1$
b $2n - 1$
c $4n + 2$
d $2n^2$
e $n^2 - 1$

5 Write down the first five terms of the sequences with these nth terms.

a $n + 2$
b $4n - 1$
c $4n - 3$
d $n^2 + 1$
e $2n^2 + 1$

6 A sequence is formed by the rule $\frac{n}{2n - 1}$.

The first term of this sequence is $\frac{1}{2 \times 1 - 1} = \frac{1}{1} = 1$.

Write down the next four terms in this sequence.

7 A taxi firm uses this chart to charge for journeys of k kilometres.

k	1	2	3	4	5	6	7	8	9	10
Charge (£)	4.50	6.50	8.50	10.50	12.50	15.00	17.00	19.00	21.00	23.00
k	11	12	13	14	15	16	17	18	19	20
Charge (£)	26.00	28.00	30.00	32.00	34.00	37.00	39.00	41.00	43.00	45.00

a Using the charges for 1 to 5 km, work out an expression for the kth term.
b Using the charges for 6 to 10 km, work out an expression for the kth term.
c Using the charges for 11 to 15 km, work out an expression for the kth term.
d Using the charges for 16 to 20 km, work out an expression for the kth term.
e What is the basic charge per kilometre?

8 A series of fractions is $\frac{3}{7}, \frac{5}{10}, \frac{7}{13}, \frac{9}{16}, \frac{11}{19}, \ldots$

a Write down the nth term of the numerators.
b Write down the nth term of the denominators.
c i Work out the fraction when $n = 1000$.
ii Give the answer as a decimal.
d Will the terms of the series ever be greater than $\frac{2}{3}$?
Explain your answer.

9 This rhyme can be used for picking members of a team.

Eeny, meeny, miney, moe

Catch a tiger by the toe

If he roars, let him go

Eeny, meeny, miney, moe.

Each time you say a word, you point at a different person. The person being pointed at on the last word of the rhyme is then chosen.

Ten friends are standing in a circle: Alexander, Briony, Chris, David, Ellie, Fran, Greta, Hermione, Isabel and Jack.

Xavier stands in the middle and starts by pointing at Alexander with the first 'Eeny'.

Jack is the first person to be chosen.

There are now nine people left in the circle. The rhyme is repeated, starting with the next person in the circle (Alexander), to choose the next person, and so on.

Write down the order in which the friends are chosen, starting with Jack.

10 Look at these two sequences.

5, 11, 17, 23, 26, 32, 38, ...

1, 4, 7, 10, 13, 16, 19, ...

Will the two sequences ever have a term in common? Give a reason for your answer.

11 Look at these two sequences.

100, 96, 92, 88, 84, ...

2, 8, 14, 20, 26, ...

Write down all the terms that appear in both sequences.

12 The nth terms of two sequences are given by $106 - 4n$ and $6n - 4$.

These two sequences have several terms in common but only one of these common terms has the same position in both sequences.

Without writing out the sequences, show how you can tell using the nth terms that it is the 11th term.

21.3 Finding the nth term of a linear sequence

Homework 21C

1 Write down the next two terms and the nth term in each linear sequence.

a 5, 7, 9, 11, 13 ... **b** 3, 11, 19, 27, 35, ... **c** 6, 11, 16, 21, 26, ...

d 3, 9, 15, 21, 27, ... **e** 4, 7, 10, 13, 16, ... **f** 3, 10, 17, 24, 31, ...

2 Write down the nth term and the 50th term in each linear sequence.

a 3, 5, 7, 9, 11, ... **b** 5, 9, 13, 17, 21, ... **c** 8, 13, 18, 23, 28, ...

d 2, 8, 14, 20, 26, ... **e** 5, 8, 11, 14, 17, ... **f** 2, 9, 16, 23, 30, ...

3 For each sequence **a** to **f**, work out:

i the nth term

ii the 100th term

iii the term closest to 100.

a 5, 12, 19, 26, 33, ... **b** 9, 11, 13, 15, 17, ... **c** 2, 7, 12, 17, 22, ...

d 2, 6, 10, 14, 18, ... **e** 5, 13, 21, 29, 37, ... **f** 6, 7, 8, 9, 10, ...

4 A physiotherapist uses the formulae below for charging for a series of n sessions, when they are paid for in advance.

For $n \leqslant 5$, cost is £$(35n + 20)$

For $6 \leqslant n \leqslant 10$, cost is £$(35n + 10)$

For $n \geqslant 11$, cost is £$35n$

a How much will the physiotherapist charge for 8 sessions booked in advance?

b How much will the physiotherapist charge for 14 sessions booked in advance?

c One client paid £220 in advance for a series of sessions.

How many sessions did she book?

d A runner gets a leg injury and is not sure how many sessions it will take to treat.

He books four sessions. After the sessions, he starts to run in races again. The leg injury returns and he has to book three more sessions.

How much more does he pay for his treatment than he would do if he booked all the sessions at the same time?

5 A fraction sequence is formed by the rule $\frac{n+1}{2n+1}$.

Show that in the first eight terms, only one of the fractions is a terminating decimal.

Hints and tips You could set this up on a spreadsheet.

21.4 Special sequences

Homework 21D

1 a is an odd number, b is an even number.

State whether each expression is odd or even.

a $a + b$ **b** $a + 1$ **c** $a + 2$ **d** $b + 1$

e $b + 2$ **f** a^2 **g** b^2 **h** $a^2 + b^2$

2 The nth term of the cube numbers is given by n^3. This gives the sequence 1, 8, 27, 64, 125, ...

a What is the 10th cube number?

b Give the nth term of these sequences.

i 2, 9, 28, 65, 126, ...

ii 2, 16, 54, 128, 250, ...

iii 0.5, 4, 13.5, 32, 62.5, ...

3 p is an odd number, q is an even number.

State whether each expression is odd or even.

a $p + 5$ **b** $q - 3$ **c** $2p$ **d** q^2

e pq **f** $2(p + q)$ **g** $p^2 + q$ **h** $q(p + q)$

4 Write down the next two lines of this number pattern.

$0 + 1 = 1 = 1^2$

$1 + 3 = 4 = 2^2$

$3 + 6 = 9 = 3^2$

$6 + 10 = 16 = 4^2$

5 p is a prime number, q is an odd number and r is an even number.

State if the following are always odd (O), always even (E) or could be either (C).

a $p + 2$ **b** $p + q$ **c** $pr + q^2$ **d** $(p + q)(p + r)$

6 The powers of 3 are $3^1, 3^2, 3^3, 3^4, \ldots$

This gives the sequence 3, 9, 27, 81, ...

a Continue the sequence for another three terms.

The nth term of the powers of 3 is given by 3^n.

b Give the nth term of these sequences.

i 2, 8, 26, 80, ...

ii 6, 18, 54, 162, ...

21.5 General rules from given patterns

Homework 21E

1 The tables at a conference centre can each take three people. The tables are always put together to sit people as shown.

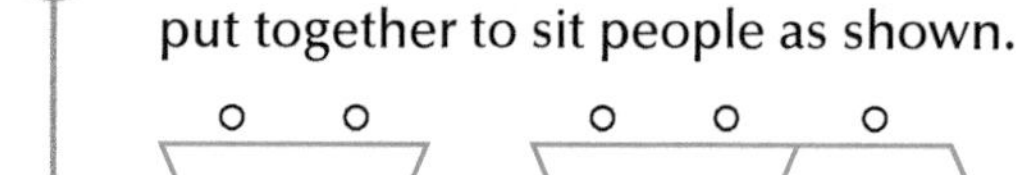

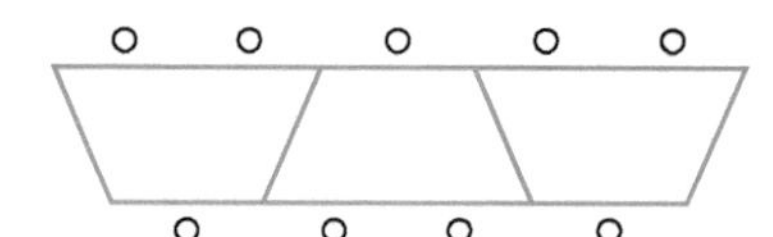

a How many people could be seated at four tables?

b How many people could be seated at n tables?

c How many tables does the conference centre need to set out for 50 people?

2 This pattern of shapes is built up from matchsticks.

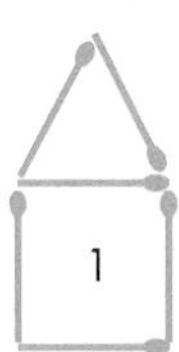

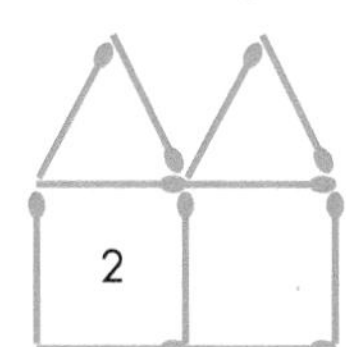

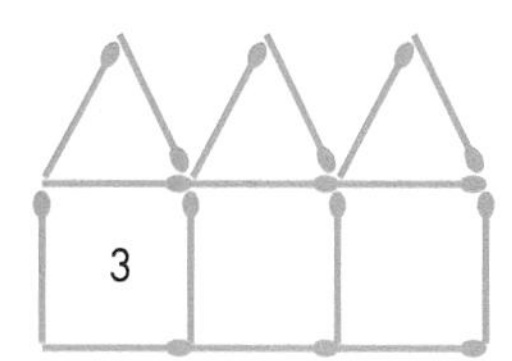

a Draw the fourth diagram.

b How many matchsticks are in the nth diagram?

c How many matchsticks are in the 25th diagram?

d Diagram 3 is the biggest diagram that you could make with 20 matchsticks. What is the biggest diagram you could make with 200 matchsticks?

3 This pattern of hexagons is built up from matchsticks.

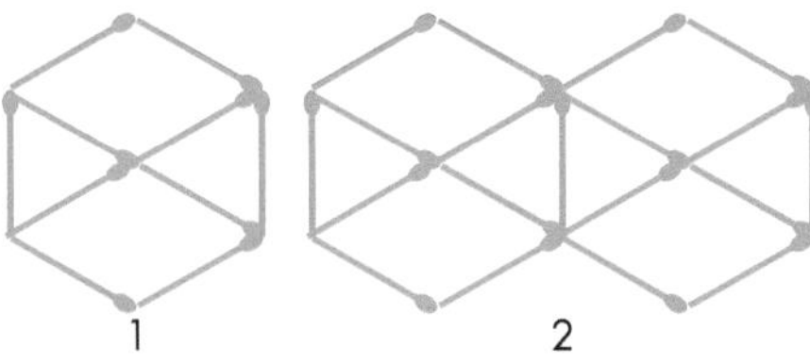

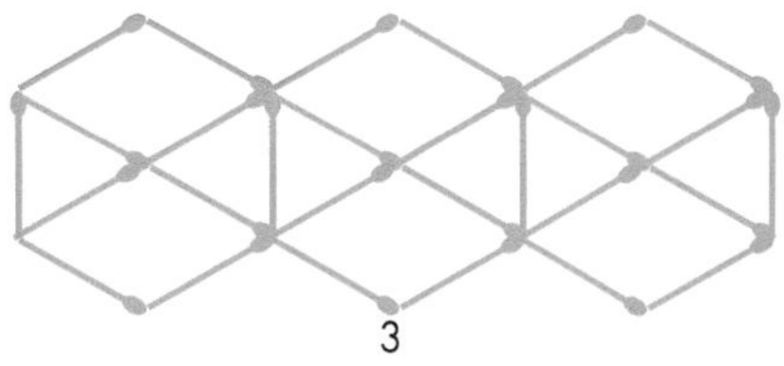

a Draw the fourth set of hexagons in this pattern.

b How many matchsticks would you need for the nth set of hexagons?

c How many matchsticks would you need for the 60th set of hexagons?

d If you only have 100 matchsticks, what is the largest set of hexagons you could make?

4 Debbie, Kim and Danielle are making patterns with building bricks. Their steps are shown in the diagram below.

	Step 1	Step 2	Step 3	Step 4
Debbie				
Kim				
Danielle				

Altogether they have 100 bricks.

How many steps in their pattern can they reach before running out of bricks?

22 Geometry and measures: Right-angled triangles

22.1 Pythagoras' theorem

Homework 22A

1. **a** Construct a triangle with sides 7 cm, 24 cm and 25 cm.
   b Measure the size of the largest angle.
   c Square the length of each side (7^2, 24^2, 25^2).
   d What is the connection between the squares?

2. **a** Construct a triangle with sides 8 cm, 15 cm, and 17 cm.
   b Measure the size of the largest angle.
   c Square the length of each side (8^2, 15^2, 17^2).
   d What is the connection between the squares?

3. **a** Construct a triangle with sides 9 cm, 12 cm and 15 cm.
   b Measure the size of the largest angle.
   c Square the length of each side (9^2, 12^2, 15^2).
   d What is the connection between the squares?

Homework 22B 

Hints and tips For any right-angled triangle:

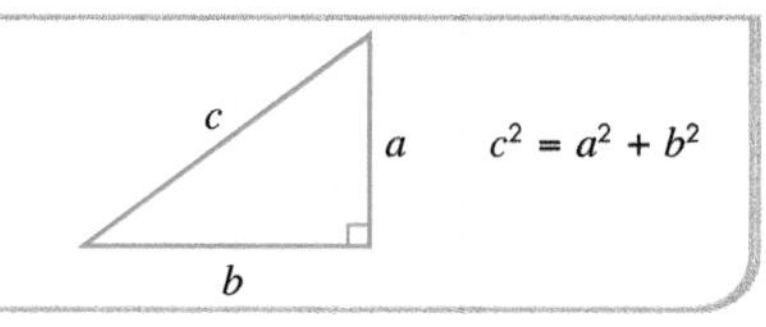

$c^2 = a^2 + b^2$

Note: The triangles in this exercise are not drawn to scale.

1. Calculate the length of the hypotenuse, x, for each triangle.
   Give your answers correct to one decimal place.

a

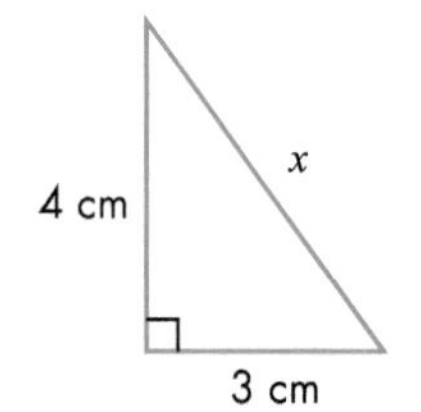

b

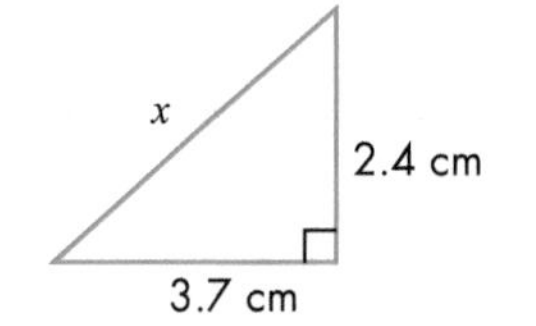

c

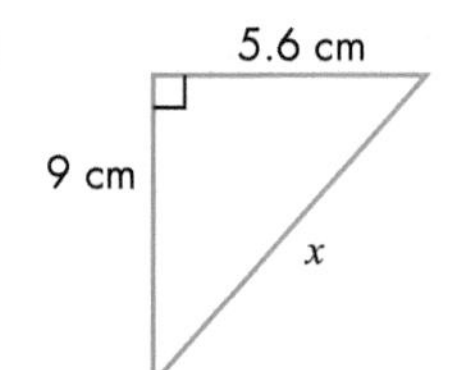

d

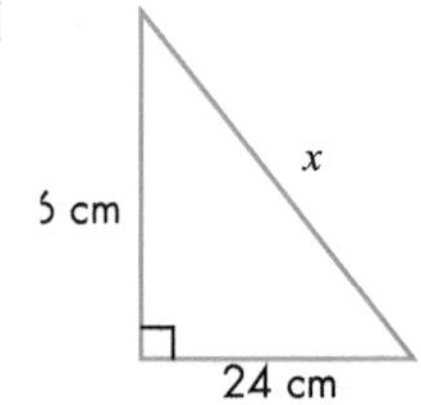

2 State which of these are right-angled triangles.

a
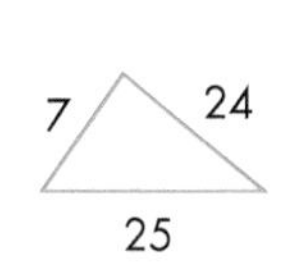

b
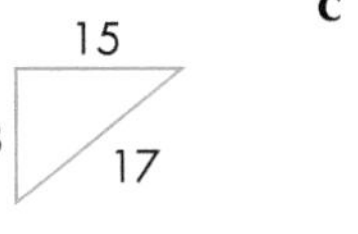

c
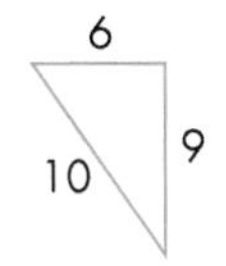

d
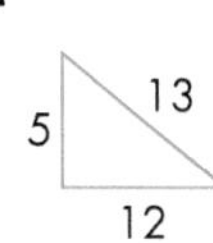

e
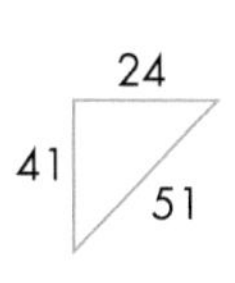

f
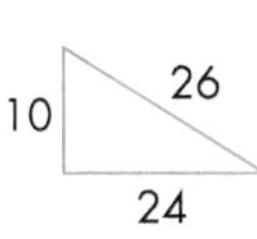

g
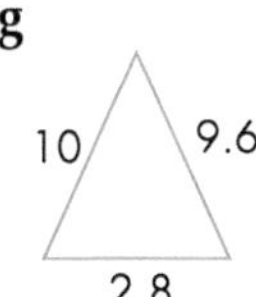

h
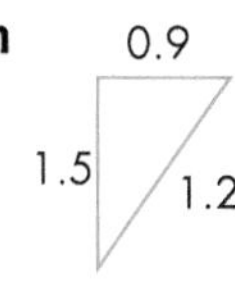

3 The diagonal of a square has length 20 cm. What is the perimeter of the square?

4 The diagonal of a square is 8 cm.
Jo says, 'The area of the square is 32 cm^2.
Explain how she worked this out.

22.2 Calculating the length of a shorter side

Homework 22C 

Hints and tips For any right-angled triangle:

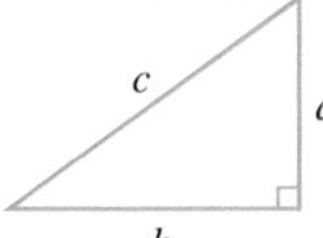

$a^2 = c^2 - b^2$ and $b^2 = c^2 - a^2$

Note: The triangles in this exercise are not drawn to scale.

1 Calculate the length of x for each triangle. Give your answers correct to one decimal place.

a
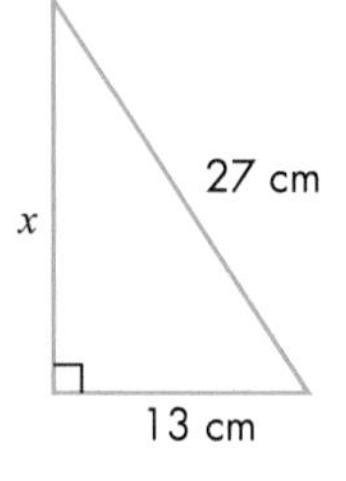

b
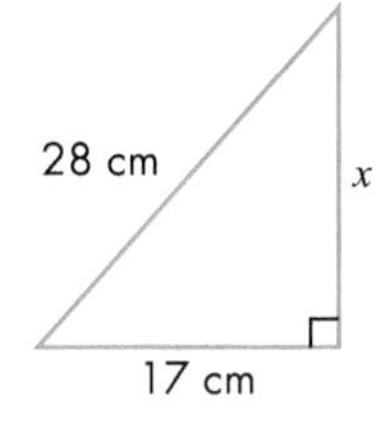

c
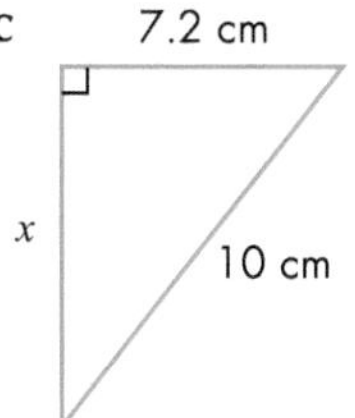

d
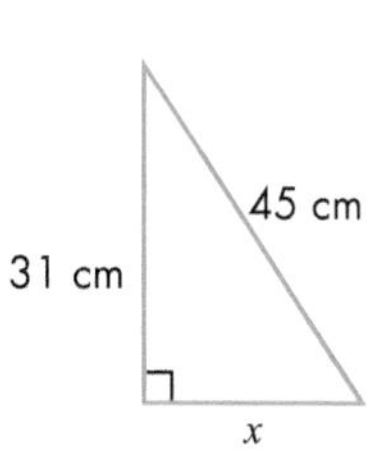

e
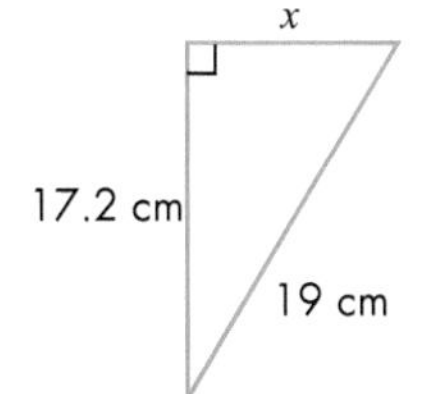

f
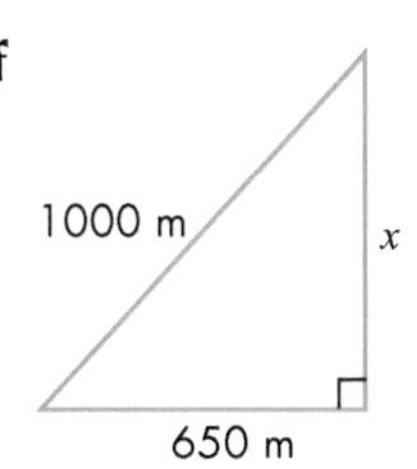

g
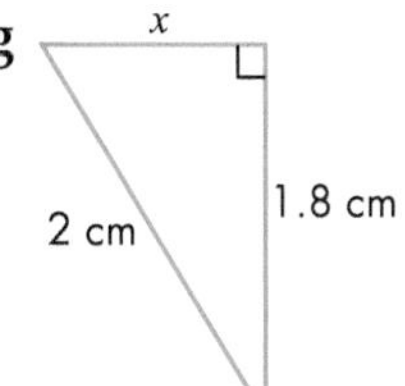

h
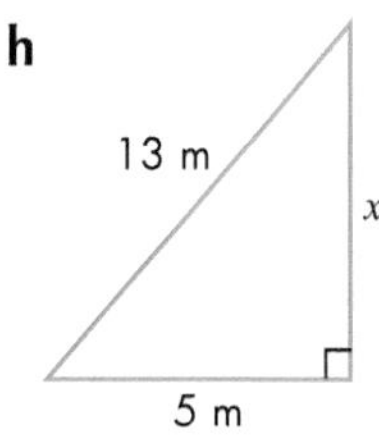

2 Calculate the length of x in each triangle. Give your answers correct to one decimal place.

a

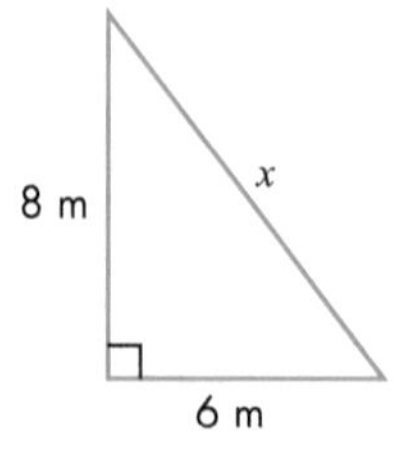

b

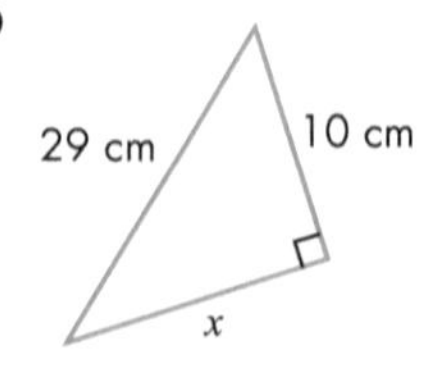

c

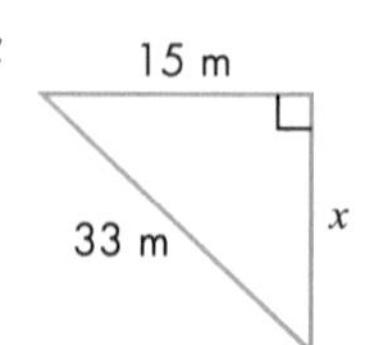

d

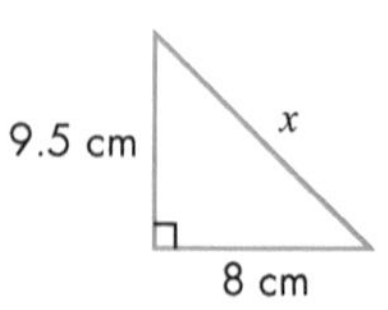

3 The diagram shows the end view of a building. Calculate the length AB.

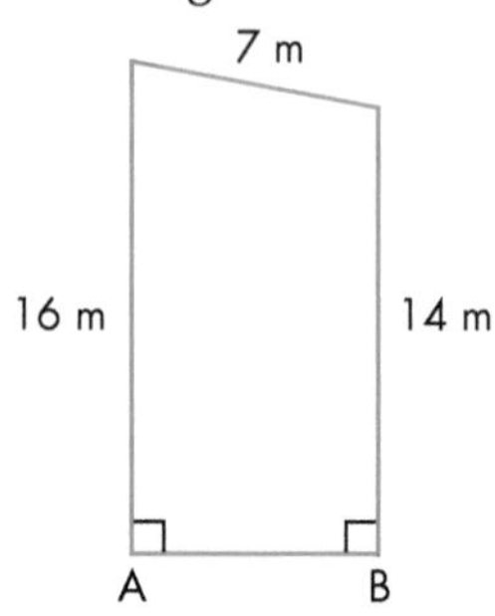

4 A pilot flies for 300 km and finds himself 200 km north of his original position. How far east has he travelled?

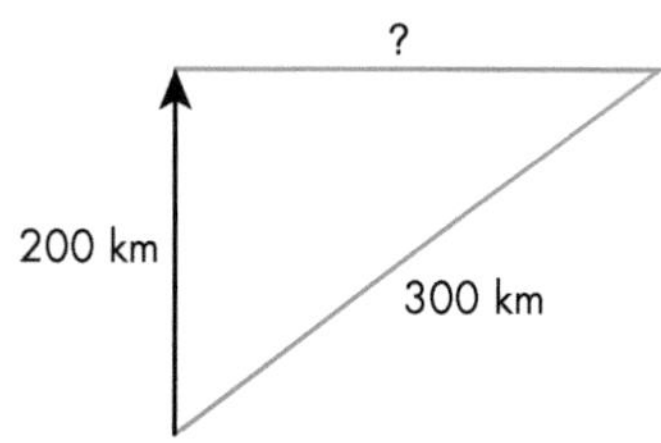

5 The diagram shows the area of a semicircle on each side of a right-angled triangle. What can you say about the areas of the semicircles?

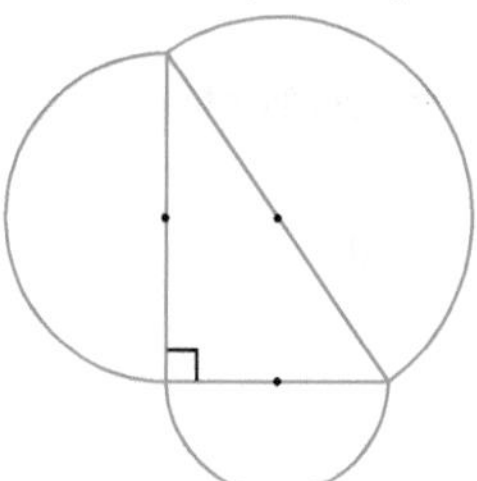

6 Ella calculates CD to be 4 cm. Evaluate her answer.

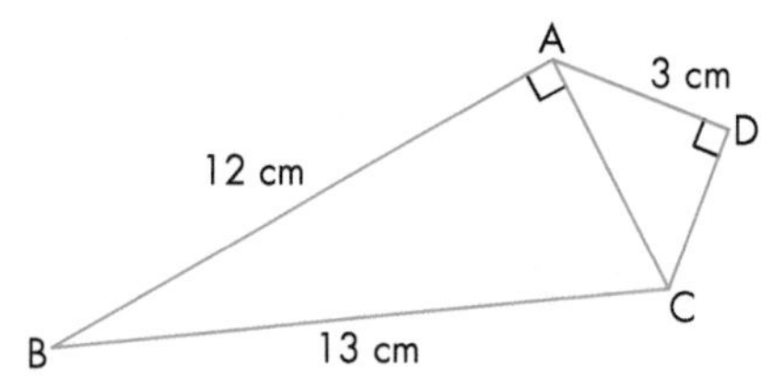

22.3 Applying Pythagoras' theorem in real-life situations

Homework 22D

1 A ladder, 15 m long, leans against a wall. The ladder must reach 12 m up the wall. How far away from the wall should the foot of the ladder be placed?

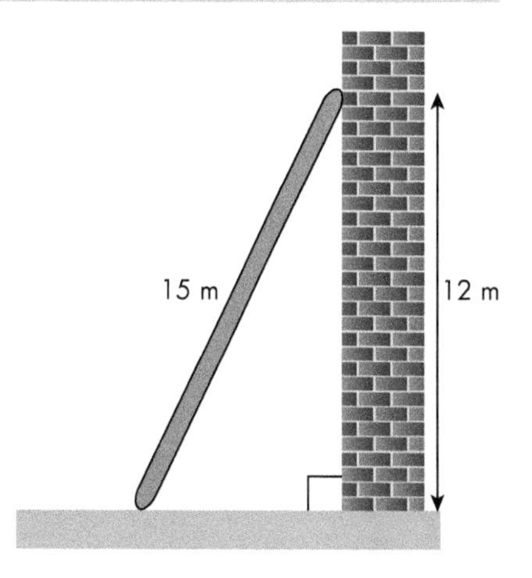

2 A rectangle is 3 m long and 1.2 m wide. How long is the diagonal? Give your answer correct to one decimal place.

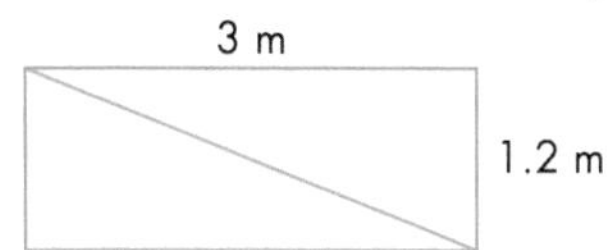

3 How long is the diagonal of a square with a side of 10 m?

4 A ship leaves port and sails 8 km east and 6 km north to a lighthouse. What is the direct distance between the port and the lighthouse?

5 The diagram shows three towns, A, B and C, joined by two roads. The council wants to build a road that runs directly from A to C.

How much shorter will the new road be than the two existing roads?

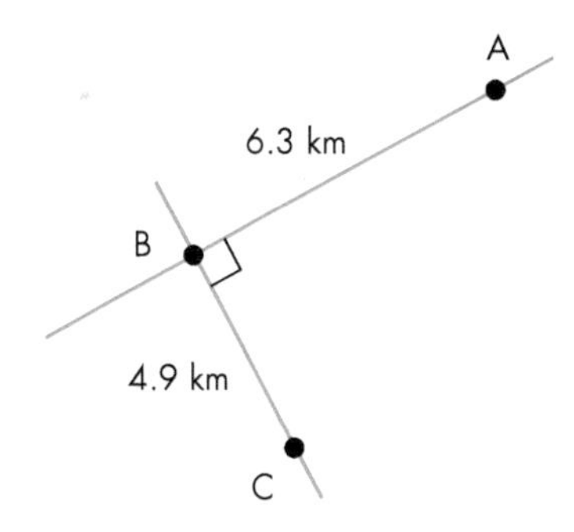

6 A ladder, 8 m long, is put up against a wall.

a How far up the wall will it reach when the foot of the ladder is 1 m away from the wall?

b When it reaches 7 m up the wall, how far is the foot of the ladder away from the wall?

7 A and B are two points on a coordinate grid. They have coordinates (1, 3) and (2, 2). What is the length of the line that joins them? Give your answer to one decimal place.

8 A rectangle is 4 cm long. The length of its diagonal is 5 cm. What is the area of the rectangle?

9 Is the triangle with sides 9 cm, 40 cm and 41 cm a right-angled triangle? Give a reason for your answer.

10 A and B are two points on a coordinate grid. They have coordinates (–3, –7) and (4, 6). Show that the line that joins them has length 14.8 units.

11 A helicopter takes off vertically from the ground. When it reaches a height of 200 m, it flies due north for 300 m then turns and flies due east for 500 m.

What is the direct distance between the helicopter and its starting point?

12 A 13-cm pencil fits exactly diagonally in a rectangular pencil tin.

The dimensions of the rectangular base are integer values. If the pencil is 13 cm in length, what are the values of the length and width of the tin?

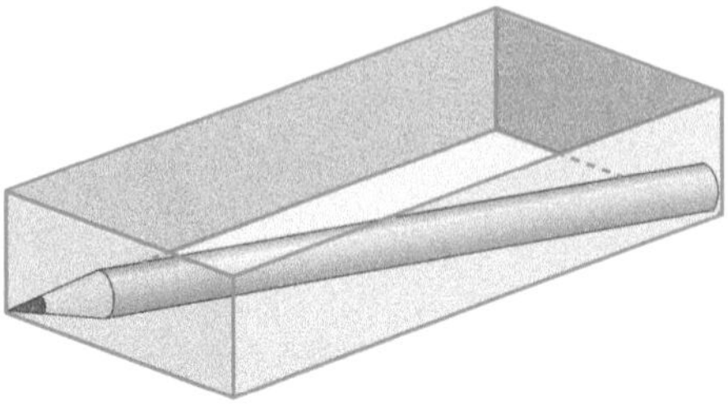

22.4 Pythagoras' theorem and isosceles triangles

Homework 22E

1 Work out the length of the hypotenuse in each diagram.

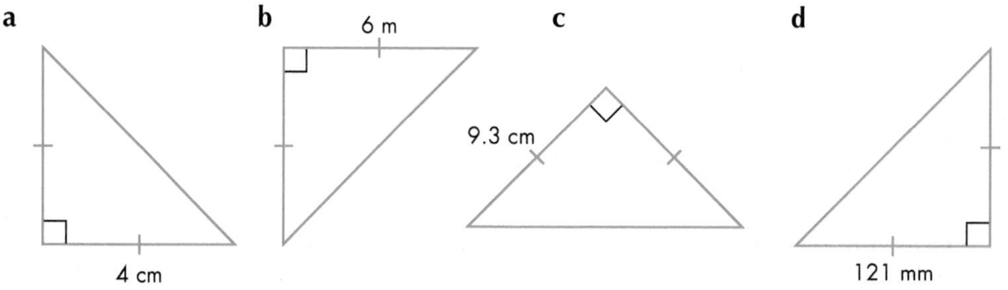

2 Calculate the length of the missing sides in each isosceles triangle.

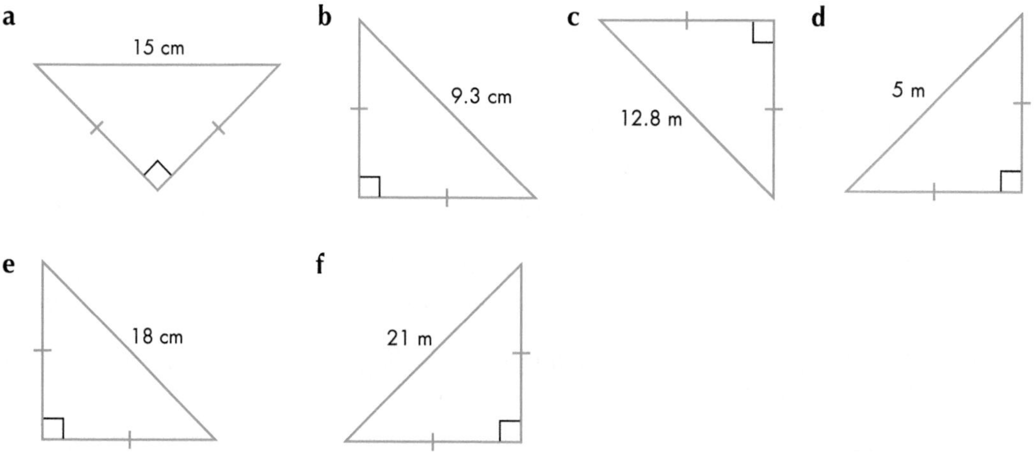

3 Calculate the areas of these isosceles triangles.

a

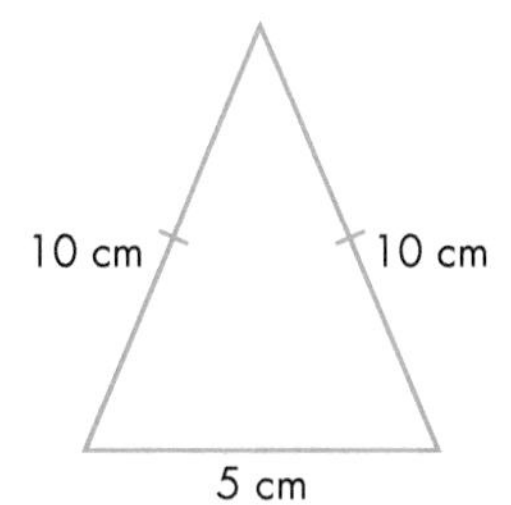

b

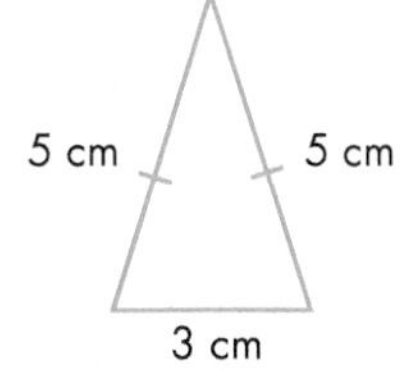

c

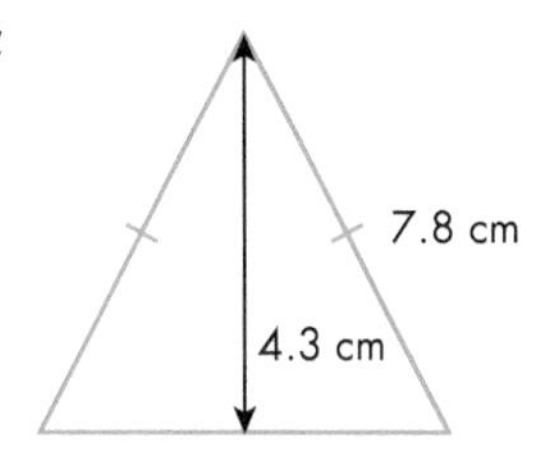

4 Calculate the area of an equilateral triangle of side 8 cm.

22.5 Trigonometric ratios

Homework 22F 

1 Choose an angle between 20° and 50°.

a Draw five different right-angled triangles that also include your chosen angle.

b Measure and label the length of all three sides of each triangle and record the information in a copy of the table below. Complete your table by calculating the required values for the last three columns, correct to three decimal places.

> **Hints and tips** Remember, the hypotenuse is the longest side.

Triangle	Opposite	Adjacent	Hypotenuse	$\frac{\text{opposite}}{\text{hypotenuse}}$	$\frac{\text{adjacent}}{\text{hypotenuse}}$	$\frac{\text{opposite}}{\text{adjacent}}$
A						
B						
C						
D						
E						

c What do you notice about the values in the last three columns?

Homework 22G

Example

Write down the trigonometric ratios for this triangle.

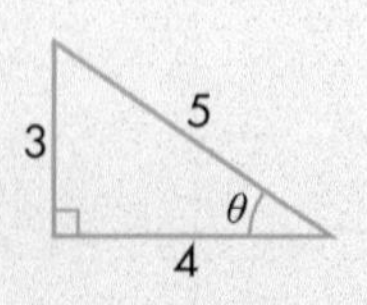

$\sin\theta = \frac{\text{opp}}{\text{hyp}} = \frac{3}{5}$ $\cos\theta = \frac{\text{adj}}{\text{hyp}} = \frac{4}{5}$ $\tan\theta = \frac{\text{opp}}{\text{adj}} = \frac{3}{4}$

For each triangle in questions **1** to **9**, make a copy and:

a label the hypotenuse (H) and the sides that are opposite (O) and adjacent (A) to θ

b write down the trigonometric ratios for each triangle.

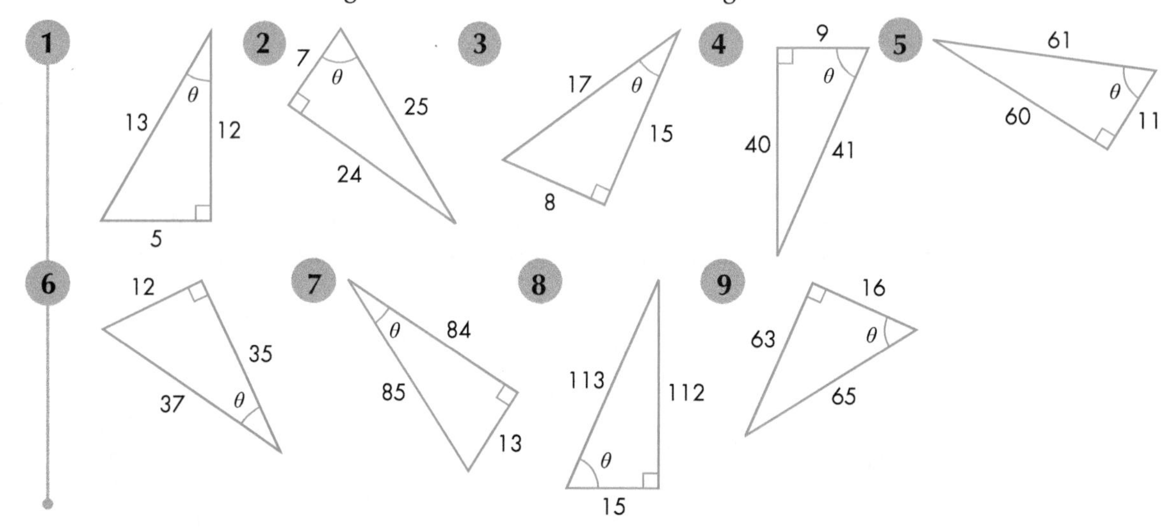

Homework 22H

1 Work out these values. Give your answers to three significant figures.

a sin 45° **b** sin 23° **c** sin 11° **d** sin 90°
e sin 270° **f** sin 180° **g** sin 67° **h** sin 56°

2 Work out these values. Give your answers to three significant figures.

a cos 34° **b** cos 87° **c** cos 90° **d** cos 180°
e cos 270° **f** cos 101° **g** cos 29° **h** cos 146°

3 Work out these values. Give your answers to three significant figures.

a 5 tan 34° **b** 6 tan 72° **c** 5 tan 0° **d** 2 tan 11°
e 3 tan 30° **f** 5 tan 90°

22.6 Calculating lengths using trigonometry

Homework 22I

Give your answers to questions **1** to **3** to three significant figures.

1 Use the tangent ratio to calculate the lengths of the lettered sides.

a

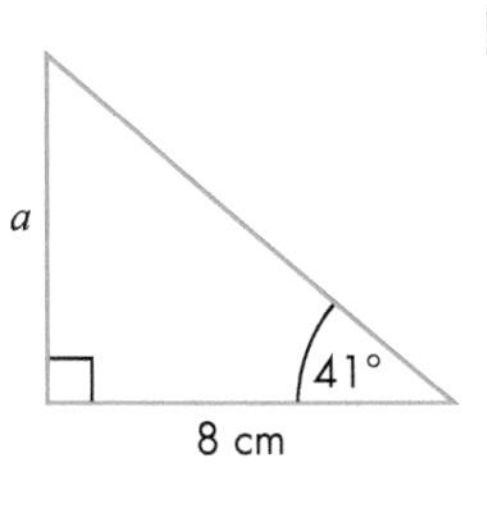

b

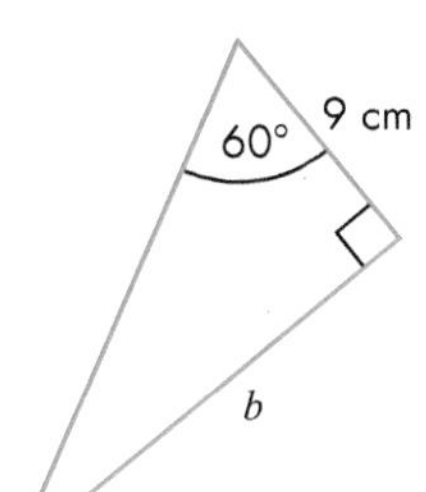

c

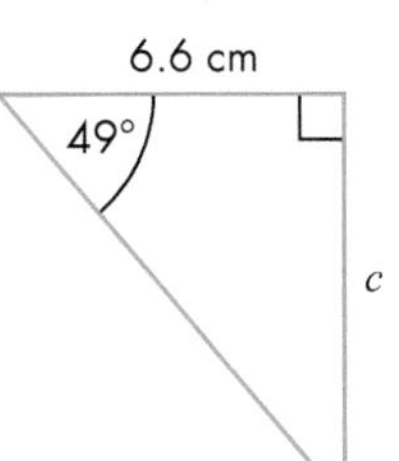

d

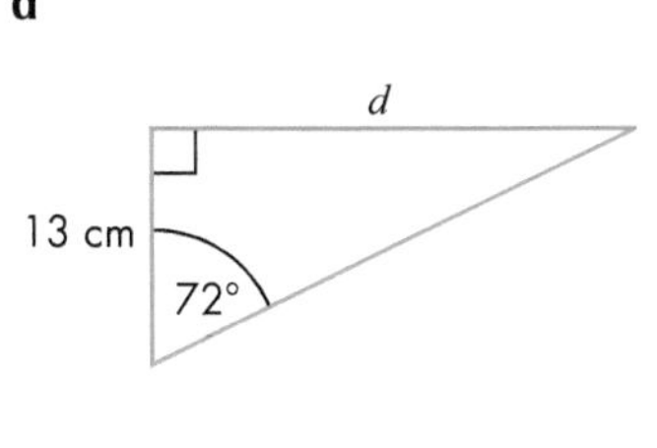

2 Use the sine ratio to calculate the lengths of the lettered sides.

a

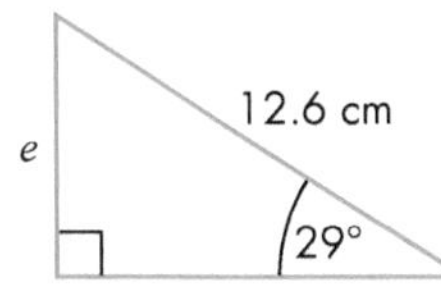

b

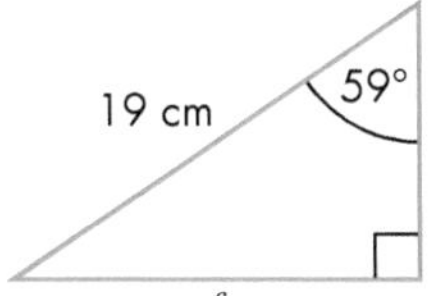

c

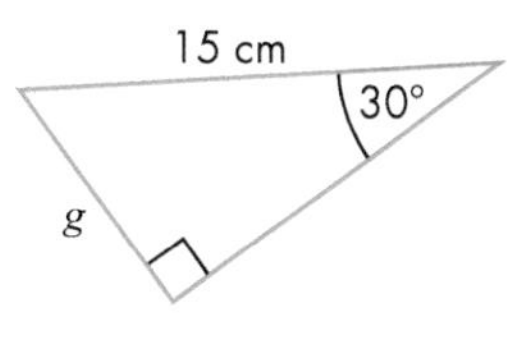

d

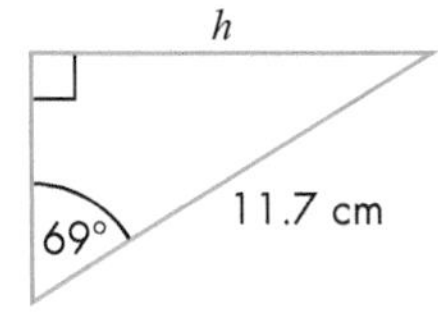

3 Use the cosine ratio to calculate the lengths of the lettered sides.

a

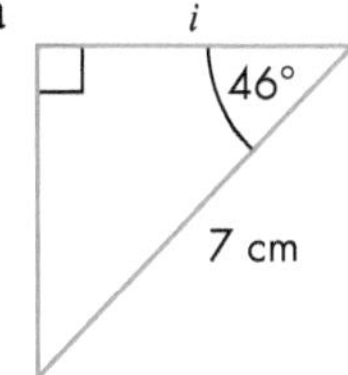

b

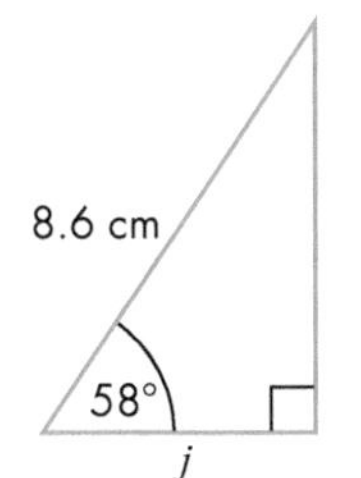

c

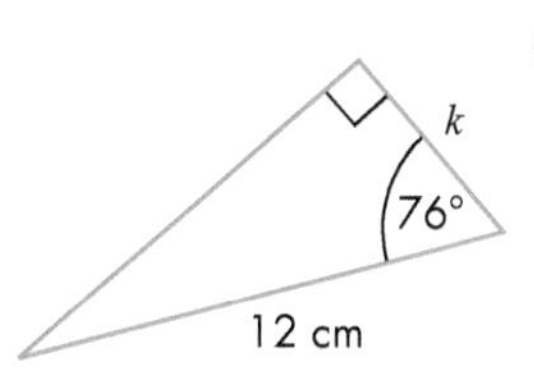

d

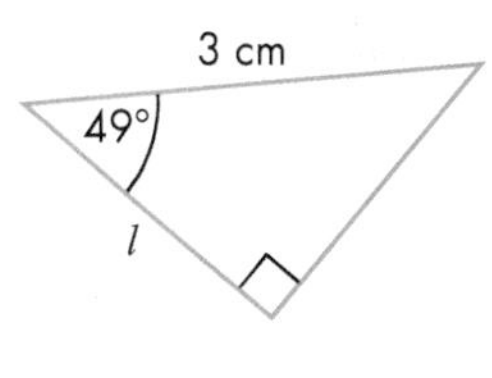

4 Calculate the lengths represented by letters. Give your answers correct to three significant figures.

a

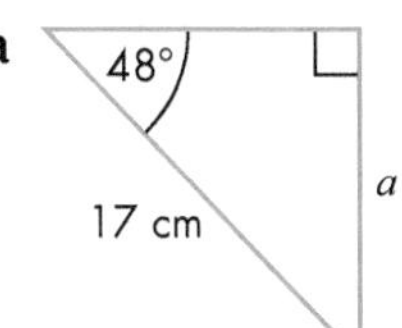

b

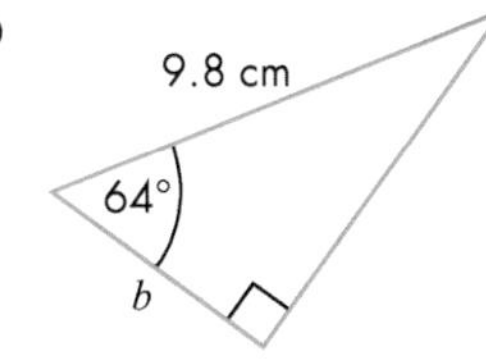

c

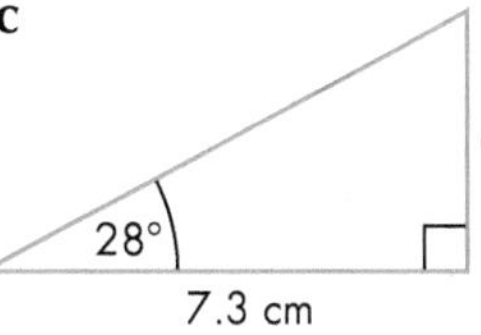

d

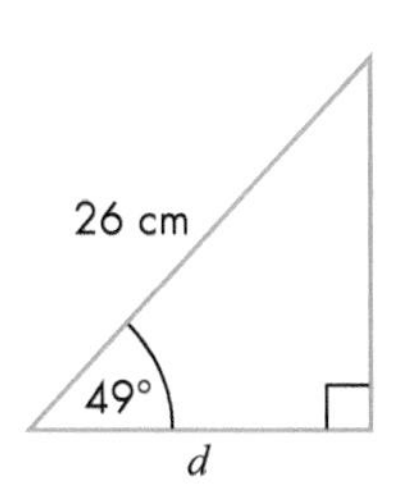

e

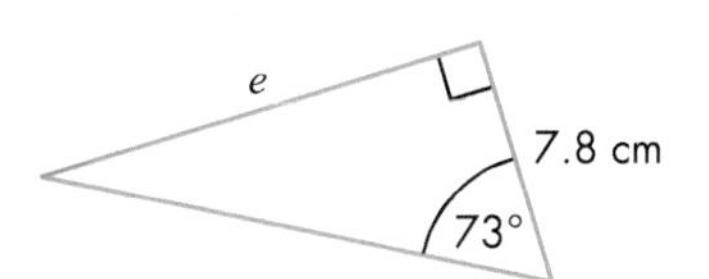

f

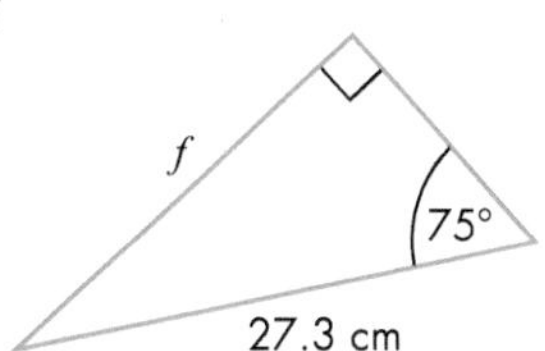

5 Calculate the length of x in each triangle.

> **Hints and tips** Labelling the known side and the side to be found (O, A or H) will help you to identify the correct formula to use.

a $67°$, x, 15 cm

b x, $72°$, 12 cm

c x, 8 cm, $65°$

d

e

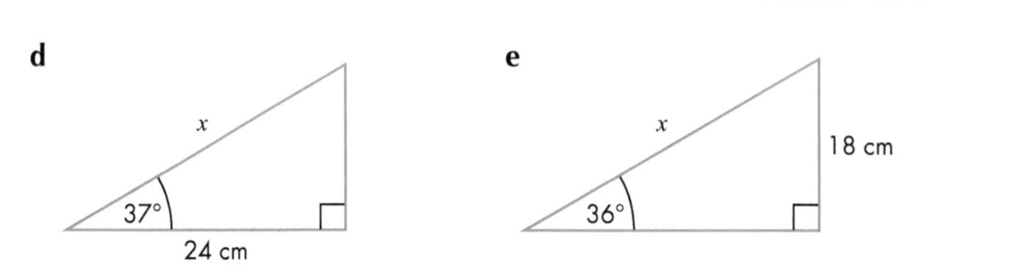

6 A slide is 10 m long and makes an angle of 37° with the playground.

How high above the ground is the top of the slide?

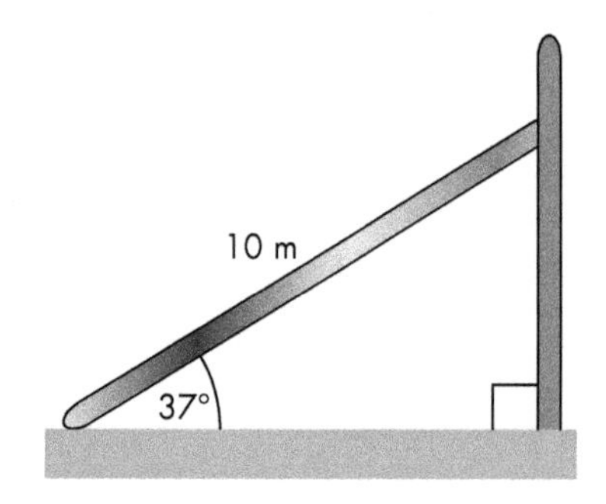

22.7 Calculating angles using trigonometry

Homework 22J

1 Work out the size of angle x.

a $\sin x = 0.8$ **b** $\sin x = 0.654$ **c** $\sin x = \frac{2}{3}$

d $\cos x = 0.623$ **e** $\cos x = 0.21$ **f** $\cos x = \frac{2}{5}$

g $\tan x = 1.25$ **h** $\tan x = 0.55$ **i** $\tan x = \frac{3}{2}$

2 Calculate the value of the lettered angle in each triangle.

a 18 cm, 11 cm, a

b

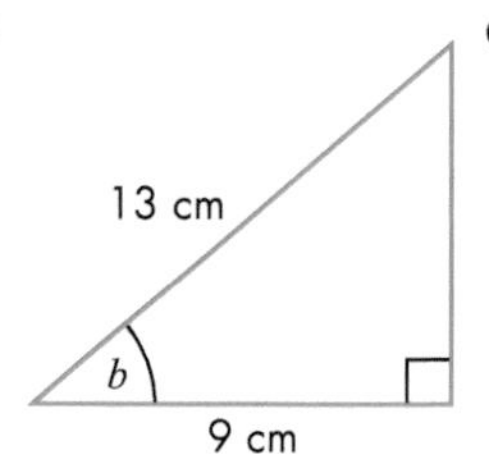

c

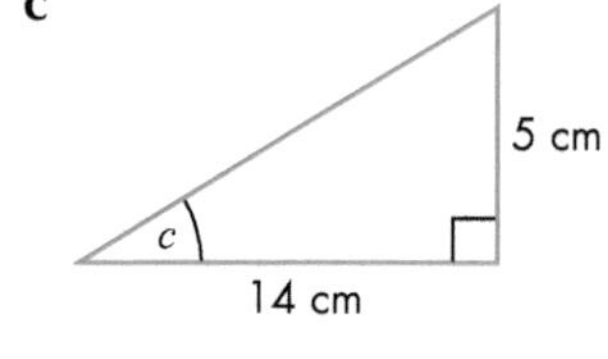

d

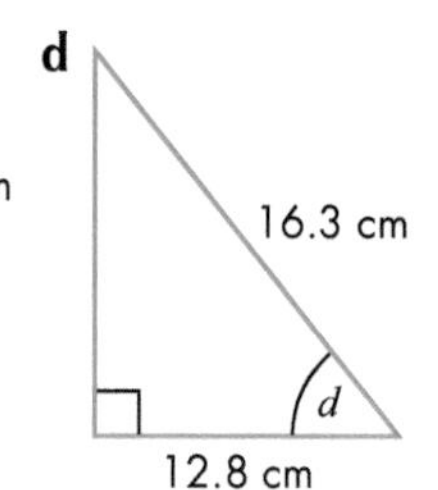

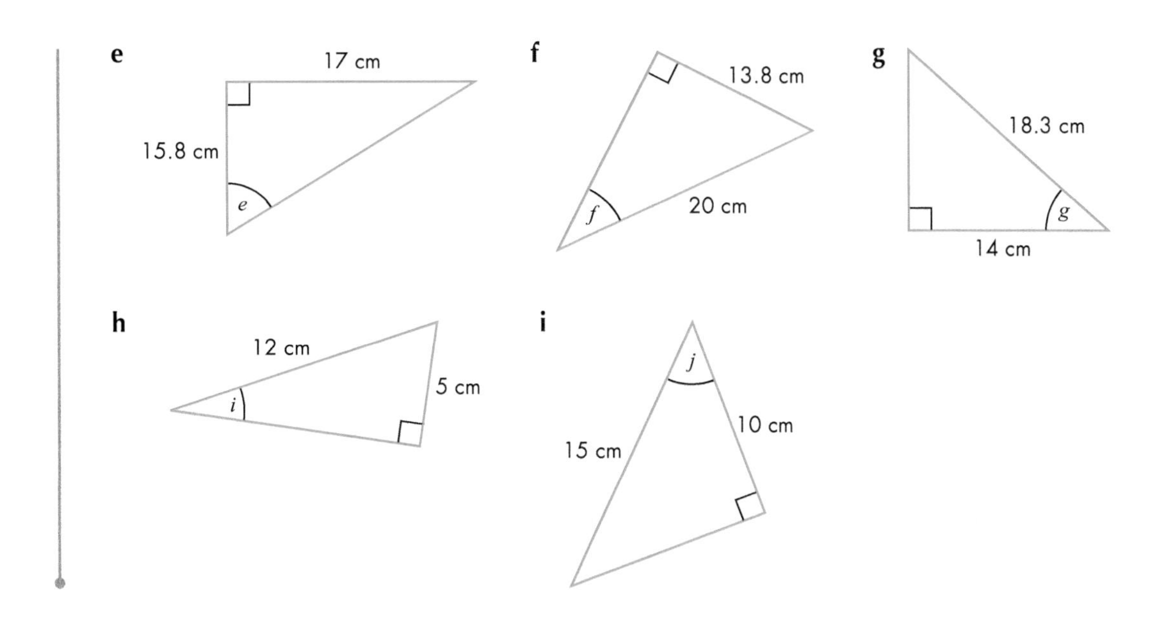

22.8 Trigonometry without a calculator

Homework 22K 

1 **a** Use these diagrams to show that $\cos 60 = \frac{1}{2}$.

b Show some other facts using these diagrams.

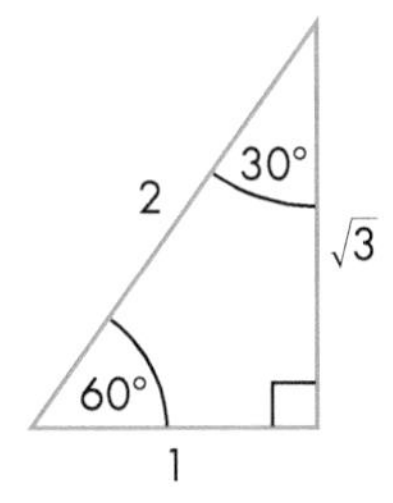

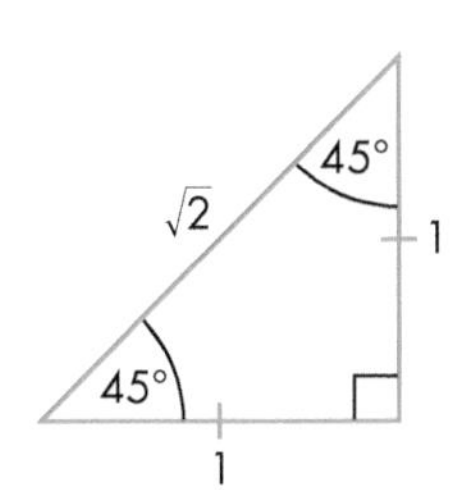

2 Calculate the length marked x in each triangle.

a

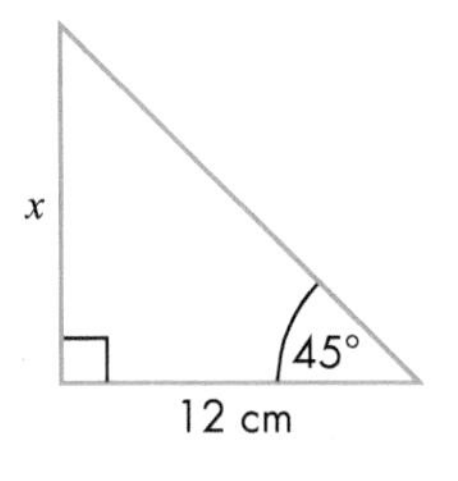

b

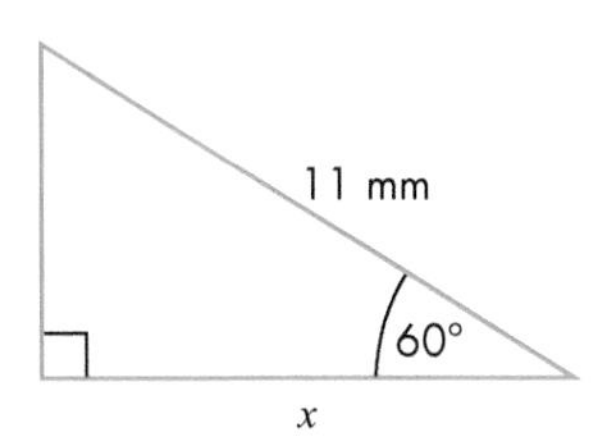

3 Calculate the length marked x in each diagram.

a

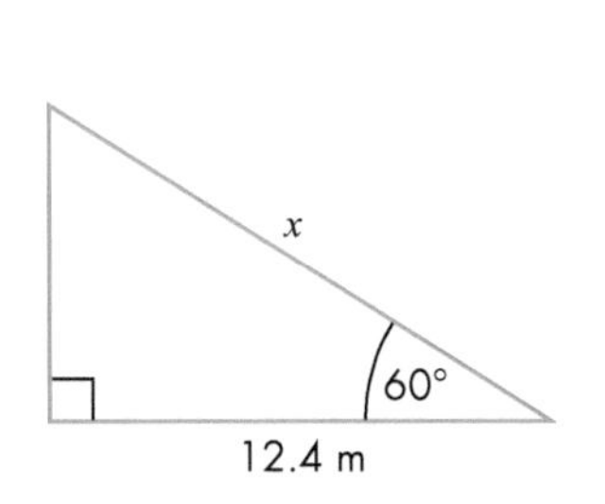

b

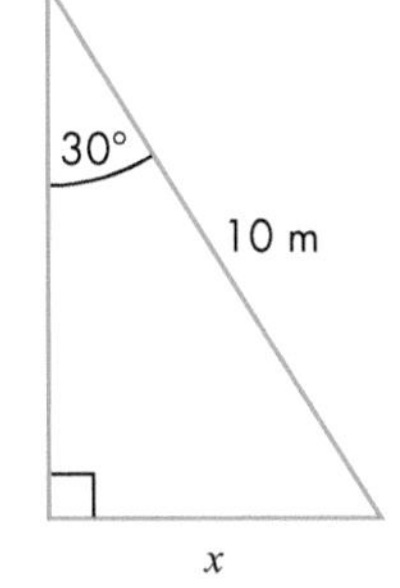

c

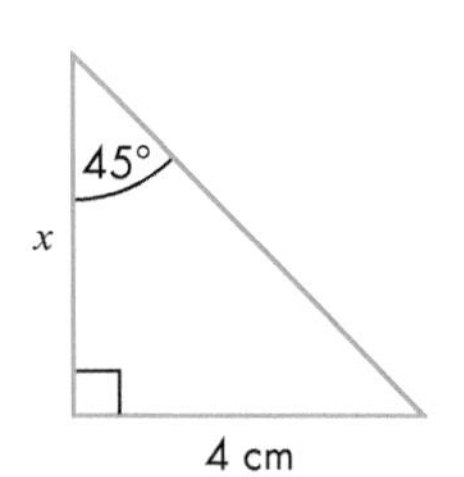

22.9 Solving problems using trigonometry

Homework 22L

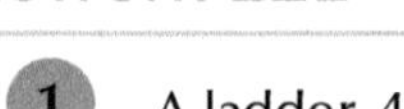

1. A ladder, 4 m long, rests against a wall. It makes an angle of 22° with the wall. Calculate:

   a how high up the wall the ladder reaches

   b the distance between the foot of the ladder and the wall.

2. A boy was flying his kite and had let out 36 m of string when it got stuck in a tree. When the string is pulled tight, it reaches the ground 27 m from the base of the tree. Calculate:

   a the angle the string makes with the ground

   b how high above the ground the kite is.

3. Calculate:

   a the height of the tree

   b the distance XY.

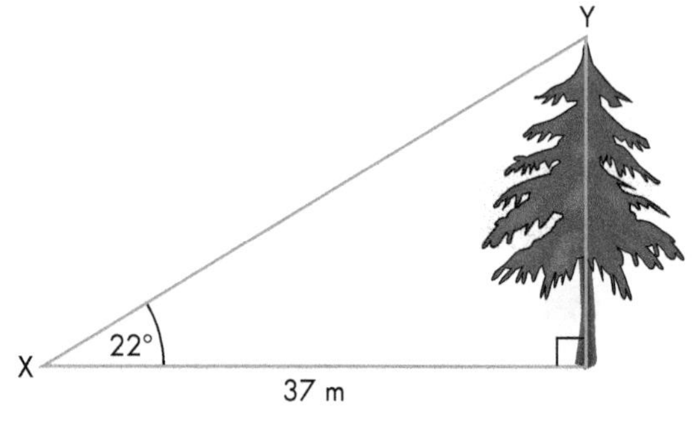

4. ABCD is a rectangular sheet of paper. AC = 21 cm and AD = 10 cm. Calculate:

   a the angle BAC

   b the length of AB in centimetres, correct to one decimal place.

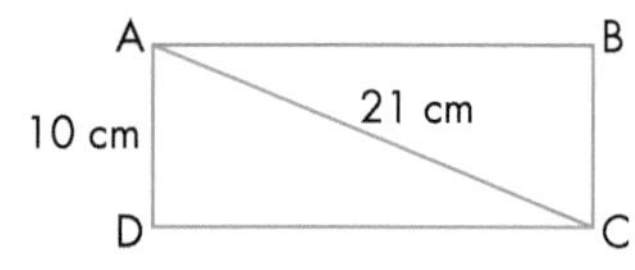

5. The string of a kite makes an angle of 72° with the horizontal. The kite is 51 m vertically above a point T. Calculate:

   a the length of string (x)

   b the horizontal distance between the start of the string and T (y).

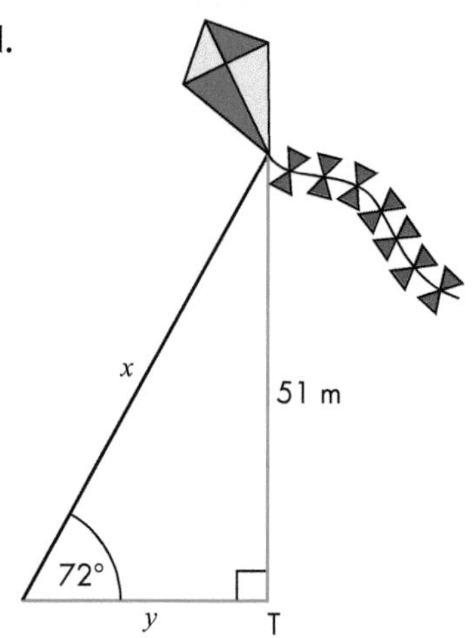

Homework 22M 

In this exercise, give any angles to one decimal place and any distances to two decimal places.

1. A surveyor is standing at the base of a hill. She measures the angle of elevation to the top of the hill as 35°. She then walks up 250 m to the top of this hill. Calculate the vertical height of the hill.

2. A rescue manager is at the top of a lighthouse. He measures the angle of depression to a boat as 35°. The lighthouse is 17 m above sea level. Calculate the distance between the base of the lighthouse and the boat.

3. John wants to measure the height of a tree. He walks exactly 50 feet from the base of the tree and looks up. The angle of elevation from the ground to the top of the tree is 23°. How tall is the tree?

4. An aeroplane is flying 1 mile above the ground. The distance along the ground from the aeroplane to the airport is 15 miles. What is the angle of depression from the aeroplane to the airport?

22.10 Trigonometry and bearings

Homework 22N

In this exercise, give your answers correct to three significant figures, where appropriate.

1. The beach (B) is 20 km from the airport (A) on a bearing of 064°. How far east of the airport is the beach?

2. A plane (P) flies 450 km on a bearing of 050° from the airport (A).

   a How far north has it travelled?

   b How far east has it travelled?

3. A helicopter leaves its base (B) and flies 63 km on a bearing of 285° to P.

   a How far west of B is P?

   b How far north of B is P?

4. A fishing boat leaves port (P) and sails on a straight course. After 2 hours it is 24 km south of P and 7 km east of P.

   On what bearing did it sail?

22.11 Trigonometry and isosceles triangles

Homework 22O

1 Work out the value of x in each triangle.

a

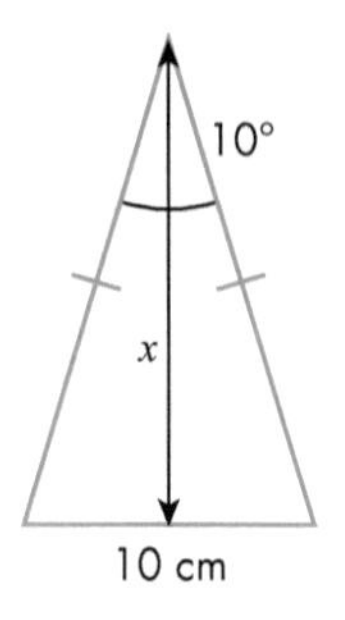

b

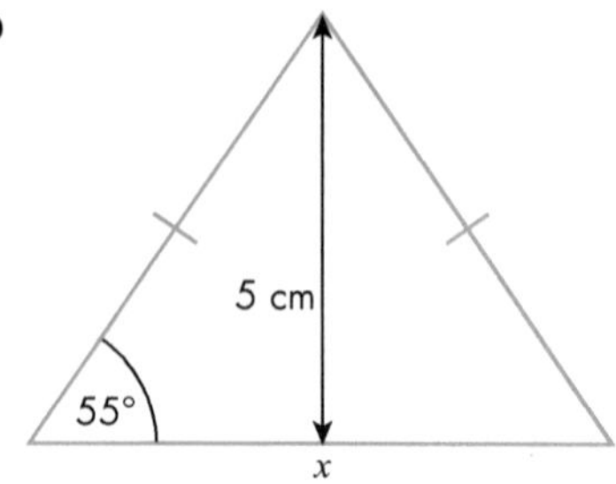

c

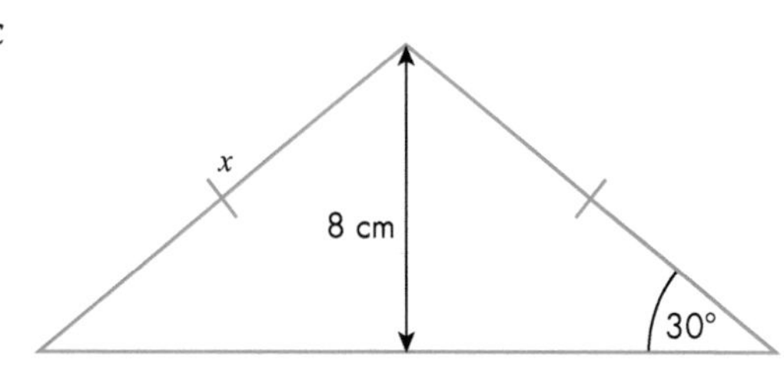

d

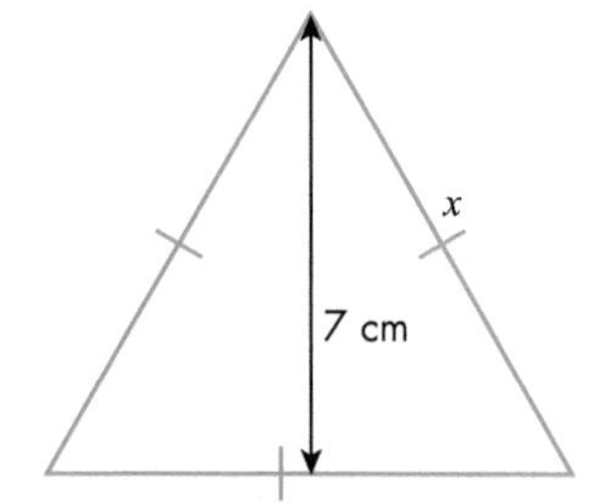

2 Calculate the area of each triangle.

a

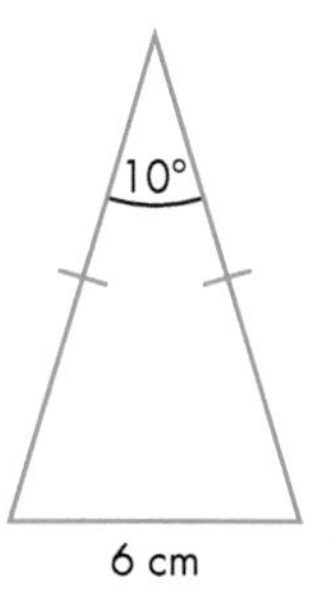

b

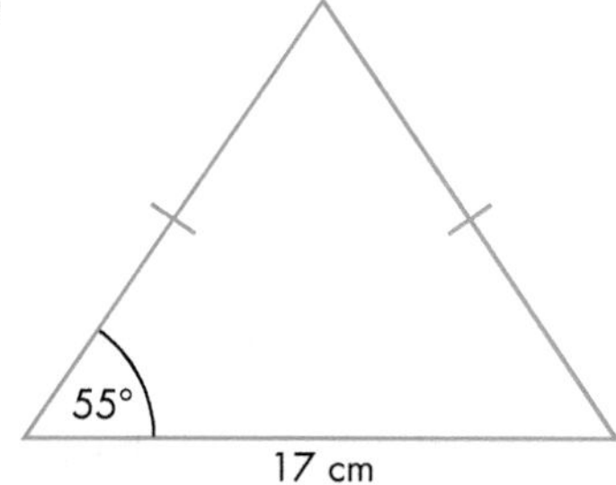

c

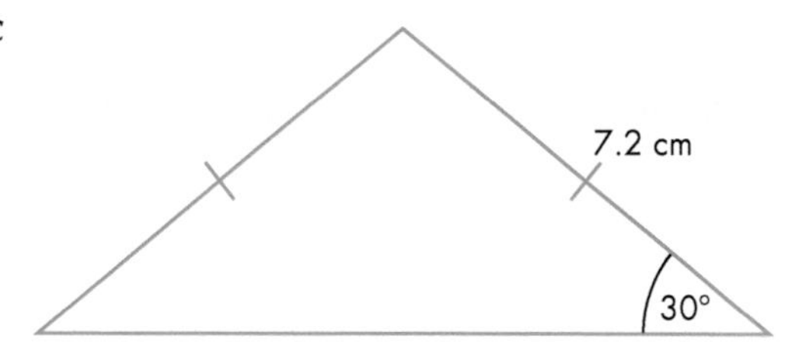

d

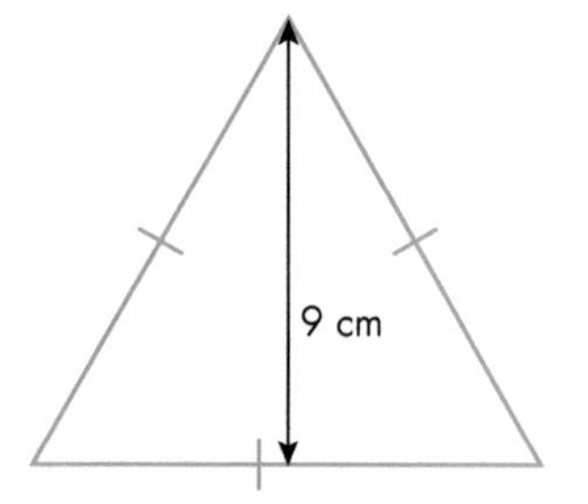

23 Geometry and measures: Congruency and similarity

23.1 Congruent triangles

Homework 23A

1 Write down:

a the pairs of congruent triangles

b the condition that shows they are congruent.

A
50°
50°
8 cm

B
50°
8 cm
50°

C
4 cm
70°
37°

D
8 cm
40°
6 cm

E
37°
4 cm
70°

F
8 cm
6 cm
40°

2 Show that triangle ABC is congruent to triangle DEF.

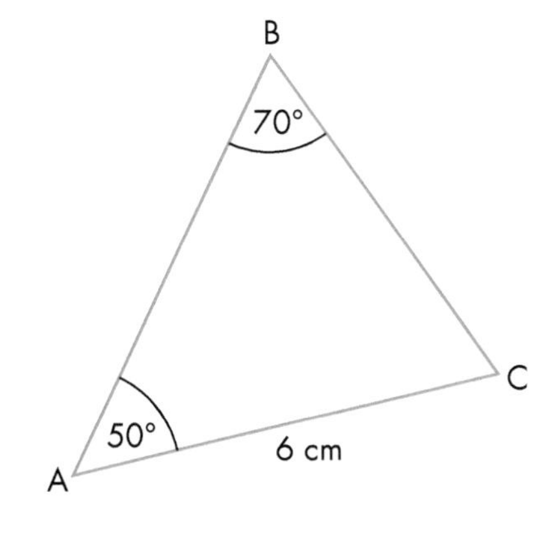

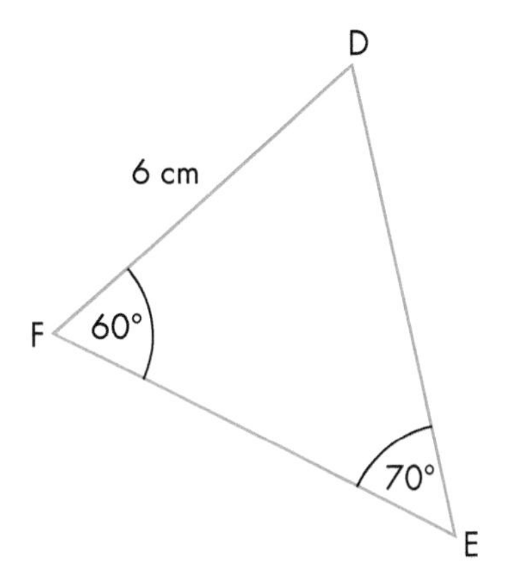

3 Show that triangle PQR is congruent to triangle STU.

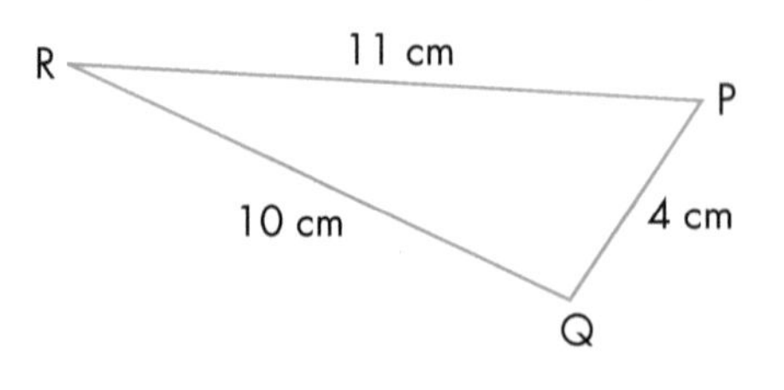

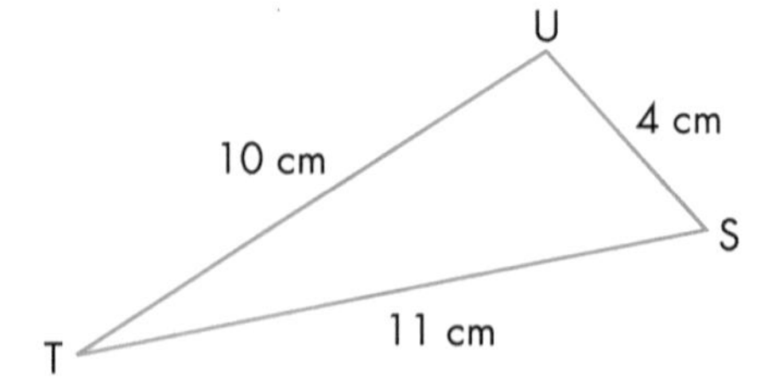

4 The diagram shows kite ABCD. The diagonals AC and BD intersect at X.

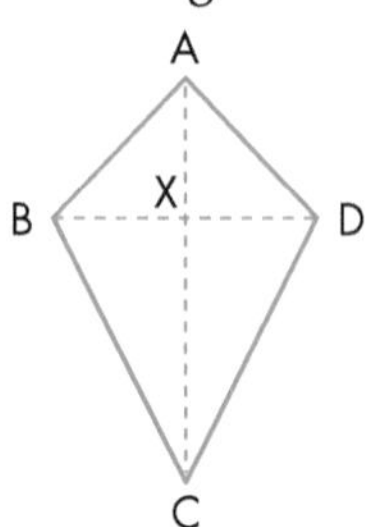

Which of the following statements are true?

a Triangle ABC is congruent to triangle ACD.

b Triangle ABD is congruent to triangle BCD.

c Triangle XBC is congruent to triangle XCD.

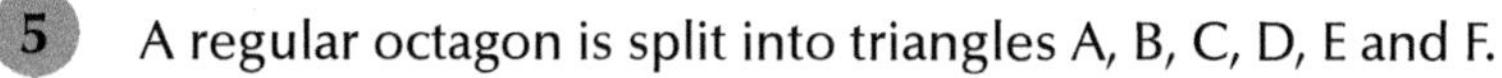

5 A regular octagon is split into triangles A, B, C, D, E and F.

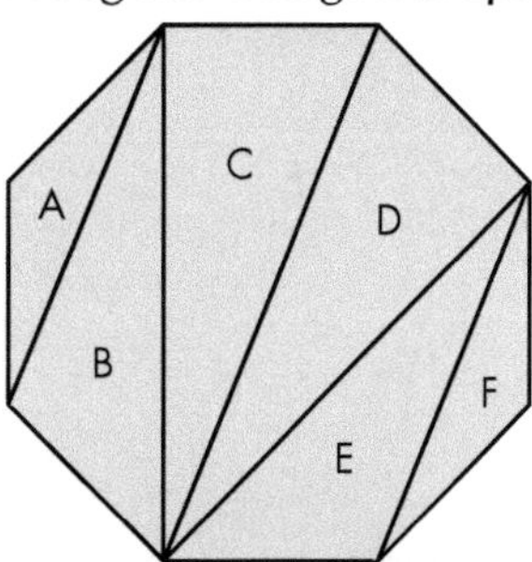

Write down the pairs of congruent triangles.

6 While she was sitting in a doctor's waiting room, Katrina looked at this simple wooden design that was built into the wall.

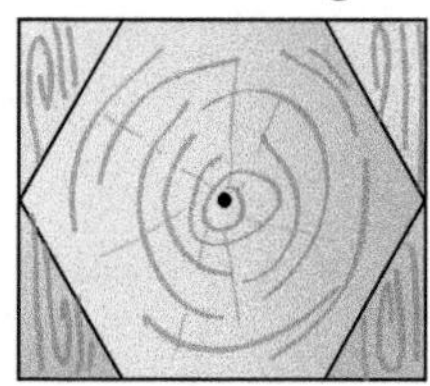

a What is the name of this regular six-sided shape?

b If you were to label the vertices ABCDEF and the centre O, how many other shapes inside the shape are congruent to:

i triangle ABC

ii quadrilateral ABCD

iii triangle ACE

iv rectangle ABDE

v triangle AOB

vi triangle ABD

vii pentagon ABCDE?

23.2 Similarity

Homework 23B

1
- **a** Construct one triangle with sides 2 cm, 3 cm and 4 cm and a second triangle with sides 6 cm, 9 cm and 12 cm.
- **b** Measure and label each angle.
- **c** What do you notice about the angles?
- **d** What is the scale factor of enlargement from the smaller triangle to the larger triangle?

2
- **a** Construct one triangle with sides 3 cm, 3.5 cm and 4 cm and a second triangle with sides 6 cm, 7 cm and 8 cm.
- **b** Measure and label each angle.
- **c** What do you notice about the angles?
- **d** What is the scale factor of enlargement from the smaller triangle to the larger triangle?

3
- **a** Construct one triangle with sides 4 cm, 8 cm and 10 cm and a second triangle with sides 6 cm, 12 cm and 15 cm.
- **b** Measure and label each angle.
- **c** What do you notice about the angles?
- **d** What is the scale factor of enlargement from the smaller triangle to the larger triangle?

Homework 23C

1 Which of these triangles are similar?

A

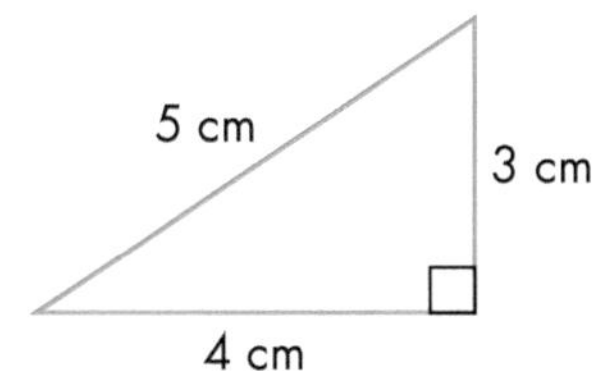

B

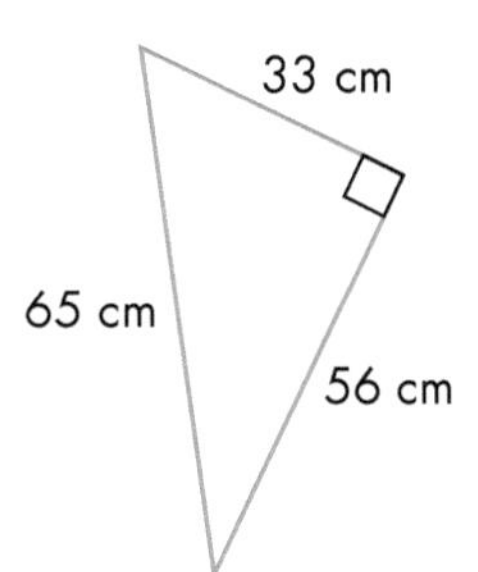

C

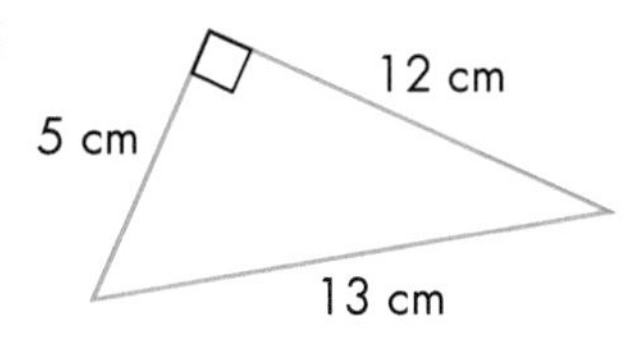

D

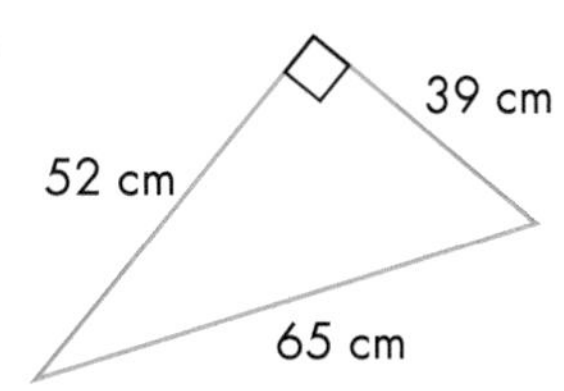

2 Are these pairs of shapes similar? If so, give a scale factor.

a

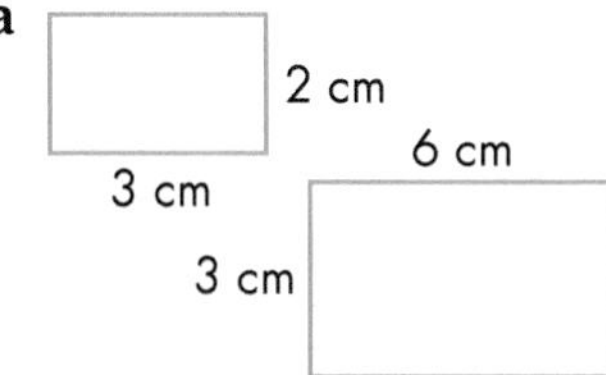

b

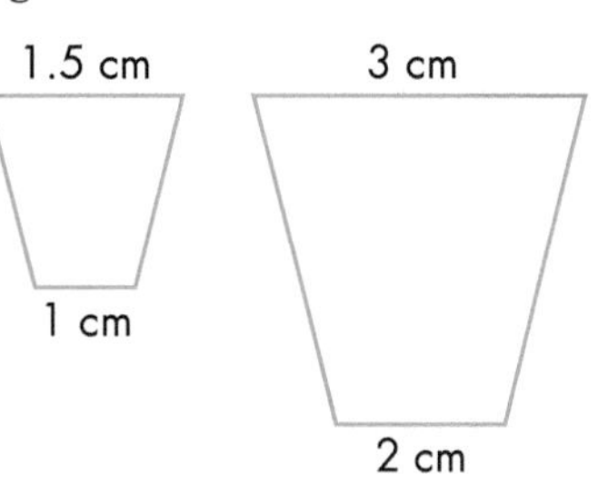

3 For each pair of similar shapes:

i state the scale factor of enlargement **ii** calculate the missing length(s).

a

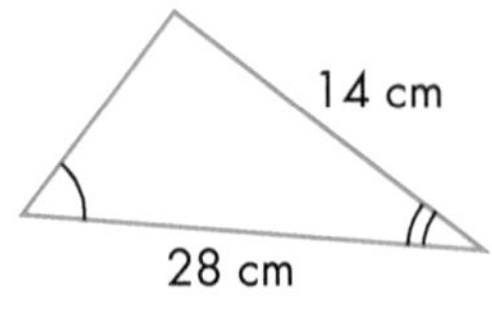

☐ cm
4 cm

b

3 cm
4 cm
11.2 cm
☐ cm

c

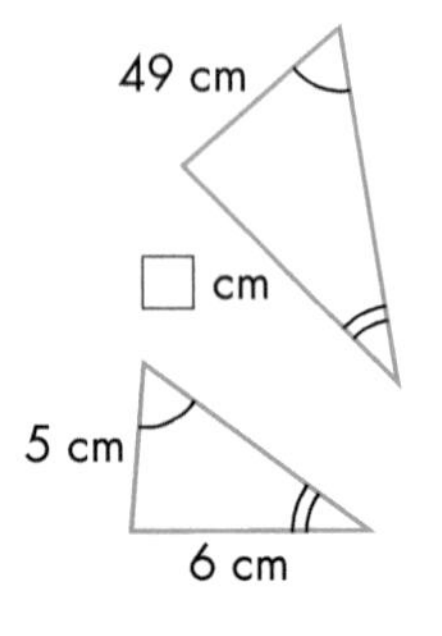

d

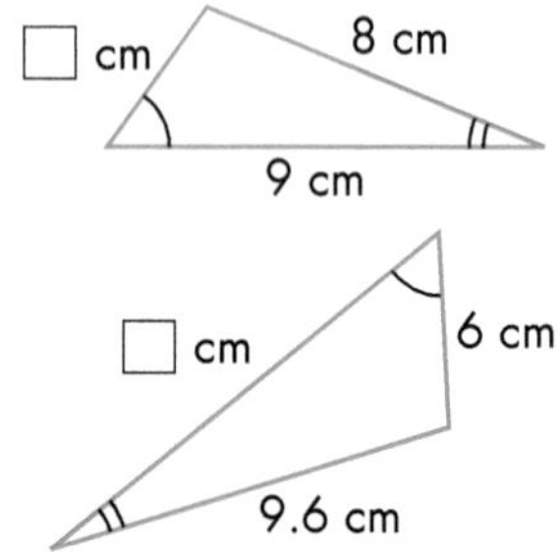

e

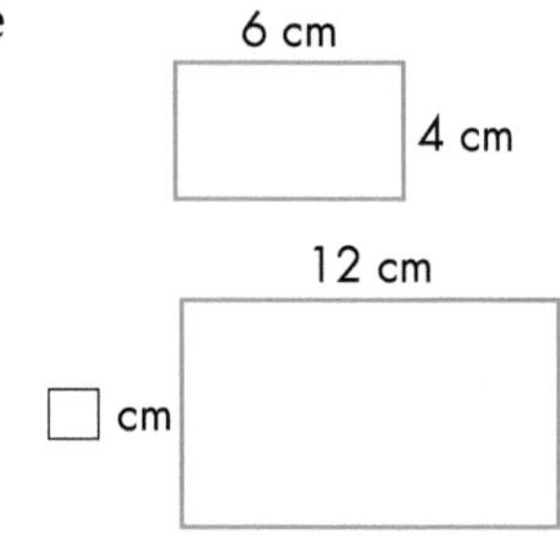

f

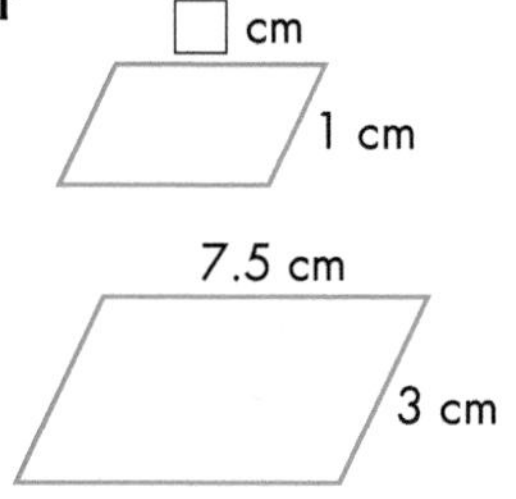

g

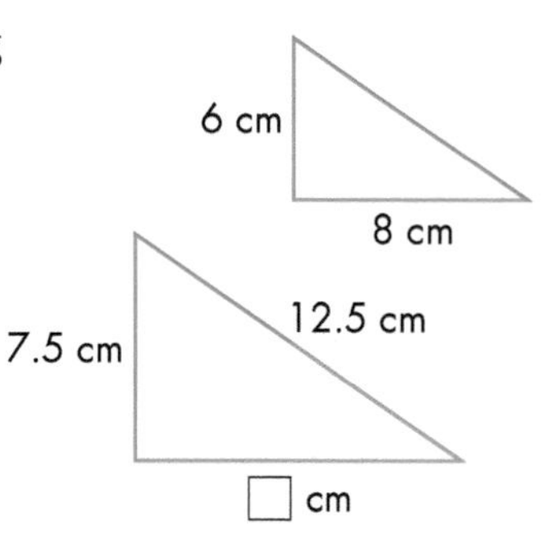

4 Calculate the missing lengths in each pair of similar shapes.

a

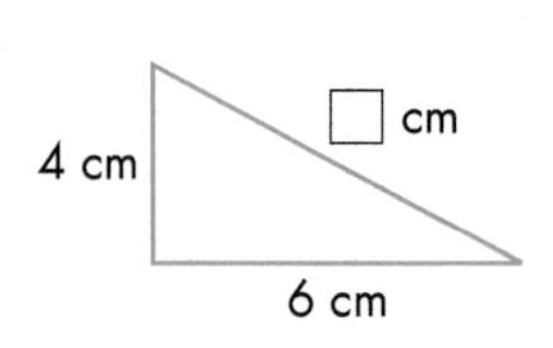

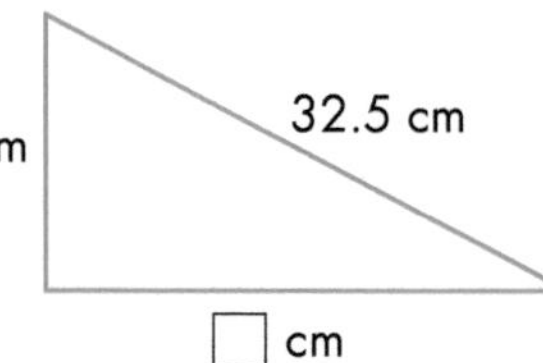

b

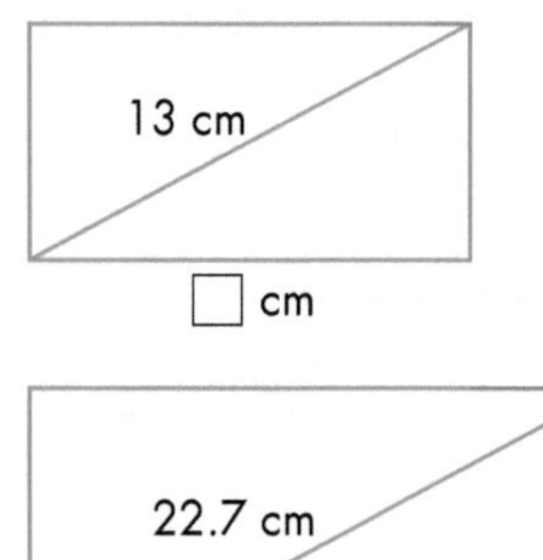

c

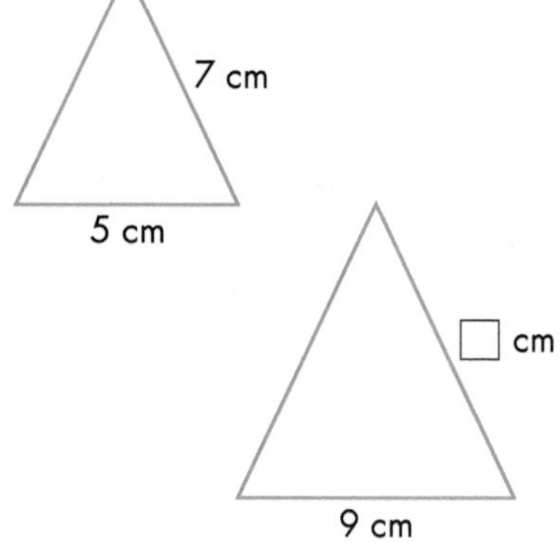

5 This pair of shapes are similar. Calculate the value of k.

Homework 23D

1 Use the information shown in the diagram to calculate the height of the statue, x.

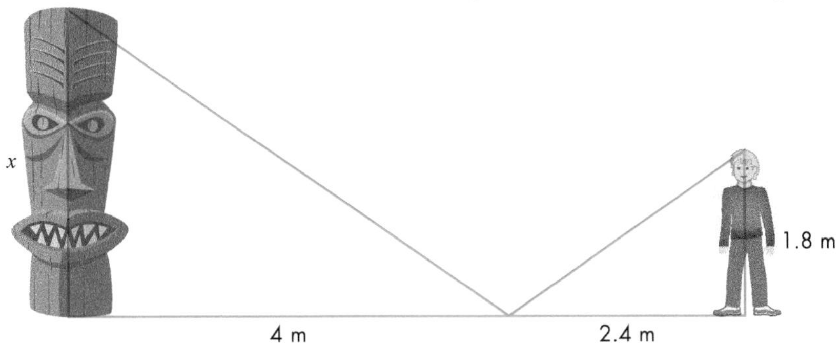

2 Ben is standing next to a tree that is 36 feet tall. The tree casts a shadow of 12 feet. Ben is 6 feet tall. How long is Ben's shadow?

3 Triangle ABC is similar to triangle XYZ.

The lengths of the sides in triangle ABC are 144 cm, 128 cm and 112 cm. The length of the shortest side of triangle XYZ is 280 cm.

What is the length of the longest side of triangle XYZ?

4 A 40-foot flagpole casts a shadow of 25 feet. Work out the length of the shadow cast by a neighbouring building that is 200 feet tall.

5 A girl, 160 cm tall, stands 360 cm from a lamp post. The light (from the lamp post) casts a 90-cm shadow of the girl on the ground. Work out the height of the lamp post.

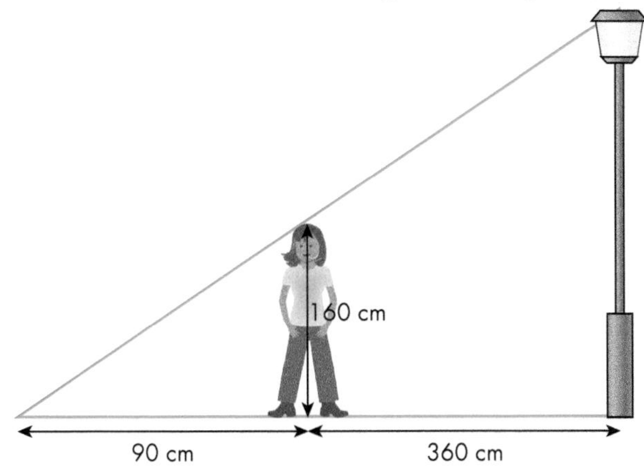

24 Probability: Combined events

24.1 Combined events

Homework 24A

1 **a** Copy and complete the sample space diagram to show all the outcomes for the total scores when two dice are thrown.

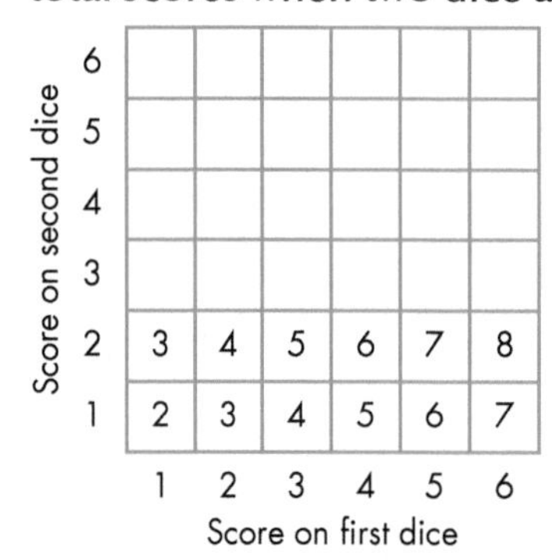

b What is the most likely total score?

c Which two total scores are least likely?

d What is the probability that a score is:

i 11 **ii** 4 **iii** greater than 9

iv an odd number **v** 4 or less **vi** a multiple of 4?

2 Complete a sample space diagram to show the outcomes when the spinner is spun and the dice is thrown.

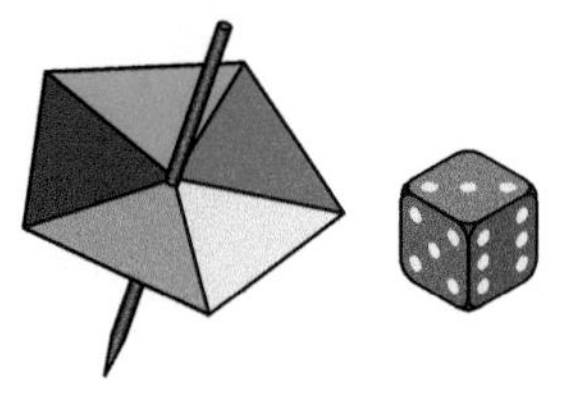

What is the probability that:

a the spinner stops on red and the score on the dice is 6

b the spinner stops on green and the score on the dice is even

c the score on the dice is 3?

3 Elaine throws a coin and spins a five-sided spinner that is numbered from 1 to 5. One possible outcome is (head, 5).

a List all the possible outcomes.

b What is the probability of getting a tail on the coin and an odd number on the spinner?

4 A bag contains five discs that are numbered 2, 4, 6, 8 and 10. Sharleen takes a disc at random from the bag, writes down the number and puts the disc back.

She then shakes the bag and takes another disc. She adds together the two numbers on the discs she has taken.

a Copy and complete the probability space diagram to show the scores for all the outcomes.

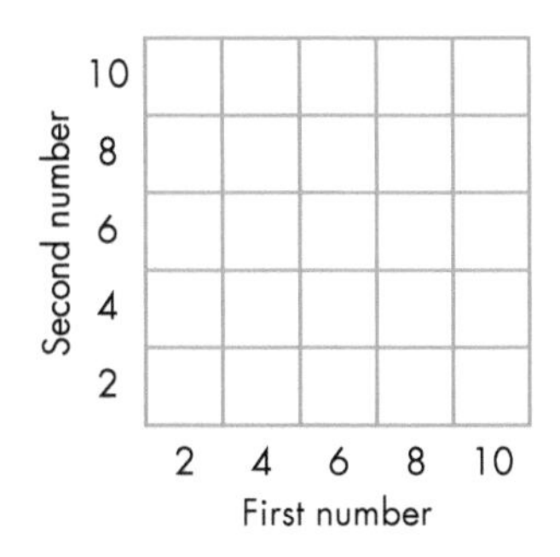

b Work out the probability that the total is:

i 12 **ii** 20 **iii** 15

iv a square number **v** a multiple of 3.

5 George throws two coins. What is the probability of each outcome?

a two heads **b** two tails **c** one head and one tail

6 Samuel is playing a board game with his sister Ellen. They take it in turns to throw two six-sided dice and add together the scores.

On his next turn, Samuel must roll a total greater than 6 to avoid losing.

On her next turn, Ellen will go to jail if she rolls a double.

a What is the probability that Samuel does not lose on his next turn?

b What is the probability that Ellen does not go to jail on her next turn?

7 Two eight-sided dice, numbered from 1 to 8, are thrown at the same time. What is the probability that the product of the two dice is a square number that is odd?

8 Isaac throws two dice and adds the two scores. He wants to work out the probability that his total score is a prime number.

Explain why a probability space diagram will help him.

24.2 Two-way tables

Homework 24B

1 Here is some information about the holiday times and destinations of a sample of 100 adults from the UK.

In July, 10 people went to Portugal, 19 went to Spain and 2 went elsewhere. In August 15 people went to Portugal, 15 went to Spain and 10 went elsewhere. In September 6 people went to Portugal, 18 went to Spain and 5 went elsewhere.

a Copy and complete the two-way table.

	Portugal	Spain	Elsewhere	Total
July				
August				
September				
Total				

b How many people went on holiday in September?

c What percentage of the sample went to Spain?

d A person is chosen at random. What is the probability they went to Portugal in August?

2 Eighty students each study one science subject. The two-way table shows some information about these students.

	Biology	Chemistry	Physics	Total
Female	18			47
Male			19	
Total		21	33	80

a Copy and complete the two-way table.

b A student is chosen at random. What is the probability that this student studies physics?

c What percentage of girls in the sample study biology?

3 The two-way table shows information about the results of candidates taking their driving test for the first time.

	Passed	Failed	Total
Male		11	
Female	17		
Total	25		50

a Copy and complete the two-way table.

b How many candidates failed their test on their first attempt?

c One driver is chosen at random. What is the probability it is a female who failed her test?

d Do a greater proportion of males or females pass their test at the first attempt? Explain your answer.

4 The two-way table shows information about the arrival times of cars operated by two taxi companies.

	On time	Early	Late	Total
Taxi4U			29	450
Cheap Eezy	374	6		
Total	784		199	1000

a Copy and complete the two-way table.

b What percentage of each taxi company's cars arrived on time?

c Were a higher percentage of Taxi4U cars late? Justify your answer.

d Cheap Eezy has 5000 bookings. For how many of these would you expect the car to arrive early?

5 USA, Germany and China topped the medals table at the world indoor athletics championships.

There were 43 silver medals won by the three countries altogether.

Germany won 18 gold and 16 silver.

USA won 10 gold and 31 bronze.

China won 9 silver and 11 bronze.

Germany won 43 medals altogether.

China won 42 medals altogether.

a Draw and complete a two-way table to show this information.

b Which country won the most medals?

c Who do you think won overall? Give a reason for your answer.

24.3 Probability and Venn diagrams

Homework 24C

1 P(A) = 0.22 and P(B) = 0.49. Write down:

a P(A′)

b P(B′).

2 ξ = {1, 2, 3, 4, 5, 6}, A = {2, 3, 5}, B = {1, 3, 5}

a Show this information in a Venn diagram

b Use your Venn diagram to work out:

i P(A)

ii P(B)

iii P(A′)

iv P(B′)

v $P(A \cap B)$

vi $P(A \cup B)$.

In Screen 3 of the cinema, 46 people had bought sweet popcorn (S), 22 had bought a drink (D), 11 had bought a drink and sweet popcorn and 10 had not bought either.

a Copy and complete the Venn diagram to show this information.

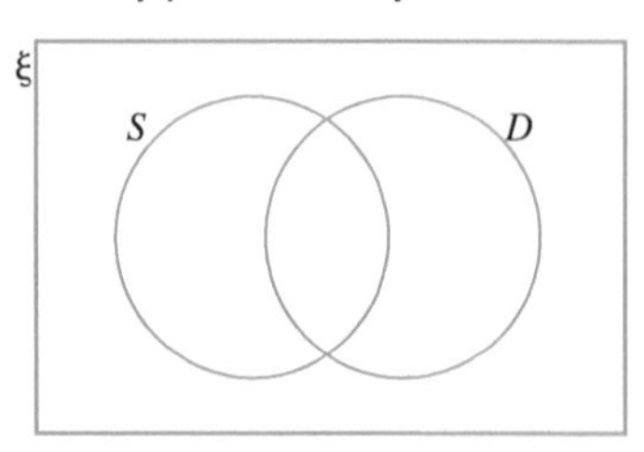

b For customers in Screen 3, what is:

i P(S)

ii $P(S \cap D)$

iii $P(S \cup D)$?

Of 100 pet owners, 34 owned a cat (C), 60 owned a cat or a dog (D) and 14 owned a cat and a dog.

a Copy and complete the Venn diagram to show this information.

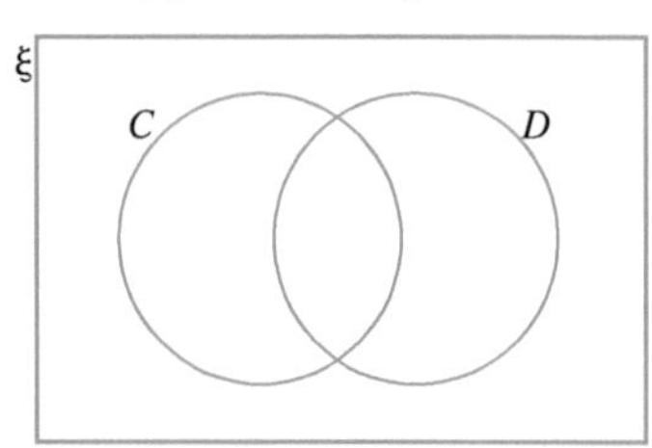

b For a pet owner chosen at random, what is:

i $P(C' \cup D')$

ii P(D)

iii $P(D \cap C')$?

In a class of 30 students, one student is chosen at random to be the class captain. There are 16 girls. Seven of the girls and six of the boys have brown eyes.

Work out the probability that the student chosen as class captain:

a has brown eyes

b is a girl with brown eyes

c is a girl but does not have brown eyes

d is a boy with brown eyes.

6 In a survey, 224 members of a health club were asked which facilities they used on their last visit. The results are shown in the Venn diagram.

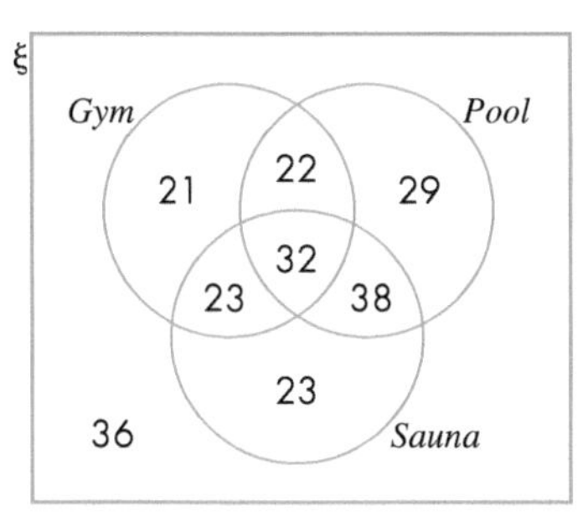

Work out the probability that a member chosen at random:

a only used the gym

b used the gym

c did not use the gym

d did not use any facilities

e used the gym and the pool but not the sauna

f used the pool and sauna, given that they did use the gym?

24.4 Tree diagrams

Homework 24D

1 A gardener had 100 customers. He is supposed to mow their lawns and strim the edges. The probability that his strimmer starts is 0.9 and the probability that his lawnmower starts is 0.6.

a Draw a frequency tree diagram for all the outcomes.

b How many customers had their lawn strimmed and mowed?

c How many customers had their lawn strimmed, mowed or both?

2 The probability that Neil wins a game is 0.4. He plays the game twice.

a Copy and complete the probability tree diagram to show the possible outcomes.

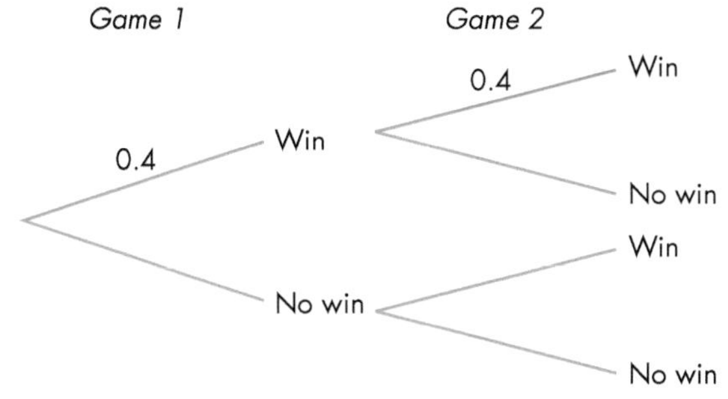

b Use the tree diagram to work out the probability that Neil:

i wins both games **ii** wins only one game.

3 Anna and Bill do not have high attendance at school. On any school day:

P(Anna attends school) = 0.6

P(Bill attends school) = 0.7

a Copy and complete the probability tree diagram to show all the outcomes.

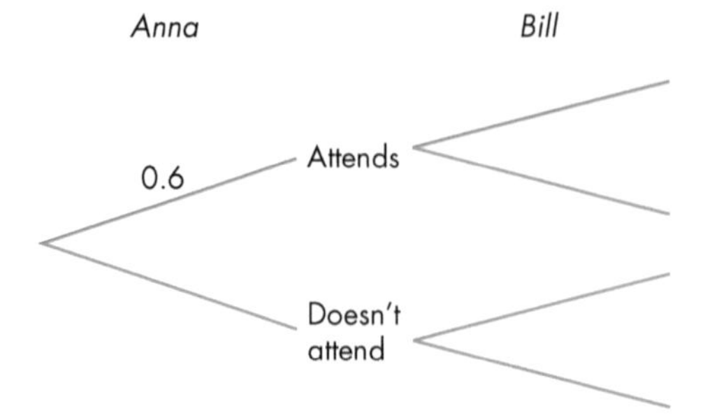

b What is the probability that:

i neither Anna nor Bill is at school

ii at least one of them is at school?

4 Chloe has to hand in homework for two subjects on Tuesday.

The probability that she has her maths homework is $\frac{2}{5}$.

The probability that she has her art homework is $\frac{7}{10}$.

Draw a probability tree diagram to show all the outcomes.

Use your diagram to work out:

a P(she doesn't have any of her homework)

b P(she has at least one piece of homework).

5 There are six red discs and three white discs in a bag. A disc is taken from the bag, its colour noted and replaced. Another disc is then taken from the bag.

Work out the probability that:

a both discs are the same colour

b at least one of the discs was red.

25 Number: Powers and standard form

25.1 Powers (indices)

Homework 25A 

Example

Work out the value of 3^5.

$3^5 = 3 \times 3 \times 3 \times 3 \times 3$

$= 243$

1 Use the power key on your calculator (or any method you prefer) to work out the value of each power term.

a 2^3 **b** 4^3 **c** 7^3 **d** 10^3 **e** 12^3

f 3^4 **g** 10^4 **h** 2^5 **i** 10^6 **j** 2^8

2 **a** Use the power key on your calculator (or any method you prefer) to work out the value of each power term.

i 11^2 **ii** 11^3 **iii** 11^4

b Describe any patterns you notice in your answers.

c Does your pattern work for other powers of 11? Give a reason for your answer.

3 A container is in the shape of a cube.

The length of the container is 30 cm.

Work out the volume of the container.

Use the formula for the volume of a cube: volume = $(\text{length of edge})^3$

4 Write each number as a power of a different number.

The first one has been done for you.

a $16 = 2^4$ **b** 64 **c** 27 **d** 36

5 **a** Using your calculator, or any method you prefer, work out the value of each power term.

i $(-2)^8$ **ii** $(-2)^7$ **iii** $(-2)^{11}$ **iv** $(-2)^{14}$

b What do you notice?

25.2 Rules for multiplying and dividing powers

Homework 25B

1 Write each expression as a single power of 2.

a $2^4 \times 2^4$ **b** $2^2 \times 2^6$ **c** $2^2 \times 2^3$ **d** 2×2^2 **e** $2^1 \times 2^9$

f 2×2 **g** $2^2 \times 2^4$ **h** $2^7 \times 2^3$ **i** $2^{12} \times 2^9$

2 Write each expression as a single power of x.

a $x^3 \times x^7$ **b** $x^5 \times x^4$ **c** $x^5 \times x^2$ **d** $x^3 \times x^2$

e $x^6 \times x^1$ **f** $x^7 \times x^5$ **g** $x^7 \times x^4$

3 Write each expression as a single power of 3.

a $3^6 \div 3^3$ **b** $3^6 \div 3^2$ **c** $3^7 \div 3^2$ **d** $3^8 \div 3^4$

e $3^2 \div 3^4$ **f** $3^{11} \div 3^2$ **g** $3^4 \div 3^2$

4 Write each expression as a single power of y.

a $y^{10} \div y^3$ **b** $y^4 \div y^3$ **c** $y^7 \div y$ **d** $y^2 \div y^2$

e $y^{20} \div y^4$ **f** $y^3 \div y$ **g** $y^4 \div y^2$

5 Simplify each expression.

a $5a^5 \times 3a^3$ **b** $3a^7 \times 3a^{-5}$ **c** $(5a^5)^3$

d $-5a^5 \times 3a^5$ **e** $-7a^3 \times -5a^5$ **f** $-5a^7 \times 5a^{-7}$

6 Simplify each expression.

a $30a^6 \div 5a^5$ **b** $15a^5 \div 3a^5$ **c** $15a^5 \div 5a$

d $18a^5 \div 3a^1$ **e** $57a^5 \div 3a^5$ **f** $30a \div 3a^5$

7 Simplify each expression.

a $5a^5b^3 \times 7a^3b$ **b** $5a^5b^7 \times 5ab^{-3}$ **c** $3a^5b^3 \times 5a^{-7}b^{-5}$

d $15a^5b^7 \div 3ab$ **e** $57a^{-3}b^7 \div 3a^5b^{-3}$ **f** $14a^7b^{-5} \div 7a^5b^3$

8 Write each of these as a single power of 7.

a $(7^5)^3$ **b** $(7^3)^5$ **c** $(7^1)^3$

d $(7^3)^{-5}$ **e** $(7^{-5})^{-3}$ **f** $(7^7)^0$

Homework 25C 

1 Write down the answers without using a calculator.

a 200×400 **b** 30×5000 **c** $(10)^3$ **d** 50×5000

2 Write down the answers without using a calculator.

a $2.5 \div 10$ **b** $20.34 \div 1000$ **c** $0.35 \div 10$ **d** $12.5 \div 10\,000$

3 Write down the value of each expression.

a 8.1×10 **b** 8.1×100 **c** 8.1×1000 **d** $8.1 \times 10\,000$

4 Write down the value of each expression.

a $8.1 \div 10$ b $8.1 \div 100$ c $8.1 \div 1000$ d $8.1 \div 10\,000$

5 Evaluate each expression.

a 2.4×10^3 b 1.24×10^5 c 6.41×10^{-3} d 4.29×10^{-2}

e 2.408×10^{-3} f 3.09×10^{-2} g 7.003×10^6

25.3 Standard form

Homework 25D

1 Write down the value of each expression.

a 12.7×0.1 b 12.7×0.01 c 12.7×0.001 d 12.7×0.0001

2 Write down the value of each expression.

a $12.1 \div 0.1$ b $12.1 \div 0.01$ c $12.1 \div 0.001$ d $12.1 \div 0.0001$

3 Write these numbers out in full.

a 2.5×10^2 b 3.12×10 c 4.32×10^{-3} d 2.43×10

e 2.0719×10^{-2} f 5.372×10^3 g 2.03×10^2 h 1.3×10^3

i 8.17×10^5 j 8.35×10^{-3} k 3×10^7 l 5.27×10^{-4}

4 Write these numbers in standard form.

a 200

b 0.305

c 40 700

d 3 400 000 000

e 20 780 000 000

f 0.000 537 8

g 2437

h 0.173

i 0.100 73

j 0.989

k 274.53

l 98.7354

m 0.0054

n 0.004 37

o 53.1045

5 Write these numbers in order of size, starting with the smallest.

3.75×10^4

37×10^3

375 000

$15 \times 2.3 \times 10^4$

6 These numbers are not in standard form. Write them in standard form.

a 53.2×10^2
b 0.03×10^4
c 34.3×10^{-2}
d 0.02×10^{-2}
e 53×10
f $2 \times 3 \times 10^5$
g $2 \times 10^2 \times 35$
h 130×10^{-2}
i 23 million
j 0.0003×10^{-2}
k 25.3×10^5
l $13 \times 10^2 \times 3 \times 10^{-1}$
m $2 \times 10^4 \times 53 \times 10^{-4}$
n $(18 \times 10^2) \div (3 \times 10^3)$
o $(53 \times 10^3) \div (2 \times 10^{-2})$

7 Work these out. Give your answers in standard form.

a $4 \times 10^4 \times 5.4 \times 10^9$
b $1.9 \times 10^4 \times 9 \times 10^4$
c $4 \times 10^4 \times 9 \times 10^4$
d $4 \times 10^{-4} \times 5.4 \times 10^9$
e $1.9 \times 10^{-4} \times 4 \times 10^4$
f $4 \times 10^4 \times 9 \times 10^{-4}$
g $7.4 \times 10^{-9} \times 4 \times 10^4$
h $(5 \times 10^9)^4$
i $(4 \times 10^{-4})^9$

8 Light travels at 299 792 458 m/s.

a Round this speed to three significant figures.
b Express this rounded speed in standard form.

9 How many seconds are there in a leap year? Give your answer in standard form.

10 The Sun's interior is heated by nuclear reactions to temperatures up to 15 million °C. Write this temperature in standard form.

11 The starship *Enterprise* travelled 3×10^{15} km at a speed of 1.5×10^3 km/s. How long did the journey take?

12 The average distance from Earth to Mars is 2.25×10^8 km. A video message is sent to Mars at a speed of 1.8×10^7 km/min.

How long will it take to arrive?

13 $x = 1.8 \times 10^7$ and $y = 4 \times 10^3$.

Express $y - 3x$ in standard form.

26 Algebra: Simultaneous equations and linear inequalities

26.1 Elimination method for simultaneous equations

Homework 26A

Example

Solve this pair of simultaneous equations by elimination.

$2x + 3y = 10 \quad (1)$

$2x + y = 8 \quad (2)$

Subtract: $(1) - (2) \quad 2y = 2$

Divide by 2: $y = 1$

Substitute $y = 1$ into (1):

$2x + 3(1) = 10$

$2x + 3 = 10$

$2x = 7$

$x = 3.5$

Substitute both values into (2) to check your solutions:

$2(3.5) + 1 = 8$

$7 + 1 = 8$

Solve each pair of simultaneous equations by the elimination method.

1 $6x + 5y = 39$
$6x + 6y = 42$

2 $6x + 10y = -22$
$6x + 6y = -6$

3 $6x + 2y = -10$
$6x + 5y = -16$

4 $4x + 5y = -32$
$4x + 6y = -36$

5 $6x + 3y = -12$
$6x + 2y = -16$

6 $12x + 4y = -4$
$2x + 4y = -4$

7. $20x + 25y = 30$
   $20x + 24y = 32$

8. $4x + 4y = -8$
   $4x + 6y = -14$

26.2 Substitution method for simultaneous equations

Homework 26B

Example

Solve this pair of simultaneous equations by substitution.

$2x + 3y = 10$ (1)

$y = 8 - 2x$ (2)

Substitute $y = 8 - 2x$ into (1):

$2x + 3(8 - 2x) = 10$

Expand and simplify:

$2x + 24 - 6x = 10$

$-4x = -14$

Divide both sides by –4:

$x = 3.5$

Substitute $x = 3.5$ into (2) to find the value of y:

$y = 8 - 2(3.5)$

$y = 8 - 7$

$y = 1$

Substitute both values into (1) to check they work:

$2(3.5) + 3(1) = 10$

$7 + 3 = 10$

Solve each pair of simultaneous equations by the substitution method.

1. $6x + 2y = 36$
   $y = 2x - 2$

2. $6x + 2y = -44$
   $y = x + 10$

3. $6x + 6y = -12$
   $y = 4x - 32$

4. $6x + 2y = 48$
   $y = 3x - 24$

5. $2x + 5y = -16$
   $y = -4x + 4$

6. $2x + 5y = 10$
   $y = 2x + 26$

7. $5x + 5y = -50$
   $y = -3x - 30$

8. $4x + 6y = -12$
   $y = -2x + 6$

26.3 Balancing coefficients to solve simultaneous equations

Homework 26C

Example

Balancing coefficients involves multiplying the equations by a constant so that the coefficient of one of the variables is the same. You can then solve by elimination.

$2x - 5y = 11$ (1)

$3x + 2y = 7$ (2)

Multiply (1) by 3 and (2) by 2 to balance the coefficients of x:

$6x - 15y = 33$ (3)

$6x + 4y = 14$ (4)

Solve by elimination:

(3) – (4) $-19y = 19$

$y = -1$

Substitute $y = -1$ into (1) to find the value of x:

$2x - 5(-1) = 11$

$2x + 5 = 11$

$2x = 6$

$x = 3$

Substitute both values into (2) to check your solutions:

$3(3) + 2(-1) = 7$

$9 - 2 = 7$

Solve each pair of simultaneous equations by balancing the coefficients.

1. $3x - 4y = 13$
   $2x + 3y = 3$

2. $3x + 7y = 26$
   $4x + 5y = 13$

3. $x + 2y = 4$
   $3x - 4y = 7$

4. $4x + 10y = 30$
   $6x - 4y = 26$

5. $6x + 4y = 56$
   $2x + 7y = 47$
6. $2f + g = 13$
   $6f - g = 3$

26.4 Using simultaneous equations to solve problems

Homework 26D

Read each situation carefully, then write a pair of simultaneous equations and solve them. Remember to give your answers in the context of the original problem.

1. Two numbers have a sum of 20 and a difference of 8. What are they?
2. I think of two numbers. When I double the first number and add the second number to the result, the answer is 17. When I treble the first number and subtract the second number from the result, the answer is 18. What are the two numbers?
3. Molly and Jenson have a combined age of 48. Three years ago Molly was double the age Jenson is now. How old are Molly and Jenson?
4. Steve and Kath have £500 between them. Steve has £75 more than Kath. How much do they each have?
5. Twelve goals are scored in a football match between Y10 and Y11. Y10 score four more goals than Y11. How many goals does each team score?
6. Two numbers have a sum of 8 and a difference of 2. What are they?

26.5 Linear inequalities

Homework 26E

1. Solve each linear inequality.

   a $x + 3 < 8$ **b** $t - 2 > 6$

   c $p + 3 \geqslant 11$ **d** $4x - 5 < 7$

   e $3y + 4 \leqslant 22$ **f** $2t - 5 > 13$

   g $\frac{x + 3}{2} < 8$ **h** $\frac{y + 4}{3} \leqslant 5$

   i $\frac{t - 2}{5} \geqslant 7$ **j** $2(x - 3) < 14$

   k $3x + 8 \geqslant 11$ **l** $4t - 1 \geqslant 29$

2 Write down the values of x that satisfy:

a $x - 2 \leqslant 3$, where x is a positive integer

b $x + 3 < 5$, where x is a positive, odd integer

c $2x - 14 < 38$, where x is a square number

d $4x - 6 \leqslant 15$, where x is a positive, odd number

e $2x + 3 < 25$, where x is a positive, prime number.

3 Frank had £6. He bought three cans of cola and lent his brother £3. When he reached home, he put a 50p coin in his piggy bank. What is the most that the cans of cola could have cost?

4 The perimeter of this rectangle is greater than or equal to 10 but less than or equal to 16.

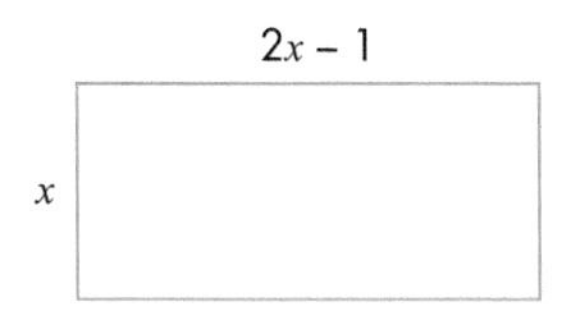

a What are **i** the smallest and **ii** the biggest values of x?

b What are **i** the smallest and **ii** the biggest values of the area?

5 A teacher asks six students to stand at the front of the class and hold up these inequality cards.

$x > 0$	$x < 2$	$x \geqslant 3$	$x = 2$	$x = 3$	$x < 9$

She writes 'TRUE' on one side of the board and 'FALSE' on the other side.

She asks the other students to call out a number, and the students holding the cards have to stand on the 'TRUE' side if their inequality card is true for the number, or on the 'FALSE' side if it isn't.

a A student calls out '2' and the students with the cards all go to the correct side.

  i Which cards are held by the students on the 'TRUE' side?

  ii Which cards are held by the students on the 'FALSE' side?

b Write down a number that would satisfy this grouping.

True			False		
$x \geqslant 3$	$x < 9$	$x > 0$	$x < 2$	$x = 2$	$x = 3$

Hints and tips Remember:

● ——▶ means $\geqslant$ ◀—— ● means $\leqslant$

○ ——▶ means $>$ ◀—— ○ means $<$

1 Write down the inequality represented by each diagram.

a

–1 0 1 2 3 4

b

–3 –2 –1 0 1 2

c

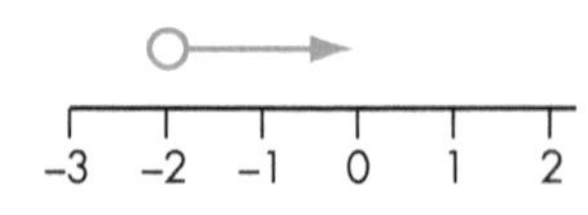

d

–3 –2 –1 0 1 2

e

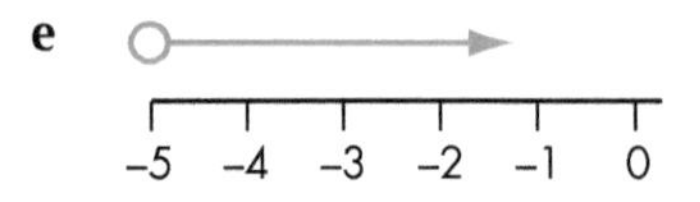

f

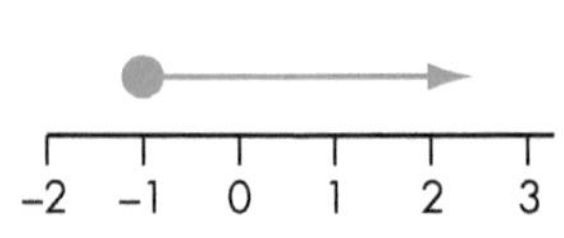

2 Draw diagrams to illustrate these inequalities.

a $x \leqslant 2$ **b** $x > -3$ **c** $x \geqslant 1$ **d** $x < 4$

e $x \geqslant -3$ **f** $1 < x \leqslant 4$ **g** $-2 \leqslant x \leqslant 4$ **h** $-2 < x < 3$

3 Mary went to the record shop with £20. She bought two CDs costing £x each and a DVD costing £9.50. When she got to the till, she found she didn't have enough money.

Mary took the DVD back and paid for the two CDs.

She counted her change and found she had enough money to buy a lipstick for £7.

a Explain why $2x + 9.5 > 20$ and solve the inequality.

b Explain why $2x + 7 \leqslant 20$ and solve the inequality.

c Show the solution to both of these inequalities on a number line.

d The price of a CD is a whole number of pounds. How much is a CD?

4 Solve these inequalities and illustrate their solutions on number lines.

a $x + 5 \geqslant 9$ **b** $x + 4 < 2$ **c** $x - 2 \leqslant 3$ **d** $x - 5 > -2$

e $4x + 3 \leqslant 9$ **f** $5x - 4 \geqslant 16$ **g** $2x - 1 > 13$ **h** $3x + 6 \leqslant 3$

i $3(2x + 1) < 15$ **j** $\frac{x+1}{2} \leqslant 2$ **k** $\frac{x-3}{3} > 7$ **l** $\frac{x+6}{6} \geqslant 1$

5 On copies of the number lines below, draw two inequalities so that only the integers {5, 6, 7, 8} are common to both inequalities.

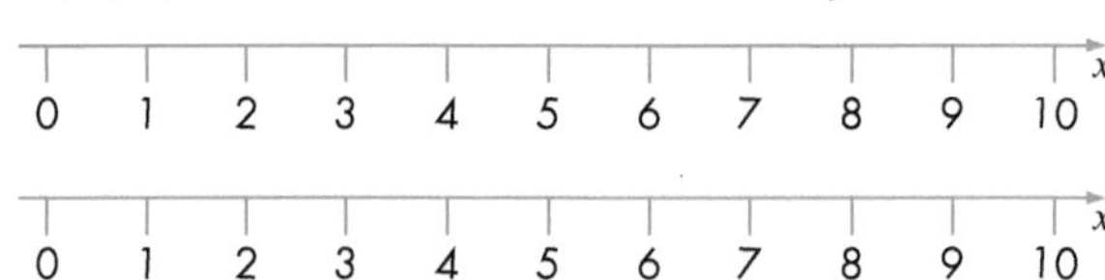

27 Algebra: Non-linear graphs

27.1 Distance–time graphs

Homework 27A

1 This distance–time graph shows Joe's car journey to meet his girlfriend. He set off from home at 9:00 pm and stopped on the way for a break.

Distance (km)
150
100
50
0
9:00 pm
10:00 pm
11:00 pm
12 midnight
Time

a At what time did he:

i stop for his break **ii** set off after his break **iii** arrive at his meeting place?

b At what average speed was he travelling:

i over the first hour **ii** over the last hour **iii** over the whole journey?

2 Jean's taxi set off from Hellaby, stopped to pick up her parents and took them all to a shopping centre. The taxi driver then travelled a further 10 km to pick up a new passenger and drove them to Hellaby.

This distance–time graph illustrates the journey.

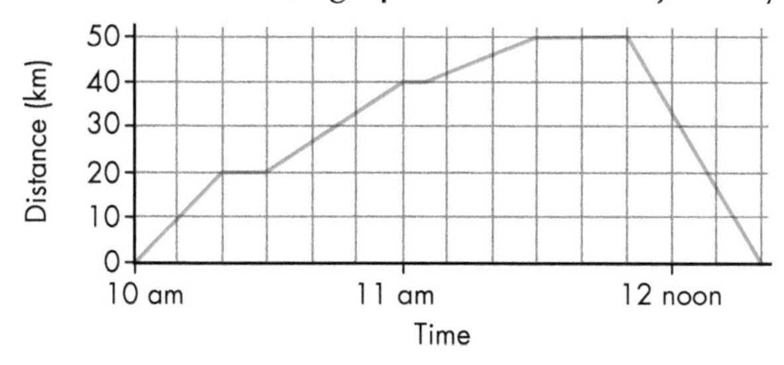

a How far from Hellaby do Jean's parents live?

b How far from Hellaby is the shopping centre?

c What was the average speed of the taxi when Jean was the only passenger?

d What was the average speed of the taxi on its return journey to Hellaby?

3. Michael took his grandchildren out for the afternoon. They set off from Norwich at 1 pm and travelled for half an hour, at an average speed of 60 km/h. They stopped to look at the sea and eat an ice cream.

   At two o'clock they set off again, travelling at a speed of 80 km/h for a quarter of an hour. They then stopped to play on the sand for 30 minutes.

   After this they returned to Norwich, at an average speed of 50 km/h.

   Draw a distance–time graph to represent this information. Use the horizontal axis for time, from 1 pm to 5 pm, and the vertical axis for distance, from 0 to 60 km.

4. A runner sets off at 8 am from point P to jog along a trail at a steady pace of 12 km/h.

   One hour later, a cyclist sets off from P on the same trail, at a steady pace of 24 km/h. After 30 minutes, the cyclist gets a puncture that takes her 30 minutes to fix. She then sets off at a steady pace of 24 km/h.

   At what time did the cyclist catch up with the runner?

   Hints and tips Drawing a distance–time graph is a straightforward method of answering this question. Remember that the cyclist doesn't start until 9 am.

Homework 27B

1. Liquid is poured at a steady rate into each container. Match each container with the correct graph.

2. Draw a graph of the depth of water in each of these containers as it is filled steadily.

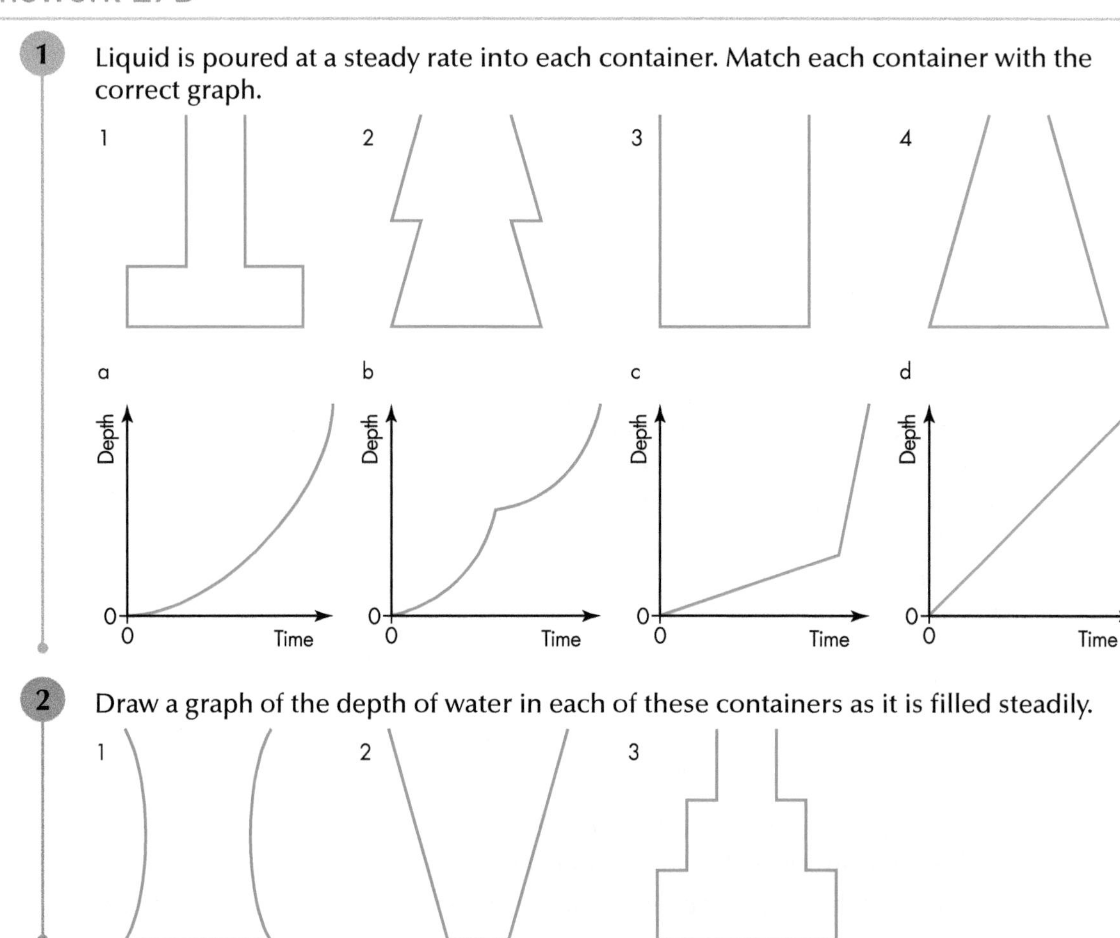

27.2 Velocity–time graphs

Homework 27C

1 Ellen has a dog. She is observing her dog's movements for a science experiment and draws a velocity–time graph.

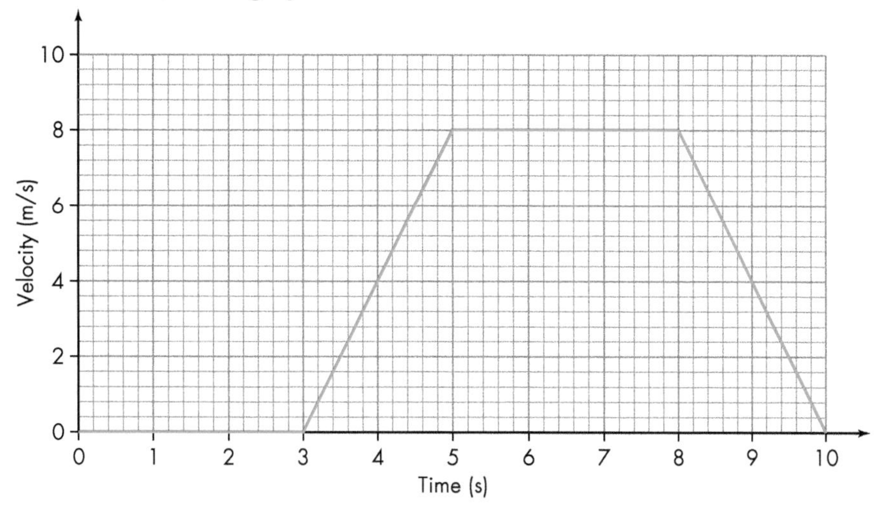

a What can you say about the velocity of the dog between:

i 0 and 3 seconds **ii** 3 and 5 seconds

iii 5 and 8 seconds **iv** 8 and 10 seconds?

b Work out the acceleration between 3 and 5 seconds.

2 **a** Describe what is happening in each section of this graph.

b Work out the acceleration for section A.

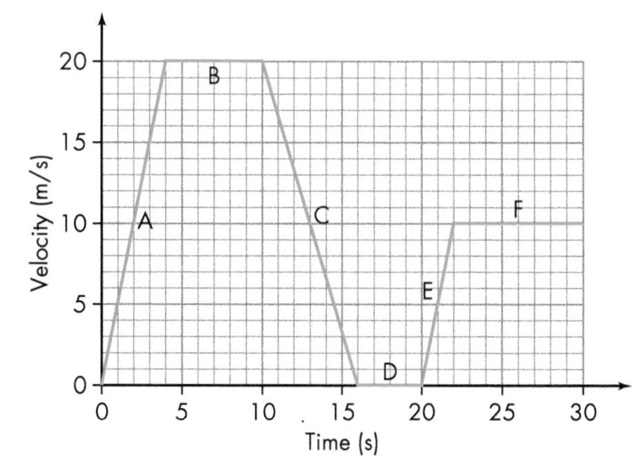

3 The velocity–time graph shows the journey of a scooter. Work out the deceleration of the scooter between B and C.

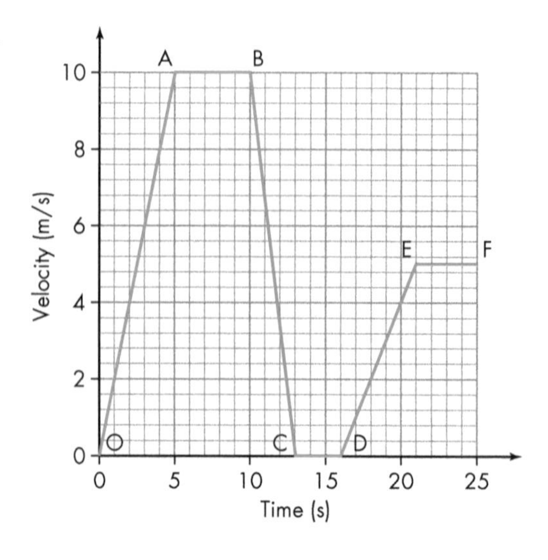

27.3 Plotting quadratic graphs

Homework 27D 

1 **a** Copy and complete the table and draw the graph of $y = 2x^2$ for values of x from –3 to 3.

x	–3	–2	–1	0	1	2	3
$y = 2x^2$	18		2			8	

b Use your graph to find the value of y when $x = –1.4$.

c Use your graph to find the values of x that give a y-value of 10.

2 **a** Copy and complete the table and draw the graph of $y = x^2 + 3$ for $–5 \leqslant x \leqslant 5$.

x	–5	–4	–3	–2	–1	0	1	2	3	4	5
$y = x^2 + 3$	28		12					7			28

b Use your graph to find the value of y when $x = 2.5$.

c Use your graph to find the values of x that give a y-value of 10.

3 **a** Copy and complete the table and draw the graph of $y = x^2 – 3x + 2$ for $–3 \leqslant x \leqslant 4$.

x	–3	–2	–1	0	1	2	3	4
$y = x^2 – 3x + 2$	20			2			2	

b Use your graph to find the y-value when $x = –1.5$.

c Use your graph to find the values of x that give a y-value of 2.5.

4 **a** Copy and complete the table and draw the graph of $y = x^2 – 5x + 4$ for $–1 \leqslant x \leqslant 6$.

x	–1	0	1	2	3	4	5	6
$y = x^2 – 5x + 4$	10	4				0	2	

b Where does the graph cross the x-axis?

c Use your graph to find the y-value when $x = 2.5$.

d Use your graph to find the values of x that give a y-value of 8.

5 Tom is drawing quadratic graphs with equations of the form $y = x^2 + bx + c$.

He notices that two of his graphs pass through the point (2, 5).

He has drawn the graphs of two of the following equations. Which two?

Equation A: $y = x^2 + 3$

Equation B: $y = x^2 + 1$

Equation C: $y = x^2 + 2x - 3$

Equation D: $y = x^2 - x + 5$

27.4 Solving quadratic equations by factorisation

Homework 27E

Example

Solve $x^2 + 10x + 16 = 0$.

First factorise. $(x + 2)(x + 8) = 0$

Then solve. Either $x + 2 = 0$ or $x + 8 = 0$

so $x = -2$ or $x = -8$

1 Solve these quadratic equations.

a $(x - 2)(x - 3) = 0$

b $(x + 2)(x + 3) = 0$

c $(x - 4)(x + 4) = 0$

d $(x + 8)(x - 2) = 0$

2 First factorise and then solve these quadratic equations.

a $x^2 + 3x + 2 = 0$

b $x^2 + 7x + 12 = 0$

c $x^2 + 8x + 16 = 0$

d $x^2 + 15x + 56 = 0$

e $x^2 + 5x - 14 = 0$

f $x^2 + 6x - 40 = 0$

g $x^2 + 2x - 63 = 0$

h $x^2 - 11x + 30 = 0$

i $x^2 - 17x - 60 = 0$

j $x^2 - 20x + 84 = 0$

3 A rectangle has width x cm. Its length is 3 cm longer than the width and its area is 40 cm^2. Work out the dimensions of the rectangle.

27.5 The significant points of a quadratic curve

Homework 27F

In question **2** of the previous exercise, you factorised the equations written below. For each equation, draw the graph and write down the coordinates of:

a the y-intercept

b the roots

c the turning point.

Can you see any connections between the equations (factorised and unfactorised) and the graphs?

1 $x^2 + 3x + 2 = 0$

2 $x^2 + 7x + 12 = 0$

3 $x^2 + 10x + 16 = 0$

4 $x^2 + 15x + 56 = 0$

5 $x^2 + 5x - 14 = 0$

6 $x^2 + 6x - 40 = 0$

7 $x^2 + 2x - 63 = 0$

8 $x^2 - 11x + 30 = 0$

9 $x^2 - 17x - 60 = 0$

10 $x^2 - 20x + 84 = 0$

Homework 27G

1 Plot each graph for $-2 \leqslant x \leqslant 3$. Then write down the coordinates of:

a the y-intercept

b the points where the curve intersects the x-axis

c the turning point.

i $x^2 + 8x + 12$

ii $x^2 + 14x + 48$

iii $x^2 + 15x + 56$

iv $x^2 - 12x + 27$

v $x^2 - 3x + 2$

vi $x^2 - x - 56$

vii $x^2 + 4x - 21$

viii $x^2 - 9x - 10$

ix $x^2 - 36$

x $4x^2 - 19x + 12$

xi $2x^2 + x - 6$

xii $4x^2 - 15x + 9$

2. For each graph, write down the coordinates of:

   a the y-intercept

   b the points where the curve intersects the x-axis

   c the turning point.

   i $4x^2 + 7x + 3$

   ii $6x^2 + 19x + 10$

   iii $2x^2 - x - 21$

   iv $10x^2 - 11x + 3$

   v $2x^2 - 10x - 28$

27.6 Cubic and reciprocal graphs

Homework 27H 

1. **a** Copy and complete the table for $y = x^3$ for $-3 \leqslant x \leqslant 3$.

x	–3	–2	–1	0	1	2	3
y	–27		–1	0		8	

   b Draw the graph of $y = x^3$.

2. **a** Copy and complete the table for $y = x^3 + 3$ for $-3 \leqslant x \leqslant 3$.

x	–3	–2	–1	0	1	2	3
y	–24		2	3		11	

   b Draw the graph of $y = x^3 + 3$.

3. **a** Draw and complete a table of values for of $y = \frac{1}{x}$ for $-3 \leqslant x \leqslant 3$.

   b Draw the graph of $y = \frac{1}{x}$.

William Collins' dream of knowledge for all began with the publication of his first book in 1819. A self-educated mill worker, he not only enriched millions of lives, but also founded a flourishing publishing house. Today, staying true to this spirit, Collins books are packed with inspiration, innovation and practical expertise. They place you at the centre of a world of possibility and give you exactly what you need to explore it.

Collins. Freedom to teach

Published by Collins
An imprint of HarperCollins*Publishers*
The News Building
1 London Bridge Street
London SE1 9GF

Browse the complete Collins catalogue at
www.collins.co.uk

10 9 8 7 6 5

ISBN 978-0-00-811388-9

A catalogue record for this book is available from the British Library

Commissioned by Lucy Rowland and Katie Sergeant
Project managed by Elektra Media Ltd
Copyedited by Marie Taylor
Proofread by Joanna Shock
Answers checked by Amanda Dickson
Edited by Caroline Green and Jennifer Yong
Typeset by Jouve India Private Limited
Illustrations by Ann Paganuzzi
Designed by Ken Vail Graphic Design
Cover design by We are Laura
Production by Rachel Weaver

Printed and bound by CPI Group (UK) Ltd, Croydon, CR0 4YY

Acknowledgements
The publishers gratefully acknowledge the permissions granted to reproduce copyright material in this book. Every effort has been made to contact the holders of copyright material, but if any have been inadvertently overlooked, the publisher will be pleased to make the necessary arrangements at the first opportunity.

The publishers would like to thank the following for permission to reproduce photographs in these pages:

Cover (bottom) Georgios Kollidas/Shutterstock, cover (top) Godruma/Shutterstock.